£3

with

(on p.21)

Functional differential equations: I. C-theory*

Pitman Monographs and
Surveys in Pure and Applied Mathematics 70

Functional differential equations: I. C^*-theory

Anatolij Antonevich and Andrei Lebedev
Byelorussian State University

Longman Scientific & Technical
Copublished in the United States with
John Wiley & Sons, Inc., New York

Longman Scientific & Technical
Longman Group UK Limited
Longman House, Burnt Mill, Harlow
Essex CM20 2JE, England
and Associated companies throughout the world.

Copublished in the United States with
John Wiley & Sons Inc., 605 Third Avenue, New York, NY 10158

First published 1994

AMS Subject Classifications: 22, 28, 35, 45, 46, 47, 58

ISSN 0269-3666

ISBN 0 582 07251 4

British Library Cataloguing in Publication Data

A catalogue record for this book is
available from the British Library

Library of Congress Cataloging-in-Publication Data

A catalog record for this book is available

Printed and bound by Bookcraft (Bath) Ltd

ВЛАДИМИРУ РЫВКИНУ – Изумительному Математику и Человеку посвящается

To VLADIMIR RYVKIN – the Wonderful Mathematician and the Man

Contents

Preface

This book is an attempt to provide the first thorough and up–to–date exposition of a relatively new field of modern analysis lying in between differential equations and C^*–algebras, and in fact deeply intersecting both of these subjects. The problems considered arise from the investigation of non–local functional differential operators containing transformation operators of an argument.

We have developed a general unified approach to the solution of these problems based on the investigation of C^*–algebras corresponding to the functional differential operators that we examine. The main objects which appear here are C^*–algebras associated with dynamical systems. Attention is principally devoted to the investigation of invertibility and Fredholmity of these operators and their spectral properties. The monograph contains a description of the structure of the C^*–algebras introduced and in particular, as a result of this description, a symbolic calculus for these kinds of problems is obtained. General formulae for the spectral radius and conditions for the invertibility of elements of these algebras are given. The fundamental correlation between the invertibility of elements and hyperbolic decomposition, trajectorial decomposition and approximations of dynamical systems is brought to light. Various applications to equations associated with singular integral operators, Toeplitz operators, pseudodifferential operators (including operators with discontinuous coefficients) and transformation operators of an argument; non–local boundary–value problems, convolution–type equations with oscillating coefficients and other equations generated by different local–type operators and transformation operators of an argument are presented.

In the first part of the book (C^*–theory), Sections 1–10 and 17–20 have been written by Antonevich; the rest is due to Lebedev.

At the end of each chapter notes and remarks concerning each section in the chapter are inserted. These short comments are probably incomplete and may contain errors. The authors ask the forgiveness of those mathematicians whose contributions to the subject have not been brought to light because of our oversight or lack of information.

The bibliography contains in essence only those sources directly related to the material considered here and does not reflect the historical development of the subject.

We wish to express our thanks to Professor Alan Jeffrey, the editor of the book, who in fact initiated the writing and has done everything to smooth the publication process.

Anatolij Antonevich　　　　Andrei Lebedev

Chapter 0: Preparative.

'Государство, держащееся на чтении книг, падет!'
A. Платонов 'Деревянное растение'*

1. Introduction

The main problems treated in this book originate in the theory of the solvability of non–local functional differential equations, that is, equations with respect to an unknown function which link values of the unknown function and its derivatives at different points of the domain (unlike 'pure' differential equations, which link values of a function and its derivatives at one and the same point). Now investigations in the theory of functional differential equations have formed an independent field of analysis which is developing intensively due to its numerous applications and its interesting internal mathematical structure.

Functional differential equations appear naturally for example when we are considering a mathematical model of a process in a certain system such that interactions between its parts do not proceed instantly but with some delays. These delays may have various origins: transport reasons – reflecting the time necessary for transportation of parts of the system or a signal passage; technological reasons – taking into account the time necessary for the implementation of technological operations, chemical reactions, etc.; information reasons – delays arising from the necessity for information processing in control systems, and so on. Problems where the influence of delays is essential arise in the theory of nuclear reactors, automatic control theory, immunology and epidemiology, mathematical economy, and various other fields of the natural sciences.

Moreover, functional differential equations often appear as a result of the transformation of 'pure' differential equations and their boundary–value problems (see Example 1.6).

In an analogous natural way there arise functional differential equation with partial derivatives containing time delays and deviations of space arguments.

---

*'A regime based on book reading will fail'

A. Platonov *Wooden plant*

We want to emphasize that in fact the ignoring of delays (even arbitrarily small delays) in a mathematical model may lead in certain situations to the violation of fundamental principles of natural philosophy. Let us recall in this connection the well–known example of an 'ideal predictor'.

Example 1.1 Consider the circuit shown in Fig. 1 where the unit [*] imposes a delay equal to –1.

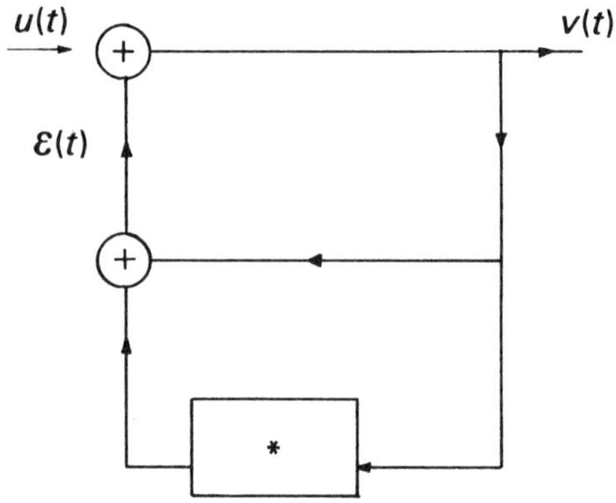

Fig. 1

The circuit works according to the equations

$$v(t) = u(t) + \varepsilon(t)$$
$$\varepsilon(t) = v(t) + v(t - 1)$$

which imply $v(t) = u(t + 1)$, that is, the output at moment t depends on the input at $t + 1$, which contradicts the causality principle.

This paradox was obtained due to the fact that the circuit does not take into account delays (which may be very small but non–zero) arising when carrying out the circuit operations.

For example, if the addition requires time equal to h then we have

$$v(t) = u(t - h) + \varepsilon(t - h)$$
$$\varepsilon(t) = v(t - h) + v(t - 1 - h)$$

and solving this system we obtain

$$v(t) = u(t - h) - u(t - 3h) + u(t - 5h) - \dots - u(t - 1 - 3h) + \\ + 2u(t - 1 - 5h) - \dots \quad .$$

Thus the output depends here on the input taken at the preceding moments of time, which agrees with the causality principle.

A vast literature is devoted to the theory of functional differential equations, including a number of monographs and detailed surveys, for example Azbelev, Rahmatullina [1], Bellman, Cooke [1], El′sgholts, Norkin [1], Hale [1], Kolmanovskij, Nosov [1, 2], Mitropolskij, Martynjuk [1], Myshkis [1, 2], and Romanenko, Sharkovskij [1].

In this book we do not expound the history and applications (non–mathematical) of the theory of functional differential equations, but refer the interested reader to the above–mentioned sources and to the bibliographies they contain.

It has to be stressed that in our opinion the structures considered in the theory of functional differential equations roughly speaking should be viewed as originating from functional equations rather than from 'pure' differential equations.

To make this statement clearer let us look through the following objects.

Example 1.2 Let $F(X)$ be some functional space over a space X (we are not specifying this space in detail because we simply want to present the idea). Consider the functional equation

$$(Lf)(x) = g(x) \tag{1.1}$$

where L is a linear operator of one of the following four types:

(1) $f(x) \overset{L}{\to} a(x)f(x)$

(2) $f(x) \overset{L}{\to} a(x)f^{(j)}(x)$

(3) $f(x) \overset{L}{\to} \sum_{i=1}^{k} a_i(x)f(\alpha_i(x))$

$$(4) \quad f(x) \xrightarrow{L} \sum_{j=1}^{n} \sum_{i=1}^{m} a_{ij}(x) f^{(j)}(\alpha_i(x)) \quad .$$

In the first case L is the operator of multiplication by some function $a(x)$; in the second case L is a 'pure' differential operator; in the third case L is a *functional operator* generated by operators of multiplication by functions a_i and transformations of the argument α_i: $X \to X$ (we call operators of the form $f(x) \to a(x)f(\alpha(x))$ weighted shift operators); in the fourth case L is a 'classical' *functional differential operator.*

If we consider the interrelations between equations of the form (1.1) with L having the form (1), (2), (3), or (4) then in the diagram of interrelations (Fig. 2) the arrows between (1) and (2) and (3) and (4) are 'short', while the

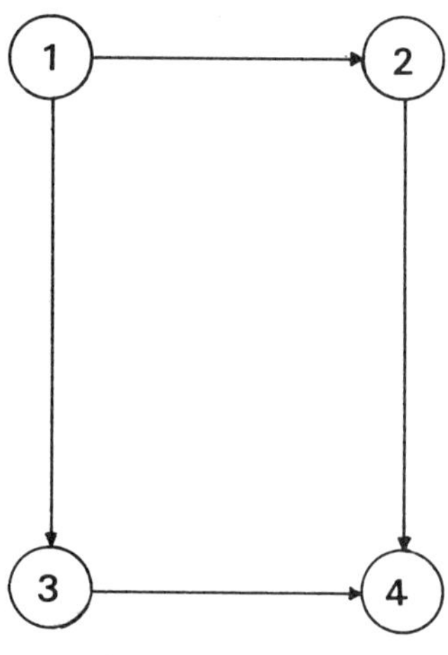

Fig. 2

arrows between (1) and (3) and (2) and (4) are 'long'.

What do we mean by this? If we consider two operators L of type (1) then they commute and (if $F(X)$ is, say $L^2(X)$) L commutes with its adjoint L^* ($(L^* f(x) = \bar{a}(x)f(x))$. Thus these operators are normal and they are invertible if and only if (in the case of continuous $a(x)$ and compact X) $a(x) \neq 0$, $x \in X$.

If we consider two operators of type (2) then in general they do not commute but they commute modulo the operators of smaller order and (if we are, say, considering Sobolev spaces $H^k(X)$) L commutes with its adjoint modulo the operators of smaller order. Moreover equation (1.1) with L of type (2) satisfies the Fredholm alternative if and only if $a(x) \neq 0$, $x \in X$.

Thus when considering problems concentrated around Fredholm theory we would say that equations generated by operators of type (1) and (2) have the same 'nature'. On the other hand, if we are considering operators of type (3) then they are very far from being normal and generate a highly non–commutative algebra, thus in fact having almost nothing in common with operators of type (1). The conditions of invertibility for these operators are rather non–trivial and one of the main purposes of this book is to obtain a general approach for finding them.

Now if we consider operators L of type (4) then it is reasonable to divide this class of operators into the subclasses:

$$\text{(4a)} \quad f(x) \xrightarrow{L} a_n(x)f^{(n)}(x) + \sum_{j=1}^{n-1} \sum_{i=1}^{m} a_{ij} f^{(j)}(\alpha_i(x))$$

$$\text{(4b)} \quad f(x) \xrightarrow{L} \sum_{j=1}^{n} \sum_{i=1}^{m} a_{ij} f^{(j)}(\alpha_i(x))$$

where in (4b) the transformations α_i in the sum $\sum_{i=1}^{m} a_{in} f^{(n)}(\alpha_i(x))$ are not trivial. In the last case equation (1.1) is called a *functional differential equation of neutral type.*

If we consider the case (4a) then the corresponding equation (1.1) satisfies the Fredholm alternative when $a_n(x) \neq 0$, $x \in X$, thus this case is analogous to 'pure' differential equations.

On the other hand, if we consider the case (4b) then the conditions for equation (1.1) to satisfy the Fredholm alternative coincide with the conditions for the functional operator

$$f(x) \rightarrow \sum_{i=1}^{m} a_{in} f(\alpha_i(x))$$

(of type (3)) to be invertible (some particular cases of this statement are proved in Proposition 10.3 and Theorem 10.4 and the general situation will be considered in the second part of the book). All this make it possible to assert that equations generated by operators L of types (3) and (4) have the same 'nature', but different from the 'nature' of equations generated by operators of type (1) and (2).

It is strange, but a lot of functional differential equation theory is developed without taking into account this observation, thus ignoring the essential features of the objects. In fact one of the main goals of this book is to apply this idea to various classes of functional differential and other types of equations having a similar internal mathematical structure.

Let us emphasize one more characteristic feature of the objects examined in this book, in which they differ greatly from 'pure' differential equations. This feature is also deeply correlated to the foregoing remark.

If we are considering, say, classical elliptic problems for partial differential equations then their algebraic structure is brought to light under the Fourier transform which establishes some sort of isomorphism between these problems and the so–called symbols of pseudodifferential operators. These symbols (which are in fact matrix functions) actually show the local structure of the initial problems and give us an opportunity to use the whole scale of finite dimensional linear algebra methods.

It may be worth mentioning that we can extend the class of pseudodifferential operators by considering the so–called Fourier integral operators (see, for example, Treves [1]) and a weighted shift operator can be viewed as a Fourier integral operator with the corresponding phase function. But problems immediately arise when we consider the sum of two Fourier integral operators having different phase functions (corresponding to different shifts). The sum of such operators is unfortunately no longer Fourier integral operator.

Thus, returning to the investigation of functional differential equations, can we find anything which, for these objects, will play a role analogous to that played by the operator symbol in the theory of pseudodifferential operators?

The answer is yes! – and the corresponding objects – namely C^*–algebras generated by 'weighted shift' operators (which have deep internal interrelations with C^*–algebras associated with automorphism groups – the so–called cross–products) are investigated in the second chapter of this book. The role of the Fourier transform (which is, in fact, non–commutative and infinite dimensional) here is played by the isomorphism theorem (Corollary 12.17) which is one of the most important technical C^*–algebra results of this book, having numerous applications. The thorough analysis of algebras arising in this way will bring to light the mathematical essence of the various classes of equations considered here, and will give us an opportunity to obtain a unified approach to different solvability problems.

Now let us provide a few examples where functional differential equations appear naturally but perhaps not in the most obvious and traditional way.

Example 1.3 Possibly one of the simplest examples of equations which do not contain a transformation of an argument in the initial statement, but which in the end are reduced to the investigation of operators with shifts, is the equation

$$U(x) + e^{ixh} \int_{-\infty}^{+\infty} K_1(x-y)U(y)dy + \int_{-\infty}^{+\infty} K_2(x-y)U(y)dy = W(y) \tag{1.2}$$

in $L^2(\mathbb{R})$ with $K_i \in L^1(\mathbb{R})$.

After implementing the Fourier transform this equation becomes an equation containing a weighted shift operator:

$$[1 + \hat{K}_2(\xi)]\hat{U}(\xi) + K_1(\xi)\hat{U}(\xi - h) = \hat{W}(\xi) \quad .$$

Equation (1.2) belongs to the class of so–called convolution–type equations with oscillating coefficients.

Example 1.4 It is worth mentioning that in sufficiently complicated systems variables can be not only of spatial or temporal type but also of other physical or geometrical types, such as variables characterizing energy, directions, frequencies, etc. For example, emission spread in space can be described by a function $f(x, w)$ $x \in \mathbb{R}^3$, $w \in S^2$ defining the energy flow distribution at a point x in a direction $w \in S^2$. If the emission reflects from a mirror–like surface Γ then the function $f(x\ w)$, when $x \in \Gamma$, should satisfy the reflection condition

$$f(x, \alpha(w)) = K(x)f(x, w) \quad , \qquad \langle w, V(x)\rangle > 0$$

where $K(x)$ is a reflection coefficient in $x \in \Gamma$ and $\alpha(w) = w - 2\langle w, V(x)\rangle V(x)$ is the mirror image of w, $V(x)$ is the normalized vector normal to the surface Γ in x.

In the simplest variant this boundary–value condition is adjoined to the transfer equation

$$\sum_{k=1}^{3} w_k \frac{\partial f}{\partial x_k} = 0 \quad .$$

Example 1.5 Now we present functional equations which naturally arise in the problem of filtration in probability theory.

Let ξ be a continuous random variable having a probability density u, let η be a discrete random variable taking a finite set of values $h_1, \ldots, h_n$ with probabilities $p_1, \ldots, p_n$. In this case the sum $\xi + \eta$ is an absolutely continuous random variable and its density v is linked with u by the equation

$$bu \equiv \sum_{k=1}^{n} p_k \, u(x - h_k) = v(x) \quad . \tag{1.3}$$

If one considers η as a noise then the problem of filtration, that is, the problem of determining u if v is a known function is reduced to the problem of determining the solution of the functional equation (1.3).

A more complicated functional equation arises in the next situation. Let a random variable ξ having a probability density u be measured by means of some instrument which may be in a finite set $\{S_k\}_{k=1}^{n}$ of n states. Being in a state S_k, the instrument receiving a value x at entry gives us $\varphi_k(x)$ where φ_k is some given differentiable function with $\varphi_k'(x) \neq 0$. Let us denote the inverse to the function φ_k by ψ_k. The instrument randomly changes its states and the probability of its being in a state S_k is equal to p_k. In this situation the output values determine some absolutely continuous random variable ζ whose density v is linked with u by the equation

$$\sum_{k=1}^{n} p_k \ \psi_k'(x)u(\psi_k(x)) = v(x) \quad . \tag{1.4}$$

The problem of restoring the density u by the measured density v is reduced now to the problem of solving equation (1.4).

Now the previous problem (reduced to equation (1.3)) can be considered as a particular case of this one if we set $\varphi_k(x) = x + h_k$. Another particular case, the case when ζ is the product of a random variable ξ and a discrete random variable η, can also be included in this scheme if we set $\varphi_k(x) = h_k \cdot x$. This is the so–called multiplicative noise case, and in fact is realized if the amplification coefficient of the measuring instrument changes randomly. In this situation equation (1.4) will have the form

$$\sum_{k=1}^{n} \frac{p_k}{h_k} \, u\left(\frac{x}{h_k}\right) = v(x) \quad . \tag{1.5}$$

If a multiplicative and an additive noise appear simultaneously then

$$\varphi_k(x) = a_k \cdot x + b_k$$

and equation (1.4) takes the form

$$\sum_{k=1}^{n} \frac{p_k}{a_k} \, u\left(\frac{x - b_k}{a_k}\right) = v(x) \quad . \tag{1.6}$$

The difference between equations (1.3), (1.5) and (1.6) is severe, and is due to the fact that the mappings $\varphi_k(x) = a_k x + b_k$ generate a non–commutative and non–compact group.

Example 1.6 Here we present a class of boundary–value problems for 'pure' differential equations which can be naturally reduced to the problem of the solvability of some functional differential equations.

To begin with we recall that the general solution of the string oscillation equation

$$\frac{\partial^2 u}{\partial t^2} = \frac{\partial^2 u}{\partial x^2} \tag{1.7}$$

has the form

$$u(x, t) = \varphi_1(x - t) + \varphi_2(x + t) \tag{1.8}$$

where φ_1, φ_2 are arbitrary functions. Because of this, boundary–value problems for the string oscillation equation are reduced to systems of functional differential equations with respect to φ_1 and φ_2. If we substitute $u(x, t)$ into the boundary conditions on curves (not coinciding with $t = 0$) the functions φ_1 and φ_2 will undergo some transformations of arguments.

In the case of the Cauchy problem and some mixed boundary–value problems the resulting system can be solved explicitly.

If the boundary condition is given on the whole of the boundary of some domain $\mathscr{D} \subset \mathbb{R}^2$ then in general one will obtain a system with a highly complicated structure which is inaccessible for complete investigation.

Let us present a class of problems for which the corresponding system for φ_1 and φ_2 is non–trivial but which can be reduced to the investigation of one weighted shift operator, thus giving us an opportunity to make progress in the investigation.

It is interesting also that the problem we shall consider now does not include any argument transformation operators in its starting position, while they arise naturally during the investigation.

So let us consider equation (1.7) and set the following boundary conditions

$$u\Big|_{t=h_1(x)} = g_1(x) \qquad -\infty < x < +\infty \tag{1.9}$$

$$\left[a_1(x)\frac{\partial u}{\partial t} + a_2(x)\frac{\partial u}{\partial x}\right]_{t=h_2(x)} = g_2(x) \qquad -\infty < x < +\infty \tag{1.10}$$

where h_1, h_2, a_1, a_2, g_1, g_2 are given functions.

In the case when $h_1(x) = h_2(x)$ this problem is called the generalized Cauchy problem.

The physical meaning of problem (1.7), (1.9) and (1.10) is rather visual: we have to find the position of the string if at a point x the position of the string is given (measured) at some moment $h_1(x)$ while the 'generalized speed' value, $a_1(x) \frac{\partial u}{\partial t} + a_2(x) \frac{\partial u}{\partial x}$, is measured at some other moment $h_2(x)$.

The problem (1.7), (1.9) and (1.10) can be regarded also as an analogue of the two–point problem for an ordinary differential equation.

Posed in this way, the problem (1.7), (1.9) and (1.10) cannot always be resolved. Thus we have to find the conditions on h_1, h_2, a_1, a_2 under which the problem (1.7), (1.9) and (1.10) can be resolved with respect to any given right–hand parts in (1.9) and (1.10) (in the corresponding spaces).

If we consider the domain $\mathscr{D} = \{(x, t): h_1(x) < t < h_2(x)\}$ then the problem (1.7), (1.9) and (1.10) can be considered as a problem with the conditions set on the whole of the boundary of $\mathscr{D}$.

The first example of the correct problem with boundary conditions set on the whole of the boundary was constructed by Sobolev [1] who obtained conditions on the coefficients of the directional derivative problem for which the problem is correct.

We show now that the problem (1.7), (1.9) and (1.10) can be reduced to an equation with a weighted shift operator and, using the properties of weighted shift operators, we shall later investigate this and some analogous problems completely.

Substituting the general solution (1.8) into the conditions (1.9) and (1.10) we have

$$\varphi_1(x - h_1(x)) + \varphi_2(x + h_1(x)) = g_1(x) \tag{1.11}$$

$$a_1(x)[-\varphi_1'(x - h_2(x)) + \varphi_2'(x + h_2(x))] + a_2(x)[\varphi_1'(x - h_2(x) + \\ + \varphi_2'(x + h_2(x))] = g_2(x) \quad . \tag{1.12}$$

Differentiating equation (1.11) we obtain

$$[1 - h_1'(x)]\varphi_1'(x - h_1(x)) + [1 + h_1'(x)]\varphi_2'(x + h_1(x)) = g_1'(x) \ . \tag{1.13}$$

Thus we arrive at the system (1.12) and (1.13) of functional differential equations with respect to the functions φ_1', φ_2'. This system establishes the relation between values of these functions on the set of points linked with each other by the mappings

$$\alpha_1^{\pm}(x) = x \pm h_1(x)$$

$$\alpha_2^{\pm}(x) = x \pm h_2(x) \quad .$$

In general (when h_1 and h_2 are arbitrary) the system (1.12) and (1.13) is very complicated. Let us consider (1.12) and (1.13) under an additional condition which makes it possible to simplify this system. Suppose that

$$|h_i'(x)| < 1 \qquad i = 1, 2 \tag{1.14}$$

or

$$|h_i'(x)| > 1 \qquad i = 1, 2 \quad . \tag{1.15}$$

The geometrical meaning of these conditions is that the curves $t = h_1(x)$ and $t = h_2(x)$ are not tangential at any point to the characteristics of equation (1.7).

Let us consider the case when (1.14) is fulfilled (the case when (1.15) is implemented can be reduced to (1.14) if one exchanges the variable x and t).

In view of (1.14) the mappings $\alpha_1^{\pm}$ and $\alpha_2^{\pm}$ are invertible. Denote their inverses by $\beta_1^{\pm}$ and $\beta_2^{\pm}$. Implementing the transformation of the argument $y = \alpha_1^{-}(x)$, $x = \beta_1^{-}(y)$ we pass from (1.13) to an equation where $\varphi_1'(y)$ is expressed in terms of φ_2':

$$\varphi_1'(y) = f_1(y)\varphi_2'(\alpha_1^{+} \circ \beta_1^{-}(y)) + f_2(y) \tag{1.16}$$

where

$$f_1(y) = -\frac{1 + h_1'(\beta_1'(y))}{1 - h_1'(\beta_1^-(y))}$$

$$f_2(y) = \frac{g_1'(\beta_1^-(y))}{1 - h_1^-(\beta_1^-(y))} \quad .$$

Substituting (1.6) into (1.12) and denoting $z = \alpha_1^+(\beta_1^-(\alpha_2^-(x)))$, $\varphi_2'(z) = \varphi(z)$ we obtain

$$a(z)\ \varphi(z) + b(z)\ \varphi(\alpha(z)) = g(z) \tag{1.17}$$

where

$$\alpha(z) = \alpha_2^+(\beta_2^-(\alpha_1^-(\beta_1^+(z))))$$

is an invertible mapping and

$$a(z) = (\hat{a}_2 - \hat{a}_1)(z)\ f_1(\alpha_1^-(\beta_1^+(z)))$$

$$b(z) = (\hat{a}_2 + \hat{a}_1)(z)$$

$$g(z) = g_2(\beta_2^-(\alpha_1^-(\beta_1^+(z)))) - f_2(\alpha_1^-(\beta_1^+(z))[\hat{a}_2 - \hat{a}_1](z)$$

$$\hat{a}_i(z) = a_i(\beta_2^-(\alpha_1^-(\beta_1^+(z)))) \qquad i = 1,\ 2 \quad .$$

The mapping α arising in equation (1.17) can be easily interpreted geometrically (Fig. 1.3).
Here

$$x_1 = \beta_1^+(z_0) = \beta_1^-(z_1) \quad , \qquad \text{that is,} \qquad \begin{aligned} z_0 &= x_1 + h_1(x_1) \\ z_1 &= x_1 - h_1(x_1) \end{aligned}$$

$$x_2 = \beta_2^-(z_1) = \beta_2^+(\alpha(z_0)) \quad , \qquad \text{that is,} \qquad \begin{aligned} z_1 &= x_2 - h_2(x_2) \\ \alpha(z_0) &= x_2 + h_2(x_2) \quad . \end{aligned}$$

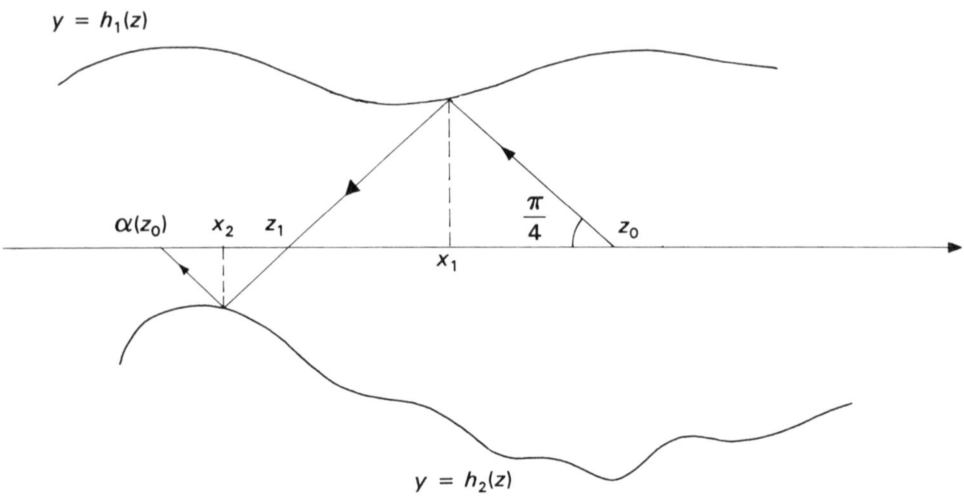

Fig. 1.3

Thus $\alpha(z_0)$ is the final point of the trace of characteristics emitted from z_0 and which reflects from the curves $y = h_1(z)$, $y = h_2(z)$.

In particular if $h_1(x) = h_2(x)$ then the point $z = x + h_1(x)$ is a fixed point for α and all the fixed points are of this kind. Thus if $h_1(x) \equiv h_2(x)$ then all the points z are fixed and equation (1.17) does not contain transformations of an argument.

If $h_1(x) = h_1 \equiv \text{const}$ and $h_2(x) = h_2 \equiv \text{const}$ then $\alpha(z) = z + h$ where $h = 2(h_2 - h_1)$.

We hope that the foregoing examples, together with the traditional examples of functional differential equations, show that the appearance of problems leading to the investigation of equations containing weighted shift operators may arise almost anywhere.

On the other hand, weighted shifts are interesting not only in connection with a wide range of applications but also as objects having an extremely rich and complicated intrinsic mathematical structure. Many different counterexamples in analysis can be constructed with the help of them (see for example Section 6). Particular representatives of the class of these operators have been treated in connection with the invariant subspace problem (see Davie [1], Gromov [2, 4], Nordgren, Rosental, Wintrobe [1]). Any automorphism of a semisimple commutative Banach algebra and any isometric operator can be written in weighted shift operator form in various functional spaces. Thus if we try to investigate a weighted shift operator in its general form we can hardly hope to succeed. On the other hand, as a rule, when dealing with concrete mathematical problems we do not meet arbitrary weighted shifts but have some correlation between weights and shifts and if this correlation is 'reasonable' then the analysis of the object can be rather fruitful. In this book we try to distinguish the class of weighted shifts which are on the one hand rather general (almost arbitrary) but on the other hand are still accessible for investigation.

Various problems of the theory of weighted shift operators appeared to be naturally and essentially connected with the investigations of algebras generated by them. These algebras, in turn, play a substantial role in the general theory of Banach algebras (and perhaps it is not surprising in view of the foregoing observations).

The investigation of such algebras originates in the celebrated work by Murray and Von Neumann [1] and Von Neumann [1]. These algebras have been used extensively in the classification of C^* and W^*–algebras, and with the help of them the problem of the isomorphism of dynamical systems was treated, as well as different problems in the spectral theory for automorphism groups, etc. Since the above–mentioned papers of Murray and Von Neumann the amount of work and the names of those who have kept in touch with the subject is immense. We refer here to the monograph by Pedersen [1] where the bibliography includes papers up to the late 1970s.

We want to emphasize that, though the object which will play a major role in this book (C^*–algebras associated with automorphism groups) should be supposed to be very well known to C^*–algebraists, the problems concerning it that are considered in this book are *non–traditional* C^*–algebra problems since they originate in the theory of the solvability of functional differential equations. To be specific, first of all we are interested in obtaining conditions of invertibility for *particular* elements in these algebras, discovering their spectral and Fredholm properties, describing compact elements, etc. We shall obtain a description of these algebras up to

*–algebraic (but not spatial) isomorphism. The last remark should be noted especially – the thing is that when constructing the symbolic calculus related to the treatment of the Fredholmity of different classes of equations we have to wipe off the initial space (when passing to the factor algebra modulo compact operators). Wiping off a space is also of extreme value in connection with the problem of the coincidence of spectra of operators given by the same formula in different functional spaces.

A number of methods and mathematical structures which are introduced and treated in the book take their origin in different fields of analysis (not initially having C^*–algebra character), such as dynamical systems, hyperbolicity theory, dichotomy, stability, ergodic theory, information theory, etc. Their usage is quite reasonable in view of the essence and nature of the problems considered. On the other hand the depth and high development of modern C^*–theory give us an opportunity to raise these methods up to the level where they can be used in a unified and effective manner for the investigation of various sorts of functional differential equation problems.

We point out that the reader will not find in this book any consideration of classical C^*–algebra theory problems and objects, such as, say, the classification of C^*–algebras, Von Neumann algebras theory, decomposition theory, spectral theory, and so on.

In fact the material of the book presents a synthesis of modern functional differential problems and methods, with modern C^*–algebra apparatus. Briefly, we are treating non–traditional objects in the classical theory of differential equations (but they are traditional in the theory of functional differential equations) with the help of non–traditional methods of classical C^*–algebra theory (but corresponding to the objects we consider).

Let us now say a few words about the structure of the book. It consists of two parts.

The first part is theoretical, where we provide the main C^*–apparatus for the class of objects considered. In the first chapter we introduce the class of operators (which we call weighted shifts) playing the role of the initial bricks of the theory. We obtain the spectral radius formula for them, describe some spectral properties and discuss different examples and problems, the general theory of which is developed in later chapters.

The second chapter contains one of the most important C^* results of the whole theory – the isomorphism theorem (in fact, a number of isomorphism theorems). The third chapter deals with different C^* results concerning the invertibility of elements in the algebras considered, and Chapter 4 describes the C^* scheme of the so–called localization method – a method of investigation of various types of functional differential equations with discontinuous coefficients. It seems to us that this method could not be applied to its full potential without an understanding of its internal operator algebraic nature provided here.

The second part presents applications and related topics. First of all we develop the symbolic calculus and Fredholmity conditions for various classes of functional differential equations with continuous coefficients. Then we proceed to abstract boundary–value problems and functional differential equations with discontinuous coefficients. After this is done we treat a number of non–Hilbert (and thus non–C^*–algebra) problems concerning functional operators in L^P–spaces, spaces of smooth functions, and so on. We finish the second part with a consideration of functional operators generated by non–invertible transformations of an argument and obtain a corresponding isomorphism theorem which gives us again an opportunity to obtain analogous results to those of Chapters 2 and 3, to construct the symbolic calculus, and to investigate the problem of the solvability of the examined equations.

2. Preliminaries. Miscellaneous results

In this book we shall use a number of notions and facts from several mathematical disciplines. So, for the readers convenience we present here the main definitions and statements (without proofs) which are necessary for the understanding of the text. The supposed level of mathematical maturity of the reader is approximately equal to a standard university course of algebra, analysis (including functional analysis), differential equations and boundary value problems (namely elliptic problems theory).

Normed spaces and linear operators

2.1 Let E, F be normed vector spaces. By $L(E, F)$ we denote the set of linear bounded operators A: $E \to F$ with domain the whole of E. If F is a Banach space then $L(E, F)$ with the norm

$$\|A\| = \sup_{\|\xi\|=1} \|A\xi\| \quad , \qquad \xi \in E$$

is also a Banach space.

If $E = F$ we denote $L(E, F)$ by $L(E)$.

2.2 A set M in a metric space is called *precompact* if any sequence of points of M contains a Cauchy subsequence.

An operator $A \in L(E, F)$ is called *compact* if the image of any bounded set of E is precompact in F. The set $K(E, F)$ of compact operators is a closed (in norm) vector subspace of $L(E, F)$.

$K(E)$ is a (left and right) ideal in $L(E)$, that is, if $K \in K(E)$ and $A \in L(E)$ then $KA, AK \in K(E)$.

2.3 The set of linear bounded functionals on E (that is $L(E, \mathbb{C})$) is called the *dual space* for E and is denoted by E^*.

The *–weak topology on E^* is the topology generated by the system of seminorms $\{p_u : p_u(f) = |f(u)|, u \in E\}$.

The closed unit ball in E^* is *–weakly compact.

An operator $A^* \in L(F^*, E^*)$ is called the *adjoint* to A if

$$(A^*f)(u) = f(Au) \quad , \qquad f \in F^* \quad , \quad u \in E \quad .$$

It is known that

$$\|A^*\| = \|A\| \quad .$$

2.4 The *kernel* of $A \in L(E, F)$ is a closed vector subspace

$$\text{Ker } A = \{u \in E : Au = 0\} \quad .$$

The *image* of A is the vector space

$$\text{Im } A = \{v \in F : \exists u \in E, \quad v = Au\} \quad .$$

2.5 An operator $A \in L(E, F)$ is said to be *invertible* if there exists an operator $B \in L(F, E)$ such that

$$AB = I_F \quad , \qquad BA = I_E$$

where $I_F(I_E)$ is the identity operator in $F(E)$.

The operator B is called the *inverse* of A and is denoted by A^{-1}.

The set of all invertible operators is an open set in $L(E, F)$.

A is invertible iff A^* is also invertible.

Theorem 2.6 (Banach's theorem on the inverse operator) *If* E *and* F *are Banach spaces and* Ker A = 0 *and* Im A = F *then* A *is invertible.*

Lemma 2.7 *If* $A \in L(E)$ *is invertible and* $B \in L(E)$ *satisfy the condition*

$$\|A^{-1}\| \; \|B\| < 1$$

then $(A - B)$ *is invertible and*

1. $$\|(A - B)^{-1}\| \le \frac{\|A^{-1}\|}{1 - \|A^{-1}\| \; \|B\|}$$

2. $$\|(A - B)^{-1} - A^{-1}\| \le \frac{\|B\| \cdot \|A^{-1}\|^2}{1 - \|A^{-1}\| \; \|B\|} \quad .$$

2.8 An operator $A \in L(E, F)$ is called *Fredholm* if

$$\dim \operatorname{Ker} A < \infty \qquad \text{and} \qquad \dim \operatorname{Ker} A^* < \infty$$

and Im A is closed.

The integer

$$\operatorname{ind} A = \dim \operatorname{Ker} A - \dim \operatorname{Ker} A^*$$

is called the *index* of the Fredholm operator A. (dim H is the dimension of a finite dimensional vector space H.)

The main properties of Fredholm operators are the following.

If A and B are Fredholm operators then $A \cdot B$ (if it is defined) is also Fredholm and

$$\operatorname{ind}(A \cdot B) = \operatorname{ind} A + \operatorname{ind} B \quad .$$

If A is Fredholm and K is compact then $A + K$ is Fredholm and

$$\operatorname{ind}(A + K) = \operatorname{ind} A$$

(stability under compact perturbations). In particular $I + K$ is Fredholm with zero index.

If A is Fredholm then there exists $\delta > 0$ such that for any operator B with $\|A - B\| < \delta$ we have:

$$B \text{ is Fredholm} \qquad \text{and} \qquad \operatorname{ind} B = \operatorname{ind} A$$

(stability under sufficiently small in norm perturbations).

If A_0 and A_1 are Fredholm operators from $L(E, F)$ which are homotopically equivalent, that is, there exists a continuous in norm set $\{A(t)\}_{0 \le t \le 1}$ of Fredholm operators, depending on t, such that $A(0) = A_0$ and $A(1) = A_1$, then

$$\operatorname{ind} A_0 = \operatorname{ind} A_1$$

(homotopic stability of the index).

A *regularizer* of $A \in L(E, F)$ is an operator $R \in L(F, E)$ such that

$$AR = I_F + K_1 \ , \qquad RA = I_E + K_2$$

where K_1 and K_2 are compact operators.

An operator A is Fredholm iff there exists a regularizer of it.

2.9 Let $A \in L(E)$ where E is a Banach space. The *spectrum* of A is the set $\sigma(A)$ of complex numbers λ such that $\lambda I - A$ is not invertible.

The *Fredholm spectrum* is the set $\sigma_F(A)$ of complex numbers λ such that $\lambda I - A$ is not Fredholm.

In some cases it is reasonable to consider the set $\sigma_{F_0}(A)$ of complex numbers λ such that $\lambda I - A$ is not Fredholm or $\operatorname{ind}(\lambda I - A) \neq 0$.

All the sets $\sigma(A)$, $\sigma_F(A)$, $\sigma_{F_0}(A)$ are not empty; they are closed and lie in the circle $|\lambda| \le \|A\|$. In fact

$$\sigma_F(A) \subset \sigma_{F_0}(A) \subset \sigma(A) \quad .$$

Lemma 2.10 *If λ_0 is a boundary point for σ(A) and λ_0 is not an isolated point of σ(A) then $\lambda_0 \in \sigma_F$(A).*

2.11 The *spectral radius* of $A \in L(E)$ is the number $r(A)$ defined by

$$r(A) = \max\{|\lambda|: \lambda \in \sigma(A)\} \quad .$$

The *Fredholm spectral radius* is the number $r_F(A)$ defined by

$$r_F(A) = \max\{|\lambda|: \lambda \in \sigma_F(A)\}$$

and in a corresponding way the number $r_{F_0}(A)$ can be introduced.

It is known that

$$r(A) = \lim_{n\to\infty} \|A^n\|^{\frac{1}{n}} \quad .$$

2.12 The spectrum is *upper semicontinuous*, that is, for a fixed A and any $\varepsilon > 0$ there exists a $\delta > 0$ such that the inequality $\|A - B\| < \delta$ implies that $\sigma(A)$ belongs to the ε–neighbourhood of $\sigma(A)$.

On the other hand, the spectrum is not continuous under small perturbations, that is, the spectrum may diminish under small perturbations of an operator.

The spectral radius is upper semicontinuous, that is, if $A_n \to A$ is norm then

$$\overline{\lim_{n\to\infty}}\, r(A_n) \le r(A) \quad .$$

2.13 The *direct sum* of a finite number of *Banach spaces* $E_1, \ldots, E_n$ is the Banach space E consisting of vectors $e = (e_1, \ldots, e_n)$, $e_k \in E_k$, with the norm $\|e\| = \max\|e_k\|$ (or any other equivalent norm). The direct sum is denoted by $\overset{n}{\underset{1}{\oplus}} E_k$.

If $E = \overset{n}{\underset{1}{\oplus}} E_k$ and $F = \overset{n}{\underset{1}{\oplus}} F_k$ then to an operator $A \in L(E, F)$ there corresponds the matrix $[A_{kj}]$ where $A_{kj} \in L(E_j, F_k)$ such that

$$(Ae)_k = \sum_j A_{kj} e_j \quad .$$

An operator A is said to be decomposed into the direct sum of operators A_k (with respect to fixed direct sums $E = \overset{n}{\underset{1}{\oplus}} E_k$ and $F = \overset{n}{\underset{1}{\oplus}} F_k$) if the corresponding matrix $[A_{kj}]$ is diagonal and $A_{kk} = A_k$. In this case we write $A = \overset{n}{\underset{1}{\oplus}} A_k$.

Theorem 2.14 (Riesz spectral decomposition) *Let* σ_1 *and* σ_2 *be two isolated and mutually supplementary parts of* $\sigma(A)$, $A \in L(E)$. *Let* $\mathscr{D}$ *be an open set of the complex plane for which*

$$\mathscr{D} \cap \sigma(A) = \sigma_1 \quad , \qquad [\mathbb{C} \setminus \overline{\mathscr{D}}] \cap \sigma(A) = \sigma_2$$

and $\partial\mathscr{D}$ *be its oriented boundary (consisting of a finite number of closed rectifiable curves). Then the operator*

$$P = \frac{1}{2\pi i} \int_{\partial\mathscr{D}} (\lambda I - A)^{-1} d\lambda$$

is a projection (the Riesz projection) and the subspaces $E_1 = \mathrm{Im}\, P$ *and* $E_2 = \mathrm{Ker}\, P$ *are both invariant under* A *and for the corresponding decomposition of* A *into the direct sum* $A = A_1 \oplus A_2$ *we have*

$$\sigma(A_1) = \sigma_1 \quad , \qquad \sigma(A_2) = \sigma_2 \quad .$$

2.15 Let H be a Hilbert space with an inner product $\langle\ ,\ \rangle$. For a subset $S \subset H$ we denote by $S^{\perp}$ the orthogonal complement of S, that is

$$S^{\perp} = \{\xi \in H : \langle \xi, h \rangle = 0,\ h \in S\} \quad .$$

The *cokernel* of $A \in L(H)$ is the set coker $A = (\mathrm{Im}\ A)^{\perp}$. This set coincides with Ker A^* (in this case we denote by A^* the operator on H defined by the equality

$$\langle Ah, \xi \rangle = \langle h, A^* \xi \rangle \quad , \qquad h, \xi \in H\ ) \quad .$$

Since for any bounded linear functional $f \in H^*$ there exists a unique $h_f \in H$ such that

$$f(\xi) = \langle \xi, h_f \rangle \quad , \qquad \xi \in H$$

we obtain the (antilinear) isomorphism $H^* \cong H$. That is why in the case of a Hilbert space we can assume that the adjoint operator deals in the same space H (cf. 2.3).

In a Hilbert space for every $A \in L(H)$ we have

$$\|AA^*\| = \|A^*A\| = \|A\|^2 \quad .$$

References for 2.1–2.15 are Banach [1], Dunford, Schwartz [1], Palais [1], Riesz, Sz.–Nagy [1], Rudin [1].

Measures, invariant measures, ergodicity

2.16 Throughout the book we shall use only σ–additive measures.

Let X be a topological space. The minimal σ–algebra of subsets of X containing all the open subsets of X is called the *Borel algebra.* A measure μ is called a *Borel measure* if it is defined on the Borel algebra.

Let μ and ν be two measures defined on the same set Y. ν is said to be *absolutely continuous with respect to* μ if every μ–measurable set is ν–measurable and $\mu(\omega) = 0$ implies $\nu(\omega) = 0$.

Theorem 2.17 (Radon–Nikodym derivative) *If* ν *is absolutely continuous with respect to* μ *then there exists a measurable function* ρ *such that for* *every measurable set $Z \subset Y$ we have*

$$\nu(Z) = \int_Z \rho(y)\, d\mu(y) \quad .$$

The function ρ is called the Radon–Nikodym derivative of ν with respect to μ and is denoted by $\rho = \frac{d\nu}{d\mu}$.

2.18 Let $\alpha : X \to X$ be a measurable mapping. A measure μ is said to be *invariant* with respect to α if $\mu(\alpha^{-1}(\omega)) = \mu(\omega)$ for any measurable set ω.

An invariant probability (that is, $\mu(X) = 1$) measure is called *ergodic* if $\mu(\omega) = 0$ or $\mu(\omega) = 1$ for any measurable set ω such that $\mu(\omega \,\Delta\, \alpha^{-1}(\omega)) = 0$ (here $\omega_1 \,\Delta\, \omega_2 = (\omega_1 \backslash \omega_2) \cup (\omega_2 \backslash \omega_1)$ is the symmetric difference of the sets ω_1 and ω_2).

If X is a topological space then the *support* of measure μ (denoted by supp μ) is the smallest closed set F such that $\mu(X \backslash F) = 0$.

If X is a compact space then the dual space (see 2.3) to $C(X)$ is the set of all the finite complex Borel measures on X.

Theorem 2.19 (Birkhoff–Hinchin ergodic theorem) *Let μ be an invariant probability measure with respect to α: $X \to X$. Then for every $f \in L^1_\mu(X)$ we have*

$$\lim_{n\to\infty} \frac{1}{n} \sum_{k=1}^{n-1} f(\alpha^k(x)) = \bar{f}(x)$$

where the limit exists μ–almost everywhere and in the $L^1_\mu(X)$ sense.

The function $\bar{f}(x)$ has the properties:

(*i*) $\bar{f}(\alpha(x)) = \bar{f}(x)$,

(*ii*) $\bar{f} \in L^1_\mu(X)$ *and* $\int_X \bar{f}(x)d\mu = \int_X f(x)d\mu$.

In particular if μ is ergodic we have

$$\bar{f}(x) \equiv \int_X f(x)d\mu \quad .$$

Lemma 2.20 *Let* μ *be an element of the set* Λ_0 *of all the invariant (with respect to* α*) probability measures. Then* μ *is an extremal point in* Λ_0 *iff it is ergodic.*

Theorem 2.21 (Choquet–Bishop–de Leeuw ergodic decomposition theorem) *Let* X *be a compact space and* μ *be an element of the set* Λ *of all the invariant (with respect to* α*) measures. Then there exists a probability measure* m *on the set* Λ *of all the ergodic measures such that*

$$\int_X f \, d\mu = \int_\Lambda \nu(f) dm(\nu) \quad .$$

References for **2.16–2.21** are Bourbaki [1], Dunford, Schwartz [1], Halmos [1], Kornfel′d, Sinaj, Fomin [1], Krengel [1], Phelps [1], Naimark [1], Walters [1].

Groups, *G*–spaces, dynamical systems

2.22 Let G be a *topological group*, that is,

(1) G is a group;
(2) G is a topological space;
(3) the group operations are continuous (with respect to the given topology).

2.23 We shall say that there is defined an *action of a group* G *on a set* X if there is defined the mapping

$$\Phi : G \times X \to X$$

such that

$$\Phi(e, x) = x \ , \qquad x \in X \ , \qquad e \text{ is the identity of } G$$
$$\Phi(gh, x) = \Phi(g, \Phi(h, x)) \ , \qquad g, h \in G \ , \quad x \in X \quad .$$

If the action of G is fixed then for a given $g \in G$ the mapping

$$x \to \Phi(g, x) \qquad x \in X$$

is denoted by $g(x)$. Thus Φ defines a homomorphism of G into the group of transformations of X:

$$g(h(x)) = (g \cdot h)(x) \quad .$$

If G is a topological group and X is a topological space then it is assumed that the mapping $\Phi\colon G \times X \to X$ is continuous.

If X is a space with a measure then it is assumed that all the mappings $g(x)$, $g \in G$, are measurable.

If X is a normed vector space and $g(x)$ are linear bounded operators then the group action is called the *linear representation* of G. The operators $g(x)$ in this situation depend continuously on g in the strong topology. The linear representation of G is a homomorphism of G into the group of invertible elements of $L(X)$.

If X is a Hilbert space and all $g(x)$ are unitary operators then this representation is called a *unitary representation.*

2.24 A *character* of a group G is a homomorphism of G into the group S^1 of all complex numbers λ such that $|\lambda| = 1$.

If G is a commutative group then by $\hat{G}$ is denoted its *dual group*, the set of all continuous characters of G equipped with the compact open topology.

Theorem 2.25 (Pontryagin duality theorem) *Let* G *be a locally compact abelian group. Then*

$$\mathrm{G} \cong \hat{\hat{\mathrm{G}}}$$

and the isomorphism is given by $\mathrm{g} \to \mathrm{f}_{\mathrm{g}}$ *where*

$$\mathrm{f}_{\mathrm{g}}(\chi) = \chi(\mathrm{g}) \quad , \qquad \mathrm{g} \in \mathrm{G} \quad , \quad \chi \in \hat{\mathrm{G}} \quad .$$

2.26 A Borel measure μ on a topological group G is said to be left invariant (right invariant) if all the mappings $g \to hg (g \to gh)$. $h \in G$ preserves μ.

Theorem 2.27 *Let* G *be a locally compact group. Then there exists a left invariant measure* μ *on* G *which is defined up to a constant multiplier.*

This measure is called the (left) Haar measure *on* G.

If G is compact then $\mu(G) < \infty$ and μ is defined uniquely by the condition $\mu(G) = 1$.

If G is a discrete group then the Haar measure is defined on finite subsets of G and for each such subset M, $\mu(M) = |M|$, the number of element in M.

The same statements are true for the right Haar measure.

A right Haar measure is linked with a left Haar measure by the existence of a continuous homomorphism Δ (the *modular function*) of G into the positive real numbers such that $\mu(\Omega g) = \Delta(g)\ \mu(\Omega)$ (where μ is a left Haar measure).

References to **2.22–2.27** are Bredon [1], Bourbaki [2], Karnfel′d, Sinaj, Fomin [1], Hewitt, Ross [1, 2], Pontryagin [1].

Vector bundles

2.28 A complex (real) *vector bundle* over a topological space X is a triple (E, X, p), where E and M are topological spaces and p: $E \to X$ is a continuous surjective mapping satisfying the following axioms:

(i) the inverse image $p^{-1}(x)$ for every $x \in X$ has the structure of a finite dimensional vector space over the field $\mathbb{C}$ (or $\mathbb{R}$);

(ii) for every point x there exists a neighbourhood U such that $p^{-1}(U)$ is homeomorphic to $U \times \mathbb{C}^N (U \times \mathbb{R}^N)$ and the homeomorphism φ: $p^{-1}(U) \to U \times \mathbb{C}^N\ (U \times \mathbb{R}^N)$ commutes with the projection p and is linear on every $p^{-1}(x)$.

Property (ii) is called the *local triviality* of a vector bundle.

E is called the *vector bundle space*, X is the *base of the vector bundle*, and p is the projection. A vector space $p^{-1}(x)$ is called a *fibre* over x, and is denoted by E_x. Often for brevity, instead of (E, X, p), one writes simply E which is also called a vector bundle.

Let (E_1, X, p_1) and (E_2, X, p_2) be two vector bundles. A *homomorphism of vector bundles* is a mapping φ: $E_1 \to E_2$ such that $p_1 = p_2 \circ \varphi$, that is, a fibre $p_1^{-1}(x)$ is mapped into $p_2^{-1}(x)$ and the mapping of fibres φ_x: $E_{1,x} \to E_{2,x}$ generated by φ is linear.

A homomorphism φ of vector bundles is called an isomorphism if there exists an inverse homomorphism. Two bundles are said to be isomorphic if there exists an isomorphism between them. If a vector bundle is isomorphic to the product bundle $X \times \mathbb{C}^N (X \times \mathbb{R}^N)$ then it is called *trivial.*

A subbundle of (E, X, p) is a subspace $E_1 \subset E$ which is in turn a vector bundle over X with the projection p.

A bundle E is said to be decomposed into a direct sum (the *Whitney sum*) of subbundels E_1 and E_2 (this is denoted by $E = E_1 \oplus E_2$) if any $\xi \in E_x$ is decomposed uniquely into the sum $\xi = \xi_1 + \xi_2$ where $\xi_1 \in E_{1x}$, $\xi_2 \in E_{2x}$.

The direct sum of bundles (E_1, X, p_1) and (E_2, X, p_2) is a space consisting of pairs (ξ, η), $\xi \in E_1$, $\eta \in E_2$ such that $p_1(\xi) = p_2(\eta)$. The topology in this set is induced from $E_1 \times E_2$, the base is the same space X, and the projection is defined in a natural way.

A *cross–section* of a bundle (E, X, p) is a continuous mapping u: $X \to E$ such that $p(u(x)) = x$. Continuous cross–sections of E form the vector space $\Gamma(E)$. If on every fibre E_x there is defined a norm then the set of bounded cross–sections in $\Gamma(E)$ can be equipped with the natural norm

$$\|u\| = \sup_X \|u(x)\|_x \quad .$$

If E is a trivial bundle of dimension N (the dimension of a vector bundle is the dimension of its fibres) then $\Gamma(E)$ is isomorphic to $[C(X)]^N$, the space of continuous vector functions of dimension N over X.

2.29 Let (E, X, p) be a vector bundle of dimension N. The set of all homomorphisms φ: $E \to E$ forms an algebra denoted by HOM(E). If by Hom E we denote the N^2–dimensional bundle over X the fibre of which over a point x is the space Hom E_x of all the linear mappings of E_x into itself then HOM(E) is isomorphic to the algebra Γ(Hom E) of all the continuous cross–sections of Hom E. If for every fibre E_x there is given a norm continuously depending on x, and X is a compact space, then one can define the norm on HOM(E) by

$$\|\varphi\| = \sup_X \|\varphi_x\|_x$$

here $\|\varphi_x\|_x$ is the norm of the linear operator $\varphi_x \colon E_x \to E_x$.

2.30 (E, X, p) is said to be given a *Hermite metric* if on every E_x an inner product $\langle \xi, \eta \rangle_x$ is defined continuously depending on x.

2.31 Given a bundle (E, X, p) and a continuous mapping $f\colon Y \to X$ then one can construct the bundle over Y as a subspace of $Y \times E$ consisting of points (y, ξ) such that $f(y) = p(\xi)$ with the projection $\tilde{p}(y, \xi) = y$. Under this definition the fibre over y is naturally isomorphic to $E_{f(y)}$. This bundle is denoted by $f^* E$.

References for 2.28–2.31 are Atiyah [1], Dubrovin, Novikov, Fomenko [1], Husemoller [1], Mischenko [1], Palais [1], Seeley [1].

Manifolds and smooth mappings

2.32 A Hausdorff topological space X is called a *manifold* if for every point $x \in X$ there exists a neighbourhood U which is homeomorphic to an open set of $\mathbb{R}^n$.

A *map* (a local coordinate system) of X is a pair (U, φ) consisting of an open set $U \subset X$ and a homeomorphism $\varphi\colon U \to \varphi(U)$ of U onto an open set $\varphi(U)$ of $\mathbb{R}^n$.

The coordinates $\varphi(x) \in \mathbb{R}^n$ are called the *local coordinates* of x. A fixed set of maps covering the whole of X is called an *atlas*.

Let (U, φ) and (V, ψ) be two maps of X. They are said to be k–smoothly linked if the mappings

$$\varphi \circ \psi^{-1} : \psi(U \cap V) \to \varphi(U \cap V)$$

and

$$\psi \circ \varphi^{-1} : \varphi(U \cap V) \to \psi(U \cap V)$$

belong to C^k, that is, they are k times continuously differentiable.

A manifold X is said to be k–smooth (belong to class C^k) if there is given an atlas consisting of k–smoothly linked maps.

2.33 Let X and Y be C^k–class manifolds. A mapping $\alpha: X \to Y$ is called l–smooth or a *mapping of class* C^l, where $l \leq k$, if for any maps (U, φ) on X and (V, ψ) on Y the mapping $\psi \circ \alpha \circ \varphi^{-1}$ is of class C^l on its domain.

A bijection $\alpha: X \to Y$ is called a *diffeomorphism* of class C^l if both α and α^{-1} are smooth mappings of class C^l.

2.34 We say that a *tangent vector* is defined in $x_0 \in X$ if, to every map containing x_0, there corresponds n numbers $(\xi_1(U), \xi_2(U), \ldots, \xi_n(U))$ in such a way that when passing from local coordinates (U, x) to local coordinates (V, y) these numbers are transformed by the formula

$$\xi_i(V) = \sum_{j=1}^{n} \frac{\partial y_i}{\partial x_j}(x_0)\, \xi_j(U)$$

where $\left[\frac{\partial y_i}{\partial x_j}\right]$ is the Jacobi matrix of the mapping $x \to y$.

The set T_x of all the tangent vectors in x has the structure of an n–dimensional real vector space and is called the *tangent space* to X in x. The set of all the tangent spaces

$$TX = \bigcup_X T_x$$

has the natural topology and is an n–dimensional vector bundle over X which is called the *tangent bundle* of X.

2.35 A linear functional on a tangent space T_x in x is called a *cotangent vector* to X in x. The space T_x^* of all the cotangent vectors in x is an n–dimensional vector space. The set of all the cotangent vectors to X forms the *cotangent bundle* T^*X.

If one uses local coordinates then a cotangent vector to X at a point x_0 is a mapping $x_0 \in U \to (\eta_1(U), \ldots, \eta_n(U))$ such that when passing from local coordinates (U, x) to (V, y) the numbers η_i are transformed by the formula

$$\eta_i(V) = \sum_{j=1}^{n} \frac{\partial x_j}{\partial y_i}(x_0)\ \eta_j(U) \quad .$$

2.36 If f: $X \to \mathbb{R}$ is a smooth function on X then the mapping which maps a tangent vector $\xi \in T_x$ into the derivative in the direction of this vector is a linear functional on T_x, thus defining the cotangent vector. This cotangent vector is called the *differential* of f and is denoted by df.

2.37 Let X and Y be smooth manifolds and g: $X \to Y$ be a smooth mapping. Then (by the foregoing argument) the mapping dg: $T_x \to T_{g(x)}$ is defined, which in local coordinates is determined as multiplication by the Jacobi matrix. This mapping is called the *differential* of g.

2.38 A diffeomorphism g: $X \to Y$ generates the mapping ∂g: $T^*X \to T^*Y$ which is called the *codifferential* of g. It is defined by the formula

$$(\partial gf)(\eta) = f(\mathrm{d}h(\eta))$$

where $h = g^{-1}$, $f \in T_x^*$, $\eta \in T_{g(x)}$.

References for 2.32–2.38 are Bourbaki [3], Dubrovin, Novikov, Fomenko [1], Husemoller [1], Mischenko [1], Palais [1].

Singular integral operators, pseudodifferential operators

The theory of singular integral operators (which are called pseudodifferential operators in the n–dimensional case $n > 1$) is now accepted as a classical part of analysis, so we sketch some parts of it, briefly extracting only the results necessary for our purposes.

2.39 The *Sobolev space* $H^s(\mathbb{R}^n)$, $s \in \mathbb{R}$ can be defined as a completion of a space of rapidly decreasing infinitely differentiable functions with respect to the norm

$$\|f\|_s = \sqrt{\int |\hat{f}(\xi)|^2(1 + |\xi|^2)^s \mathrm{d}\xi} \tag{2.1}$$

where $\hat{f}$ is the *Fourier transform* of *f*:

$$\hat{f}(\xi) = \frac{1}{(2\pi)^{n/2}} \int e^{-i(x,\xi)} f(x)dx \quad . \tag{2.2}$$

Since for $L^2(\mathbb{R}^n)$ we have

$$\langle f, g\rangle = \int f\cdot\bar{g}\ dx = \int \hat{f}\cdot\bar{\hat{g}}\ d\xi$$

it follows that $H^0(\mathbb{R}^n) = L^2(\mathbb{R}^n)$.

It is known that $H^{s_1}(\mathbb{R}^n) \subset H^{s_2}(\mathbb{R}^n)$ when $s_1 > s_2$. In particular the spaces $H^s(\mathbb{R}^n)$ for $s > 0$ are vector subspaces of $L^2(\mathbb{R}^n)$. For $s < 0$ the space $H^s(\mathbb{R}^n)$ with the help of the inner product in $L^2(\mathbb{R}^n)$ can be identified with the space of linear bounded functionals on $H^{-s}(\mathbb{R}^n)$ since for any smooth function f we have

$$\|f\|_s = \sup_{\|g\|_{-s}\le 1} \left|\int f(x)g(x)dx\right| \quad .$$

Let $H^{+\infty}(\mathbb{R}^n) = \cap\ H^s(\mathbb{R}^n)$ be the projective limit and $H^{-\infty}(\mathbb{R}^n) = \cup\ H^s(\mathbb{R}^n)$ be the inductive limit of $H^s(\mathbb{R}^n)$.

$H^{+\infty}(\mathbb{R}^n)$ is norm dense in every $H^s(\mathbb{R}^n)$.

Let A: $H^{+\infty}(\mathbb{R}^n) \to H^{+\infty}(\mathbb{R}^n)$ be a linear operator.

It is said that the operator A has the *order* r if for every s there exists a constant C_s such that

$$\|Af\|_{s-r} \le C_s\|f\|_s \quad .$$

In this case for every s the operator A can be uniquely extended by continuity up to the bounded operator from $H^s(\mathbb{R}^n)$ into $H^{s-r}(\mathbb{R}^n)$.

In particular the operator

$$D^\alpha = \left(\frac{\partial}{\partial X}\right)^\alpha = \frac{\partial^{|\alpha|}}{\partial X_1^{\alpha_1}\ldots\partial X_n^{\alpha_n}}$$

where $\alpha = (\alpha_1,\ldots,\alpha_n)$, $\alpha_k \in \mathbb{N}^+ \cup \{0\}$ and $|\alpha| = \alpha_1 + \ldots + \alpha_n$ is an operator of order $|\alpha|$ and is not an operator of order r for $r < |\alpha|$.

Note that the operator $\Lambda^2 = I + \Delta$ where

$$\Delta = -\sum_{i=1}^{n} \frac{\partial^2}{\partial x_i^2}$$ — the Laplace operator establishes the isometry

$H^s(\mathbb{R}^n) \to H^{s-2}(\mathbb{R}^n)$.

2.40 Sobolev spaces on vector bundles are obtained from $H^k(\mathbb{R}^n)$ by transplanting.

Let (E, X, p) be a vector bundle over a C^∞ smooth compact manifold X. We shall assume that on X some Riemann metric is fixed and on E is fixed some C^∞ Hermite metric $\langle\xi, \eta\rangle_x$. The Riemann metric induces a C^∞ smooth measure $d\mu$ on X. Thus we obtain the inner product for C^∞ cross–sections

$$\langle\xi, \eta\rangle = \int_X \langle\xi(x), \eta(x)\rangle_x \, d\mu \qquad \xi, \eta \in \Gamma(E) \quad . \tag{2.3}$$

If f is some C^∞ function with support $U \subset X$ and φ is a coordinate mapping of U onto $V \subset \mathbb{R}^n$ then by f_φ we denote the mapping from the set of functions defined on X into the set of functions defined on $\mathbb{R}^n$ which is given by

$$(f_\varphi g)(y) = \begin{cases} f(\varphi^{-1}(y))\, g(\varphi^{-1}(y)) , & y \in V \\ 0 , & y \bar{\in} V \end{cases} . \tag{2.4}$$

If $s \geq 0$ then by $H^s(E)$ we denote the set (of equivalence classes) of cross–sections ξ of E such that $f_\varphi(\langle\xi, \beta_j\rangle) \in H^s(\mathbb{R}^n)$ for every local basis $\{\beta_j\}_{j=1}^m$ on the domain of φ (such a basis does exist in view of the local triviality of E) and every C^∞ smooth function f having a support in the domain of φ.

$H^s(E)$ can be given a Banach space structure if one uses an infinitely smooth partition of the identity $\sum f_i = 1$ subordinate to the finite atlas

$(U_i, \varphi_i)_{i=1}^p$ and an orthonormal base $\{\beta_j^i\}$ on the support of every φ_i

$$(\|\xi\|_s)^2 = \sum_{i,j} \|f_{i\varphi_i}(\langle\xi, \beta_j^i\rangle)\|_s^2 \quad .$$

$H^0(E) = L^2(E)$ is a Hilbert space with the inner product

$$\langle\xi, \eta\rangle = \int_X \langle\xi, \eta\rangle_x \; d\mu \quad .$$

If $s < 0$ then $H^s(E)$ is defined as the antidual space for $H^{-s}(E)$.

$$H^{+\infty}(E) = \cap \; H^s(E)$$

where $H^{+\infty}(E)$ is dense in every $H^s(E)$.

$H^{+\infty}(E)$ coincides with the space of all C^∞ smooth cross–sections of E.

The antidual to it, $H^{-\infty}(E) = \cup \; H^s(E)$, in fact consists of all the distributional cross–sections of E.

A linear mapping A: $H^{+\infty}(E) \to H^{+\infty}(E)$ is said to be an operator of order $\leq r$ if, for every s, A can be extended up to a continuous mapping $\overline{A}_s$: $H^s(E) \to H^{s-r}(E)$.

2.41 To define *pseudodifferential operators* (PDOs) on $H^s(E)$ we first have to define them on $H^s(\mathbb{R}^n)$ and then simply transplant them onto $H^s(E)$. Since we are interested mainly in the Fredholm theory of such operators we restrict ourselves to the definition of PDOs modulo compact summands and to the consideration of the so–called principal symbol of an operator (which is a generalization of the principal part of the characteristic polynomial of a differential operator). We begin with operators of order 0.

Let $a(x, \xi)$ be a C^∞ function on $\mathbb{R}^n \times [\mathbb{R}^n \backslash 0]$ which is zero degree positive homogeneous with respect to ξ. We shall assume also that there exists a constant C such that $a(x, \xi)$ does not depend on x when $|x| > C$. Thus for $|x| > C$ the function $a(x, \xi)$ coincides with some function $a(\infty, \xi)$.

With the function $a(x, \xi)$ we can associate the operator $A = a(x, D)$ acting from $C_0^\infty(\mathbb{R}^n)$ (the space of infinitely differentiable functions with compact supports) into $C^\infty(\mathbb{R}^n)$ by the formula

$$Au(x) = \frac{1}{(2\pi)^{\frac{n}{2}}} \int_{\mathbb{R}^n} e^{i<x,\eta>} a(x, \eta) \hat{u}(\eta)d\eta \quad . \tag{2.5}$$

This operator is of zero order, that is, for every s there exists a constant $C(s)$ such that

$$\|Au\|_s \leq C(s)\|u\|_s \quad .$$

Thus we can extend it by continuity up to an operator in $H^{-\infty}(\mathbb{R}^n)$.

Let $\mathscr{A}$ be the uniform closure of the algebra of operators in $L^2(\mathbb{R})$ generated by the operators of the form (2.5) and compact operators.

Let $\bar{\mathbb{R}}^n = \mathbb{R}^n \cup \{\infty\}$ be the one–point compactification of $\mathbb{R}^n$, S^{n-1} be the unit sphere, and $M = \bar{\mathbb{R}}^n \times S^{n-1}$.

For any operator A defined by (2.5) let us consider the function

$$\sigma(A)\colon M \to \mathbb{C}$$

which is the restriction of $a(x, \xi)$ onto M.

The mapping σ_0: $A \to \sigma(A)$ can be extended by continuity up to the mapping σ: $\mathscr{A} \to C(M)$ (which is called the mapping of taking a symbol) and the function $\sigma(A) \in C(M)$ is called the *(principal) symbol* of a pseudodifferential operator $A \in \mathscr{A}$.

The principal symbol has the properties:

(1) $\sigma(A_1 + A_2) = \sigma(A_1) + \sigma(A_2)$;

(2) $\sigma(A_1 \cdot A_2) = \sigma(A_1) \cdot \sigma(A_2)$;

(3) $\sigma(A) = 0$ iff A is compact ;

(4) $\sigma(A)$ is invertible iff A is Fredholm ;

(5) $\sigma(A^*) = \sigma(A)^*$.

Remarks

(1) In an analogous way one can define psuedodifferential operators of zero order in the space $L^2(\mathbb{R}^n, \mathbb{C}^m)$ of m–dimensional vector functions. In this case the symbol will be an $m \times m$ continuous matrix function on M.

(2) For every $r \in \mathbb{R}$ let us consider the operator

$$(\Lambda^r f)(x) = \frac{1}{(2\pi)^{\frac{n}{2}}} \int e^{i<x,\xi>}(1 + |\xi|^2)^{\frac{r}{2}} \hat{f}(\xi)d\xi \quad .$$

This operator has the order r and for every s establishes the isometry $H^s(\mathbb{R}^n) \to H^{s-r}(\mathbb{R}^n)$. Pseudodifferential operator of order r can be defined as an operator of the form $A = A_0\Lambda^r$, where A_0 is an operator of zero order defined by (2.5). The *principal symbol* for A is defined as a positively homogeneous of degree r function on $\mathbb{R}^n \times \{\mathbb{R}^n\backslash 0\}$ of the form

$$\sigma(A)(x,\xi) = \sigma(A_0)(x,\xi)\|\xi\|^r$$

(here we regard $\sigma(A_0)(x,\xi)$ as a positively homogeneous of zero degree function on $\mathbb{R}^n \times \{\mathbb{R}^n\backslash 0\}$ which is obtained from the above–introduced symbol of zero order pseudodifferential operators by extension by homogeneity from $\mathbb{R}^n \times S^{n-1}$).

Passing to the algebra $\mathscr{A}$ and m–dimensional vector functions, it is reasonable to define the *principal symbol of a pseudodifferential operator of order* r as a matrix function (which is convenient to consider as an operator valued function)

$$\sigma(A)(x,\ \xi):\ \mathbb{C}^m \to \mathbb{C}^m \quad , \qquad (x,\ \xi) \in \mathbb{R}^n \times \{\mathbb{R}^n\backslash 0\}$$

which is r degree positive homogeneous with respect to $\xi \in \mathbb{R}^n\backslash 0$.

In this case the principal symbol has properties analogous to the above–mentioned properties (1)–(5).

2.42 Let us describe now *pseudodifferential operators on a vector bundle* $E = (E,\ X,\ p)$.

For any coordinate mapping $X \supset U \overset{\varphi}{\to} V \subset \mathbb{R}^n$ and any C^∞ smooth function f on X having support in U we define by f^φ the mapping from the set of functions defined on $\mathbb{R}^n$ into the set of functions defined on X given by the formula

$$f^\varphi(g)(x) = \begin{cases} f(x)\ g(\varphi(x)) \quad , & x \in U \\ 0 & x \bar{\in} U \quad . \end{cases} \tag{2.6}$$

We say that $A\colon H^s(E) \to H^{s-r}(E)$ is a PDO of order r in $H^s(E)$ if

(i) for any U and any C^∞ smooth functions f, h on X having supports in U

$$fAh = f^{\varphi}A_U h_{\varphi} + fKh$$

where A_U is a standard PDO of order r on $\mathbb{R}^n$ (see Remark (2) in 2.41), $K\colon H^s(E) \to H^{s-r}(E)$ is a compact operator, and f_{φ}, f^{φ} are defined by (2.4) and (2.6).

(ii) for any C^∞ smooth functions f,h on X having non–intersecting supports the operator

$$fAh\colon H^s(E) \to H^{s-r}(E)$$

is compact.

2.43 Now we pass to the principal symbol of a pseudodifferential operator on a vector bundle.

By T_0^*X we denote the subbundle of T^*X consisting of non–zero cotangent vectors (see 2.35).

Let $\pi\colon T_0^*X \to X$ be the canonical projection and set $E_1 = \pi^*E$ (see 2.31).

The *principal symbol* of a PDO A of order r acting on E is a homomorphism

$$\sigma(A)\colon E_1 \to E_1$$

which is r degree positive homogeneous and in local coordinates is the transplantation of the principal symbol of A_U mentioned in 2.42 (i).

In particular if E is trivial then the principal symbol is given by an $m \times m$ matrix function on T_0^*X (where m is the dimension of E).

Theorem 2.44 *The principal symbol possesses the properties:*

(1) $\sigma(A_1 + A_2) = \sigma(A_1) + \sigma(A_2)$;

(2) $\sigma(A_1 \cdot A_2) = \sigma(A_1)\ \sigma(A_2)$;

(3) $\sigma(A) = 0$ *iff* A *is compact in the corresponding Sobolev spaces;*

(4) $\sigma(\mathrm{A})$ *is invertible iff* A *is Fredholm in the corresponding Sobolev spaces;*

(5) *an operator* A^* *is also a PDO of order* r (*the same as for* A) *and* $\sigma(\mathrm{A}^*) = \sigma(\mathrm{A})^*$.

Pseudodifferential operators with invertible symbols are called *elliptic* operators, so property (4) of the theorem can be formulated as

A is Fredholm iff A is elliptic.

In addition to property (4) we have

(6) *if* A_1 *and* A_2 *are elliptic and* $\sigma(\mathrm{A}_1) = \sigma(\mathrm{A}_2)$ *then*

$$\mathrm{ind}\ \mathrm{A}_1 = \mathrm{ind}\ \mathrm{A}_2.$$

Which means that the principal symbol in fact solves the main problems in the Fredholm theory of pseudodifferential operators.

2.45 Now we restrict ourselves to the case of a zero–order PDO acting on $H^0(E) = L^2(E)$.

We denote by $\mathcal{A}^0$ the uniform closure of the set of all zero–order PDOs acting on $H^0(E)$. Let S^*X be the bundle of unit vectors of T^*X, π: $S^*X \to X$ be the canonical projection, and $E_2 = \pi^*E$.

In view of the structure of the principal symbol for zero–order operators we can assume that

$$\sigma(A)\colon E_2 \to E_2$$

for every zero–order operator A.

Theorem 2.46

(i) *The above–defined mapping* $\sigma(\,\cdot\,)$ *can be extended up to the mapping*

$$\sigma\colon \mathcal{A}^0 \to \mathrm{HOM}(E_2)$$

which is onto.

(ii) *The set of compact operators* $K = K(H^0(E))$ *belongs to* $\mathcal{A}^0$.

(iii) $\sigma(K) = 0$ *and the corresponding canonical mapping*

$$\sigma\colon \mathcal{A}^0/_K \to \mathrm{HOM}(E_2)$$

establishes an algebraic **–isomorphism, that is,*

$$\sigma(A_1 + A_2) = \sigma(A_1) + \sigma(A_2)\ ; \qquad \sigma(A_1 \cdot A_2) = \sigma(A_1)\sigma(A_2)\ ,$$
$$\sigma(A^*) = [\sigma(A)]^* \qquad (A \in \mathcal{A}^0 \text{ iff } A^* \in \mathcal{A}^0)$$

(iv) $A \in \mathcal{A}^0$ *is Fredholm iff* $\sigma(A)$ *is invertible.*

(v) *Let* A_0 *and* A_1 *be elliptic operators* (see **2.44**) *in* $\mathcal{A}^0$ *and suppose there exists a homotopy* F_t, $t \in [0, 1]$, *connecting* $\sigma(A_0)$ *with* $\sigma(A_1)$ *in* $\mathrm{HOM}(E_2)$ *such that all* F_t, $t \in [0, 1]$, *are invertible. Then*

$$\mathrm{ind}\ A_0 = \mathrm{ind}\ A_1 \quad .$$

(vi) $\|\sigma(A)\| = |||A|||$ *where* $|||\cdot|||$ *is the essential norm*

$$|||A||| = \inf_{k \in K} \|A + k\| \quad .$$

2.47 In fact, though we have not yet stressed it, one of the main properties of PDOs (which makes the definition 2.42 correct and not depend on any particular choice of atlas on a manifold) is the invariance of the set of PDOs with respect to transformation of an argument.

So let us again first formulate this property with respect to PDOs on $\mathbb{R}^n$ and then transplant it onto PDOs on a vector bundle.

Consider a diffeomorphism $\alpha\colon \mathbb{R}^n \to \mathbb{R}^n$ which is linear out of some ball, that is, there exists $C > 0$ such that when $\|x\| > c$, $\alpha(x) = Lx + h$, where

L is a matrix and $h \in \mathbb{R}^n$. We denote by $U_\alpha\colon L^2(\mathbb{R}^n) \to L^2(\mathbb{R}^n)$ the operator of the form

$$(U_\alpha f)(x) = f(\alpha^{-1}(x)) \quad .$$

It is known that if A is a PDO of order r (see **2.41** Remark (2)) with the symbol $\sigma(A)(x, \xi)$ then

$$U_\alpha A U_\alpha^{-1} = B + K$$

where B is a PDO with the symbol

$$\sigma(B)(x, \xi) = \sigma(A)(\alpha(x), J_\alpha \xi)$$

and J_α is the inverse of the transpose of the Jacobi matrix of α, and K is a compact operator.

To define an anologous to U_α operator in a vector bundle case we need to introduce a certain additional structure. Let $E = (E, X, p)$ be a vector bundle over a compact manifold X and $\alpha\colon X \to X$ be a diffeomorphism. Let also be given a diffeomorphism $\theta\colon E \to E$ such that the restriction θ_x on a fibre E_x maps this fibre linearly and bijectively onto $E_{\alpha(x)}$ (in this case we say that θ is a *linear extension* of α). The operator $U_{\alpha,\theta}$ (dealing in the spaces of cross–sections of E) is defined by the formula

$$(U_{\alpha,\theta} f)(x) = \theta[f(\alpha^{-1}(x))] \quad .$$

Let $E_1 = \pi^* E$ be the bundle over $T_0^* X$ induced by the canonical projection (see 2.43). The codifferential $\partial\alpha\colon T^* X \to T^* X$ and the mapping θ generate the mapping $\tilde{\theta}_\alpha\colon E_1 \to E_1$ which in local coordinates (x, ξ, z) $x \in X$, $\xi \in T_x^*$, $z \in E_x$ is given by the formula

$$\tilde{\theta}_\alpha(x, \xi, z) = (\partial\alpha(x, \xi), \theta(z)) \quad .$$

The codifferential $\partial\alpha$ also generates the mapping $\widetilde{\partial\alpha}: S^*X \to S^*X$ by the equation

$$\widetilde{\partial\alpha}(x, \xi) = \frac{1}{\|\partial\alpha(x, \xi)\|} \partial\alpha(x, \xi) \quad .$$

Let E_2 be the restriction of E_1 onto S^*X (see 2.45) and $\tilde{\tilde{\theta}}_\alpha: E_2 \to E_2$ be the mapping given in local coordinates by the formula

$$\tilde{\tilde{\theta}}_\alpha(x, \xi, z) = (\widetilde{\partial\alpha}(x, \xi), \theta(z)) \quad .$$

Theorem 2.48 *If* A *is a PDO of order* r *on a vector bundle* E *then the operator* $\mathrm{B} = \mathrm{U}_{\alpha,\theta}\mathrm{A}\mathrm{U}_{\alpha,\theta}^{-1}$ *is a PDO of order* r *and*

$$\sigma(\mathrm{B}) = \tilde{\theta}_\alpha \circ \sigma(\mathrm{A}) \circ \tilde{\theta}_\alpha^{-1} \quad .$$

In particular if r = 0 *then*

$$\sigma(B) = \tilde{\tilde{\theta}}_\alpha \circ \sigma(\mathrm{A}) \circ \tilde{\tilde{\theta}}_\alpha^{-1} \quad .$$

2.49 In the case when a manifold X is a curve on a plane (i.e. one–dimensional) pseudodifferential operators coincide with the classical singular integral operators which are often introduced in other ways.

Let Γ be a smooth, closed curve on a complex plane. The *Cauchy singular integral* operator S in $L^2(\Gamma)$ is the operator defined by

$$(Sf)(t) = \frac{1}{\pi i} \int_\Gamma \frac{\varphi(\tau)}{\tau - t}\, d\tau \quad , \qquad t \in \Gamma \tag{2.7}$$

where, on the right, is a principal value integral.

This operator has the following properties:

(i) $S^2 = I$

(ii) $Sa - aS \in K$

where a is an operator of multiplication by a continuous function

$a(t) \in C(\Gamma)$, and K is the set of compact operators.

Property (i) is known as the Poincare–Bertrand equality.

If $\Gamma = S^1$ is the unit circle, then S is an isometry.

2.50 Let $\alpha\colon \Gamma \to \Gamma$ be a diffeomorphism of Γ such that $\alpha'(t) \neq 0$ and $\alpha'(t)$ satisfies the Holder condition

$$\sup_{\tau, t\in\Gamma} \frac{|\alpha'(t) - \alpha'(\tau)|}{|t - \tau|^{\mu}} < \infty$$

with some $0 < \mu \leq 1$.

By U_α we denote the operator in $L^2(\Gamma)$ given by

$$(U_\alpha f)(t) = f(\alpha(t)) \quad . \tag{2.8}$$

U_α is a bounded invertible operator in $L^2(\Gamma)$.

Theorem 2.51 *If α is the mapping described in* 2.50 *then*

$$U_\alpha S - \gamma\, SU_\alpha \in K$$

where $\gamma = +1(-1)$ when α preserves (changes) the orientation on Γ.

2.52 The operators in $L^2(\Gamma)$ having the form

$$A = a + bS + k \tag{2.9}$$

where a, b are operators of multiplication by continuous functions $a(t)$, $b(t) \in C(\Gamma)$ and k is a compact operator are called the (*classical*) *singular integral operators* (with continuous coefficients).

In view of the properties of S (see **2.48** (i), (ii)) singular integral operators form an algebra.

2.53 The symbol of a singular integral operator A of the form (2.9) is a 2×2 matrix function on Γ

$$\sigma(A)(t) = \begin{bmatrix} (a+b)(t) & 0 \\ 0 & (a-b)(t) \end{bmatrix} . \tag{2.10}$$

Remark. Evidently one can regard $\sigma(A)$ as a scalar function on $\Gamma \times \Gamma \cong S^*(\Gamma)$ (cf. 2.45).

Theorem 2.54 *The mapping* $\sigma(\,\cdot\,)$ *defined by* (2.10) *has the following properties*:

(i) $\sigma(A) = 0$ iff $A \in K$;

(ii) $\sigma: \left(\mathscr{A}/_K \right) \to \mathrm{Diag}_2(\Gamma)$
establishes an algebraic **–isomorphism, where* $\mathscr{A}$ *is the algebra of all the singular integral operators and* $\mathrm{Diag}_2(\Gamma)$ *is the set of all the diagonal continuous matrix functions of dimension* 2;

(iii) *if* A *is Fredholm (that is,* $\sigma(A)$ *is invertible) then*

$$\mathrm{ind}\, A = \mathrm{ind}\left[\frac{a(t) - b(t)}{a(t) + b(t)}\right]_\Gamma$$

where the index of the continuous function on Γ *is defined by the equality*

$$\mathrm{ind}[\varphi] = \frac{1}{2\pi}\, [\arg\, \varphi(t)]_\Gamma$$

where $[\ \ ]_\Gamma$ *is the increase of the function around a curved circuit traversed in the positive sense.*

2.55 Along with the Cauchy singular integral operator it is often useful to consider the projections

$$P = \frac{I + S}{2} \ , \qquad Q = \frac{I - S}{2} = I - P$$

(they are indeed projections since $S^2 = I$).

In the case when $\Gamma = S^1$ is the unit circle these projections have the elementary meaning: if $f = \sum_{-\infty}^{+\infty} a_n z^n \in L^2(S^1)$ then

$$Pf = \sum_{0}^{+\infty} a_n z^n \quad .$$

2.56 In an obvious way the definitions of 2.49–2.52 can be raised up to the case of the space $L^2_n(\Gamma)$ of n–dimensional vector functions.

The singular integral operators are defined here again by the formula (2.9) where a, b are operators of multiplication by continuous $n \times n$ matrix functions. The operator S given by (2.7) again has the properties described in 2.49 and Theorem 2.51.

The symbol of a singular integral operator A in this case is an $(n + n) \times (n + n)$ block matrix function of the form (2.10) and the statement of Theorem 2.54 here is true with the obvious changes: in (ii) instead of the set of diagonal matrices we have to consider the set of diagonal block matrices; in (iii) we have

$$\text{ind } A = \text{ind}\left[\frac{\det(a(t) - b(t))}{\det(a(t) + b(t))}\right]_\Gamma \quad .$$

References to 2.39–2.56 are Agranovich [1], Hörmander [1], Palais [1], Seeley [1], Shubin [1], Taylor [1], Treves [1]; Gahov [1], Gohberg, Krupnik [1], Muskhelishvili [1].

Banach algebras and C^*–algebras

2.57 A set A is called an *algebra* over the complex field $\mathbb{C}$ if A is a vector space over $\mathbb{C}$ with an associative multiplication depending linearly on the multipliers.

The algebra A is called a *normed algebra* if it is a normed vector space and multiplication is norm continuous with respect to the total combination of multipliers.

Any normed algebra can be given an equivalent norm with respect to which

$$\|a_1 \cdot a_2\| \leq \|a_1\| \cdot \|a_2\| \quad .$$

If A is complete as a normed space then it is called a *Banach algebra.*

A subspace $A_1 \subset A$ is called a *subalgebra* of A if $a_1, a_2 \in A_1$ implies $a_1 \cdot a_2 \in A_1$.

A norm closed subalgebra of a Banach algebra is in turn a Banach algebra. When discussing Banach algebras we shall consider only closed subalgebras.

An element $I \in A$ is called the identity if $aI = Ia = a$ for every $a \in A$.

If A contains the identity then an element a^{-1} is called the inverse of a if $a^{-1}a = aa^{-1} = I$.

If an algebra A does not have an identity then we can always *adjoin the identity* in the following way. Consider the vector space $\bar{A} = A \oplus \mathbb{C}$ and define the multiplication by

$$(a_1, \lambda)\cdot(a_2, \mu) = (a_1a_2 + \mu a_1 + \lambda a_2, \lambda\mu) \quad .$$

Set $I = (0, 1)$.

Then we have $A \cong (A, 0) \subset \bar{A}$ and I is the identity in $\bar{A}$.

If A is a normed algebra then we define the norm in $\bar{A}$ by

$$\|(a, \lambda)\| = \|a\| + |\lambda| \quad .$$

With this norm $\bar{A}$ becomes a normed algebra (a Banach algebra if A is Banach).

2.58 Let A be an algebra with identity. The *spectrum* $\sigma(a, A)$ of an element $a \in A$ is the set of all $\lambda \in \mathbb{C}$ such that $(\lambda I - a)$ does not have an inverse.

When it does not lead to misunderstanding we shall denote the spectrum of a simply $\sigma(a)$. $\sigma(a, A)$ is a non–void closed set lying in the circle $|\lambda| \le \|a\|$.

The number

$$r(a) = \max_{\lambda\in\sigma(a)} |\lambda|$$

is called the *spectral radius* of a.

It is known that

$$r(a) = \lim_{n\to\infty} \|a^n\|^{\frac{1}{n}} \quad . \tag{2.11}$$

If $A_1 \subset A$ is a subalgebra of A then in general the invertibility of a in A_1. A subalgebra A_1 is called full if for any invertible (in A) element $a \in A_1$ it follows that $a^{-1} \in A_1$.

If A_1 is a full subalgebra of A then we have $\sigma(a, A) = \sigma(a, A_1)$.

Note that the spectral radius of a does not depend on the subalgebra considered (in view of (2.11)).

The set $L(E)$ of all the linear bounded operators in a Banach space E with its natural norm (see 2.1) is a Banach algebra. The spectrum of an operator in this case coincides with its spectrum as an element of a Banach algebra.

The statement of Lemma 2.7 is true for elements of any Banach algebra.

Theorem 2.59 *Let* A *be a Banach algebra with identity and* A_1 *be its closed subalgebra containing* I. *Then for any* $a \in A_1$ *we have*

$$\sigma(a, A_1) \supset \sigma(a, A)$$

and the boundary of $\sigma(a, A)$ *contains the boundary of* $\sigma(a, A_1)$.

Corollary 2.60 *The set* $\sigma(a, A_1)$ *is a union of* $\sigma(a, A)$ *and may be some bounded connected components of* $\mathbb{C}\backslash\sigma(a, A)$.

2.61 A subset $J \subset A$ is called a (two–sided) *ideal* in A if J is a vector subspace of A and $a \in A$, $j \in J$, implies $a \cdot j \in J$ and $j \cdot a \in J$. We shall consider only closed ideals.

Any ideal J induces the equivalence relation

$$a_1 \sim a_2 \Leftrightarrow a_1 - a_2 \in J \quad .$$

The set ${}^{A}/_{J}$ of equivalence classes possesses a natural algebraic structure and is called a factor algebra.

When A is a Banach algebra and J is a closed ideal then the formula

$$\|[a]\| = \inf_{j \in J} \|a + j\|$$

defines the norm of the equivalence class $[a] \in A/_J$, containing a. With this norm $^A/_J$ becomes a Banach algebra.

The subset K of compact operators acting on a Banach space E forms a two–sided closed ideal in $L(E)$ and for any $[a] \in L(E)/_K$ we have

$$\sigma([a]) = \sigma_F(a)$$

(see 2.9).

In particular a is Fredholm iff $[a]$ is invertible.

2.62 By a *morphism* π between algebras A and B we mean an algebraic homomorphism of A into B, that is,

$$\pi\colon A \to B \quad \text{is a linear mapping}$$

and

$$\pi(a_1 \cdot a_2) = \pi(a_1) \cdot \pi(a_2) \quad .$$

If A and B contain identities then we demand that

$$\pi(I_A) = I_B \quad .$$

A morphism $\pi\colon A \to B$ is also called a *representation* of A in B.

If $B = L(E)$ then any morphism of A into B is called a *linear representation* of A. In this case E is called the *representation space*.

We say that $\pi\colon A \to B$ is a *isomorphism* if it has an inverse morphism $\pi^{-1}\colon B \to A$.

A representation $\pi\colon A \to B$ is called *faithful* if Ker $\pi = 0$.

If π is not a faithful representation then Ker π is a two–sided ideal in A.

An *isomorphism between normed algebras* A and B is an algebraic isomorphism $\pi\colon A \to B$ such that both π and π^{-1} are continuous.

A linear representation $\pi: A \to L(E)$ is called *irreducible* if $E \neq 0$ and the only invariant subspaces with respect to $\pi(A)$ are E and $\{0\}$.

Irreducible representations are the main (the simplest, indecomposable) objects in terms of which the basic plot of the Banach algebras theory is written.

Two representations $\pi_1: A \to L(E_1)$ and $\pi_2: A \to L(E_2)$ are called *equivalent* if there exists an isomorphism $U: E_1 \to E_2$ transforming π_1 into π_2.

2.63 An ideal $J \subset A$ is called a *primitive ideal* if it is the kernel of some non–zero irreducible representation of A. The set Prim A of all the primitive ideals of A can be equipped with some natural topology.

For any set $F \subset A$ by $V(F) \subset \text{Prim } A$ we denote the set of all the primitive ideals in A containing F. We have:

(i) $V(\{0\}) = \text{Prim } A$;

(ii) $V(A) = \varnothing$;

(iii) $V(\bigcup_{s\in S} F_s) = V(\sum_S F_s) = \bigcap_{s\in S} V(F_s)$.

If L is the two–sided ideal generated by F then evidently $V(F) = V(L)$ and it can be shown that for any two ideals L_1 and L_2

(iv) $V(L_1 \cap L_2) = V(L_1 L_2) = V(L_1) \cup V(L_2)$.

Properties (i)–(iv) mean that the sets $V(F)$, $F \subset A$ can be regarded as closed sets in some topology on Prim A. This topology is called the *Jacobson topology* on Prim A.

Prim A, equipped with the Jacobson topology, is a T_0–space, that is, if j_1 and j_2 are two different points in Prim A then one of these points has a neighbourhood which does not contain another point (but, in general, Prim A is not a T_1–space).

2.64 Let $\hat{A}$ be the set of equivalence classes of non–zero irreducible representations of A (see 2.62). The mapping $\pi \to \text{Ker } \pi$ is a canonical surjective mapping $\hat{A} \to \text{Prim } A$.

The set $\hat{A}$, equipped with this topology – the inverse image of the Jacobson topology, is called the *spectrum of A*.

Theorem 2.65 *If* A *contains the identity then* Prim A *and* $\hat{\text{A}}$ *are quasicompact (i.e. any open cover of* Prim A ($\hat{\text{A}}$) *contains a finite subcover).*

2.66 Let A be a commutative Banach algebra with an identity. In this case one can also consider another useful set closely connected with the structure of A.

By $M(A)$ we denote the set of all the *characters* of A, that is, nontrivial morphisms of A into $\mathbb{C}$.

It is known that every character is a continuous mapping with norm ≤ 1.

The set $M(A)$ is a subset in the dual space A^* (see 2.3) and is compact in the *–weak topology.

Theorem 2.67 *Let* A *be a commutative Banach algebra with identity. Then the mapping*

$$\pi \to \text{Ker } \pi \quad , \qquad \pi \in \text{M(A)}$$

is the bijection of M(A) *onto the set of all the maximal ideals of* A.

This is why we can identify $M(A)$ with the set of all the *maximal ideals* in A.

Obviously we can identify in this case the sets Prim A, $\hat{A}$, and $M(A)$; thus one can consider the Jacobson and *–weak topologies on $M(A)$. The Jacobson topology on $M(A)$ is weaker than the *–weak topology and in general these topologies do not coincide. When we speak of $M(A)$ we shall always have in mind the *–weak topology.

2.68 If A is a commutative Banach algebra then the morphism $G\colon A \to C(M(A))$ defined by

$$G(a)(\pi) = \pi(a) \quad , \qquad a \in A \quad , \qquad \pi \in M(A) \tag{2.12}$$

is called the *Gelfand transform.*

2.69 The *radical* R_A of a commutative Banach algebra A is the intersection of all the maximal ideals of A. The Gelfand transform induces the algebraic isomorphism $A/R_A \cong G(A) \subset C(M(A))$.

An algebra A is called *semisimple* if $R_A = 0$.

2.70 A normal algebra A is called an *algebra with an involution* * (or a *–algebra) if on A there is defined a mapping $a \to a^*$ with the properties:

(i) $(a^*)^* = a$;

(ii) $(a_1 + a_2)^* = a_1^* + a_2^*$;

(iii) $(\lambda a)^* = \bar{\lambda}\, a^*$, $\lambda \in \mathbb{C}$;

(iv) $(a_1 \cdot a_2)^* = a_2^* \cdot a_1^*$.

A Banach *–algebra A is called a C^*–algebra if it satisfies

$$\|aa^*\| = \|a\|^2 .$$

The algebra $L(H)$ of all the linear bounded operators in a Hilbert space H is a C^*–algebra.

The element a^* is called the *adjoint* to a (cf. 2.15). If $a = a^*$ then a is called *selfadjoint*. If $aa^* = a^*a$ then a is called *normal*. If A has identity then we say that $a \in A$ is *unitary* when $aa^* = a^*a = I$. An element a is called *positive* if it can be expressed as $a = xx^*$ for some $x \in A$. Every positive element is selfadjoint.

The spectrum of any selfadjoint element lies on the real axis. The spectrum of any positive element belongs to $\mathbb{R}_+ = \{t \in \mathbb{R}: t \geq 0\}$. The spectrum of any unitary element is contained in the unit circle $S^1 = \{z \in \mathbb{C}: |z| = 1\}$.

For any element a of a C^*–algebra A we denote by $|a|$ the positive square root of a^*a. It is known that $|a| \in A$. (In some cases it will be convenient for our purposes to consider also the element $\sqrt{aa^*} \in A$ (see Section 22), which we shall also denote by $|a|$ if it does not cause misunderstanding.)

2.71 We say that $\pi\colon A \to B$ is a *morphism* of two *–algebras A and B if in addition to the properties mentioned in **2.62** we have

$$\pi(a^*) = \pi(a)^* \quad , \qquad a \in A \quad .$$

A *linear representation* of a *–algebra A is a morphism of A into $L(H)$ for some Hilbert space H. H is called a *representation space* for π.

Two representations $\pi_1\colon A \to L(H_1)$ and $\pi_2\colon A \to L(H_2)$ are called *equivalent* if there exists a unitary operator $U\colon H_1 \to H_2$ which transforms π_1 into π_2.

2.72 Let $\pi\colon A \to L(H)$ be some representation. A closed linear subspace $K \subset H$ generated by $\pi(a)\xi$, $a \in A$, $\xi \in H$, is called the *essential subspace* for π. π is called *non–degenerate* if $K = H$.

K and $H \ominus K$ are invariant with respect to $\pi(A)$ and $\pi(A)'$ ($\pi(A)'$ is the commutant of $\pi(A)$). $H \ominus K$ coincides with the set of vectors $\xi \in H$ such that $\pi(a)\xi = 0$ for all $a \in A$.

2.73 For any $\xi_0 \in H$ the subspace $\overline{\{\pi(a)\xi_0\}_{a\in A}}$ is invariant with respect to $\pi(A)$. If this subspace coincides with the whole of H then ξ_0 is called a *totalizating* (or *cyclic*) vector for π. If π is irreducible then any non–zero vector is cyclic.

Theorem 2.74 *Let* A *be a Banach* **–algebra and* B *be a* C^**–algebra. Then for any morphism* $\pi\colon$ A $\to$ B *we have*

$$\|\pi(a)\| \le \|a\| \qquad a \in A \quad .$$

Thus any morphism of C^*–algebras is automatically continuous and consequently any algebraic isomorphism of C^*–algebras is an isometry.

Theorem 2.75 *Let* A *be a* C^**–algebra,* J *be a closed two–sided ideal in* A. *Then* J *is a* **–ideal (that is,* $a \in J$ *iff* $a^* \in J$*) and* ${}^{A}/_{J}$ *is a* C^**–algebra.*

Corollary 2.76 *Let* A *and* B *be* C^**–algebras and* π: A $\to$ B *be some morphism with the kernel* J. *Consider the canonical decomposition of* π

$$A \to {}^{A}/_{J} \xrightarrow{\psi} \pi(A) \to B \quad .$$

Then J *is closed,* $\pi(A)$ *is closed and* ψ *is an isomorphism of the* C^**–algebra* ${}^{A}/_{J}$ *onto the* C^**–algebra* $\pi(A)$.

Corollary 2.77 *Let* B *be a* C^**–subalgebra of a* C^**–algebra* A *and* J *be a closed two–sided ideal in* A. *Then* B + J *is a* C^**–subalgebra in* A *and* $(B + J)/_J$ *is canonically isomorphic to* $B/_{B\cap J}$.

Theorem 2.78 *Every* C^**–subalgebra* B *of a* C^**–algebra* A *is full.*

Commutative C^*–algebras can be described totally with the help of the Gelfand transform (see 2.68).

Theorem 2.79 (The first Gelfand–Naimark theorem) *Let* A *be a commutative* C^**–algebra with identity. Then the Gelfand transform establishes the* **–isomorphism*

$$A \overset{G}{\cong} C(M(A)) \quad .$$

Thus the theory of commutative C^*–algebras with identity is in fact equivalent to the theory of continuous functions on compact spaces.

Theorem 2.80 (The second Gelfand–Naimark theorem) *Let* A *be a* C^**–algebra. Then there exists an isometric representation of* A *in some Hilbert space* H.

So without loss of generality when dealing with a C^*–algebra A we can always assume that A is a C^*–subalgebra of $L(H)$.

Theorem 2.81 *Let* B *be a* C^**–subalgebra of a* C^**–algebra* A *and* ν *be some representation of* B *in a Hilbert space* K. *Then there exists a Hilbert space* H *containing* K *as a subspace and a representation* π *of* A *in* H *such that*

$$\pi(a)|_K = \nu(a) \quad , \qquad a \in B \quad .$$

If ν *is an irreducible representation of* B *then we can choose* π *to be an irreducible representation of* A.

This is a rather useful statement – it enables us to construct representations of a C^*–algebra starting from representations of a suitable subalgebra.

In fact C^*–algebras have many wonderful properties. For example any closed two–sided ideal defines the decomposition of Prim A and $\hat{A}$.

Lemma 2.82 *Let* A *be a* C^**–algebra,* J *be a closed two–sided ideal in* A, π *be a representation of* A *in a Hilbert space* H, K *be the essential subspace for* $\pi|_J$ (see 2.72), *and* B *be the strong closure of* $\pi(A)$. *Then:*

(i) $P_K \in B \cap B'$

(*where* P_K *is the orthogonal projection onto* K *and* B′ *is the commutant of* B);

(ii) *the subrepresentation of* π *defined on* $H \ominus K$ *annihilates on* J;

(iii) *if* $\bar{\pi}$ *is the subrepresentation of* π *determined by* K *then* $\bar{\pi}(J)$ *is strongly dense in* $\bar{\pi}(A)$.

Theorem 2.83 *Let* A *be a* C^**–algebra,* J *be a closed two–sided ideal in* A, *and* $\hat{A}_J$ ($\hat{A}^J$) *be the set of* $\pi \in \hat{A}$ *such that* $\pi(J) = 0$ ($\pi(J) \neq 0$). *Then:*

(i) *if for any* $\pi \in \hat{A}_J$ *we define by* π' *the representation of* ${}^A/_J$ *obtained from* π *by passing to the factor then* $\pi \to \pi'$ *is a bijection of* $\hat{A}_J$ *onto* $\left({}^A/_J\right)^\wedge$;

(ii) *the mapping* $\pi \to {}^{\pi}|_J$ *is a bijection of* $\hat{A}^J$ *onto* $\hat{J}$.

Theorem 2.84 *Let* A *be a* C^**–algebra,* J *be a closed two–sided ideal in* A, *and* $\mathrm{Prim}_J A$ ($\mathrm{Prim}^J A$) *be the set of primitive ideals containing* (*not containing*) J. *Then:*

(i) *the mapping* $F \to F/_J$ *is a bijection of* $\mathrm{Prim}_J A$ *onto* $\mathrm{Prim}\left(A/_J\right)$;

(ii) *the mapping* $F \to F \cap J$ *is a bijection of* $\mathrm{Prim}^J A$ *onto* Prim J.

2.85 The canonical bijections

$$\mathrm{Prim}_J A \to \mathrm{Prim}\left(A/_J\right) \quad , \quad \hat{A}_J \to \left(A/_J\right)^\wedge$$
$$\mathrm{Prim}^J A \to \mathrm{Prim}\ J \quad , \quad \hat{A}^J \to \hat{J}$$

(defined in Theorems 2.83 and 2.84) are homeomorphisms. The sets $\mathrm{Prim}_J A$ and $\hat{A}_J$ are closed in Prim A and $\hat{A}$. The set $\mathrm{Prim}^J A$ and $\hat{A}^J$ are open in Prim A and $\hat{A}$.

2.86 For any $a \in A$ and $x \in \mathrm{Prim}\ A$ we denote by $\breve{a}(x)$ the number

$$\breve{a}(x) = \inf_{j \in x} \|a + j\| \quad .$$

The function $\breve{a}(x)$ on Prim A (and the corresponding function $\|\hat{a}(\pi)\| = \|(\pi)(a)\|$ on $\hat{A}$) possess some semicontinuity properties with respect to the Jacobson topology on Prim A ($\hat{A}$).

Theorem 2.87 *For every* $a \in A$ *the function* $\breve{a}(x)$ ($\|\bar{a}(\pi)\|$) *is lower semicontinuous on* Prim A ($\bar{A}$) *and the sets*

$$\{x \in \mathrm{Prim}\ A:\ \breve{a}(x) \geq \alpha\}$$
$$\alpha \geq 0$$
$$(\pi \in \bar{A} : \|\bar{a}(\pi)\| \geq \alpha)$$

are quasicompact and form a basis for the Jacobson topology, which is therefore locally compact.

Lemma 2.88 *For every* $a \in A$ *the function* $\breve{a}(x)$ ($\|\bar{a}(\pi)\|$) *on* Prim A ($\bar{A}$) *attains its upper bound equal to* $\|a\|$.

2.89 Recall that the *centre* of an algebra A is the set Z of all the elements $b \in A$ such that $ba = ab$ for every $a \in A$.

The centre is a commutative algebra and in the case of a C^*–algebra its maximal ideal space $M(Z)$ (see **2.66** and Theorem 2.67) is closely related to Prim A which is stated in the following.

Theorem 2.90 (Dauns–Hofmann theorem) *Let* A *be a* C^**–algebra with identity. Then the complexification of the map* $a \to \breve{a}$ *(here* a *belongs to the set of positive elements of* A) *establishes an isomorphism between the centre* Z *of* A *and the class of continuous functions on* Prim A.

2.91 There exists a natural correlation between group theory and C^*–algebra theory. Let G be a locally compact group with a fixed Haar measure (see Theorem 2.27). In the Banach space $L^1(G)$ define the operation of *convolution* (×) and *involution* (*) by

$$(f \times \varphi)(s) = \int_G f(g)\varphi(g^{-1}s)dg$$

$$f^*(s) = \Delta(s)^{-1} \cdot \bar{f}(s^{-1}) \ .$$

With these operations $L^1(G)$ becomes a *–algebra (but not a C^*–algebra).

The *group* C^**–algebra* $C^*(G)$ is the *enveloping* algebra of $L^1(G)$, that is, the completion of $L^1(G)$ in the norm

$$\|f\| = \sup_\pi \|\pi(f)\|$$

where the supremum is taken over all representations π of $L^1(G)$ continuous in the L^1–norm.

Theorem 2.92 *There exists a bijective correspondence between the unitary representations of* G (see 2.23) *and non–degenerate representations of* $L^1(G)$ (see 2.72). *Under this correspondence any representation* π *of* $L^1(G)$ *in* L(H) *has the form*

$$\pi(f) = \int_G f(g)U(g)dg \qquad f \in L^1(G) \tag{2.13}$$

where U: G → L(H) *is the corresponding unitary representation of* G.

Theorem 2.93 *If* G *is abelian and* $\hat{G}$ *is its dual group* (see **2.24**) *then*

$$\widehat{C^*(G)} = \hat{G}$$

(*in this case* U(g) *in* (2.13) *is the character of* G).

2.94 A very important class of C^*–algebras (but not often used in this book) is the class of *Von Neumann algebras*, that is, C^*–subalgebras of $L(H)$ which are closed in the strong operator topology (in this case they are also closed in the weak operator topology).

Recall that we denote by A' the *commutant* of A, i.e. the set of all the elements in $L(H)$ which commute with every $a \in A$ and $A'' = (A')'$.

Theorem 2.95 (Von Neumann bicommutant theorem) *A subalgebra* $A \subset L(H)$ *is a Von Neumann algebra iff* $A = A''$.

2.96 Among the fundamental properties of a Von Neumann algebra is that it is generated (in the strong operator topology) by projections contained in it.

If $a = U|a|$ is the polar decomposition of an operator $a \in L(H)$ then both U and $|a|$ belong to the Von Neumann algebra generated by a.

We denote by A^1 the unit ball of A and by A_{sa} we denote the set of all the selfadjoint elements in A. $A^1_{sa} = A_{sa} \cap A^1$.

We denote by A_+ the set of all the positive elements in A. $A^1_+ = A_+ \cap A^1$.

One of the most useful results in the theory of Von Neumann and C^*–algebras is:

Theorem 2.97 (Kaplansky's density theory) *Let* A *be a* C^**–subalgebra of* L(H) *with strong closure* B. *Then the unit ball* A^1 *is strongly dense in the unit ball* B^1. *Furthermore* A^1_{sa} (*respectively* A^1_+) *is strongly dense in* B^1_{sa} (*respectively* B^1_+). *Finally if* A *contains the identity operator then the unitary group of* A *is strongly dense in the unitary group of* B.

References to 2.57–2.97 are Arveson [3], Bourbaki [2], Dixmier [1, 2], Kadison, Ringrose [1], Pedersen [1], Rickart [1], Sakai [1].

Chapter 1: Weighted shifts. Spectral radius

'Андрей Семенович: Хоть ты и математик, а, честное слово ты не умен.

Математик: Нет, умен и знаю очень много!

Андрей Семенович: Много, да только все ерунду.

Математик: Нет, не ерунду!

... возьми вот этот шарик – научную модель вселенной.' Д. Хармс*

'Because the world is round it turns me on' Lennon–McCartney *Because*

In this Chapter we introduce the object which is at the centre of the main problems considered in the book and which is thus responsible for the appearance of the book itself – an abstract version of a weighted shift operator (which we shall still call a weighted shift).

We begin with the case when weights form a commutative algebra. Our first step in the investigation of the invertibility conditions of operators generated by weighted shifts is the evaluation of the spectral radius of a weighted shift (Sections 4 and 5). The general formula arising here (Theorem 4.4) definitely has not only pure mathematical but also some aesthetic value.

Continuing to consider the spectral properties of a weighted shift we obtain an estimate for a number of connected components of its spectrum, and looking through a number of examples (Sections 6–8) we find that though a weighted shift is extremely useful as a tool for the construction of various counterexamples (see Sections 6 and 9) in many apparently different cases its spectrum has some common features (Section 8). The intrinsic reasons for this will be investigated in the subsequent part of the book (Chapters 2 and 3).

We finish the Chapter with the presentation of a number of examples of functional differential equations and discover deep interrelations between the

---

*'Andrei Semenovich: Though you are a mathematician but frankly speaking you are not clever.

Mathematician: No, I am clever and know a lot!

Andrei Semenovich: A lot, but all of it is rubbish.

Mathematician: No, not rubbish!

... take say this ball as a scientific model of the Universe.' D. Harms

problems of solvability of these equations and the conditions for invertibility of the corresponding classes of operators generated by weighted shifts.

3. Weighted shift operators. Examples

Here we introduce the objects we shall treat in detail throughout the whole of the book – the weighted shift operators.

3.1 Let A be a Banach subalgebra of the algebra $L(E)$ of bounded linear operators acting on a Banach space E and let $T: E \to E$ be an invertible isometry such that $TAT^{-1} = A$; then an operator of the form aT, $a \in A$, will be called a *weighted shift operator* (with coefficient in the algebra A).

Examples

3.2 Let $E = l^p(\mathbb{Z})$, $1 \le p \le \infty$ ($\mathbb{Z}$ is the set of integers with discrete measure $\mu(\,\cdot\,) = 1$). Let A be the algebra of operators a_φ of multiplication by a function $\varphi \in l^\infty(\mathbb{Z})$, and let us define an operator T on the space $l^p(\mathbb{Z})$ by the formula

$$(Tf)(k) = f(k-1) \qquad f \in l^p(\mathbb{Z}) \quad .$$

Then $Ta_\varphi T^{-1} = a_\psi$ where $\psi(k) = \varphi(k-1)$. So T determines an automorphism of A and $a_\varphi T$ is, by definition 3.1, a weighted shift operator.

Actually in this example $a_\varphi T$ is a *classical bilateral weighted shift* and this is the reason why the object introduced in definition 3.1 was called a weighted shift operator. For the same reason it is rather natural to call A the algebra of weights and T the shift operator.

3.3 Let (X, Ω, μ) be a space with a positive σ–finite measure μ, and $\alpha: X \to X$ be an invertible measurable mapping which preserves the equivalence class of μ (that is, $\mu(\omega) = 0 \Leftrightarrow \mu(\alpha(\omega)) = 0$, $\omega \in \Omega$).

We define an operator T on the space $L^p_\mu(X)$, $1 \le p \le \infty$, by the formula

$$[Tf](x) = [\alpha'(x)]^{\frac{1}{p}} f(\alpha(x)) \qquad \left(\text{here } \tfrac{1}{\infty} = 0\right)$$

where α' is the Radon–Nikodym derivative of the measure $\alpha(\mu)$ with respect to the measure μ. By construction T is an isometry and obviously the inverse of T is the operator

$$[T^{-1}f](x) = [\alpha^{-1\,\prime}(x)]^{\frac{1}{p}} f(\alpha^{-1}(x)) \quad .$$

Let A be the algebra of operators a_φ of multiplication by a function $\varphi \in L^\infty_\mu(X)$. Then $Ta_\varphi T^{-1} = a_{\varphi\circ\alpha}$, so T determines an automorphism of A. Thus $a_\varphi T$ is a weighted shift operator.

It is evident that example 3.2 was a special case of this one.

3.4 Let X be a topological space and let $C(X)$ be the set of bounded continuous functions with the norm $\|f\| = \sup_{x\in X} |f(x)|$. Let $\alpha: X \to X$ be a homeomorphism and T: $C(X) \to C(X)$ be the operator defined by the formula

$$(Tf)(x) = f(\alpha(x)) \quad .$$

The algebra A is the algebra of operators of multiplication by continuous bounded functions. It is evident that in this case aT, $a \in A$, is a weighted shift operator.

3.5 Let A be a uniform algebra on a compact space X (this means that A is a uniformly closed subalgebra of $C(X)$, containing constants and separating points of X). Equipped with the max norm

$$\|f\| = \max_{x\in X} |f(x)|$$

the algebra A becomes a Banach algebra. Let $T: A \to A$ be an automorphism of the algebra A and $a_\varphi: A \to A$ be an operator of multiplication by an element $\varphi \in A$. The operator $a_\varphi T$ is a weighted shift operator.

3.6 Consider the *Hardy space* H^p, $1 \le p < \infty$, that is, the space of analytic functions f on the disk $\mathscr{D} = \{z: |z| < 1\}$ such that

$$\|f\|_{H^p} = \sup_{0<r<1} \frac{1}{2\pi} \left[\int_0^{2\pi} |f(re^{i\theta})|^p d\theta \right]^{\frac{1}{p}} < \infty .$$

Let α: $\mathscr{D} \to \mathscr{D}$ be a conformal mapping (that is, $\alpha'(z) \neq 0$, $z \in \mathscr{D}$) of $\mathscr{D}$ onto $\mathscr{D}$. Since $\alpha'(z) \neq 0$ for $z \in \mathscr{D}$ and $\alpha'(\mathscr{D})$ is a domain we can choose an analytic branch of $\alpha'^{\frac{1}{p}}(z)$ and define an operator T: $H^p \to H^p$ by the formula

$$(Tf)(z) = \gamma\, \alpha'^{\frac{1}{p}}(z)\, f(\alpha(z))$$

where $|\gamma| = 1$. It is shown by Forelli [1] that only these operators are isometries in H^p, $p \neq 2$. The operator T^{-1} is constructed in an analogous way from the mapping α^{-1}: $\mathscr{D} \to \mathscr{D}$. Let A be the algebra of operators α_φ of multiplication by bounded functions which are analytic on $\mathscr{D}$ (this means that $A = H^\infty$). By the construction of T we have that $Ta_\varphi T^{-1} = a_{\varphi\circ\alpha}$, so $a_\varphi T$ is a weighted shift operator.

In the foregoing examples the weighted shift operators were actually the weighted composition operators (the operator T was always a composition operator). The following examples show that in fact weighted shift operators can have a form different from weighted composition operators.

3.7 Let A be the subalgebra of $L(L^2(\mathbb{R}))$ generated by the operators of convolution with functions in $L^1(\mathbb{R})$ and by the operator of multiplication by continuous functions which have limits when $x \to +\infty$ and $x \to -\infty$. As T we take the operator of multiplication by e^{ihx}. The operator T generates an automorphism of A. Indeed for any operator of multiplication by a function a we have $TaT^{-1} = a$, and for the operator Φ of convolution with a function φ we have

$$[T\Phi T^{-1}f](x) = e^{ihx} \int_{-\infty}^{+\infty} \varphi(x - t)e^{-iht} f(t)dt$$

$$= \int_{-\infty}^{+\infty} \varphi_1(x - t)f(t)dt$$

where $\varphi_1(x) = e^{ihx}\varphi(x)$. Thus a weighted shift operator in this example is a convolution operator with an oscillating coefficient.

Consideration of convolution operators naturally leads to the following example.

3.8 Let G be a commutative locally compact group with a Haar measure dg. In the space $L^2(G)$ consider the standard group algebra A of operators of the form

$$[af](g_0) = \int_G a(g)f(g_0 - g)dg$$

where $a(g) \in L^1(G)$, and + is the group operation symbol.

Let χ be any particular character of G and define T: $L^2(G) \rightarrow L^2(G)$ by the formula

$$(Tf)(g) = \chi(g)f(g) \quad .$$

Then we have that

$$[(TaT^{-1})f](g_0) = \chi(g_0) \int_G a(g)\chi(-g_0 + g)f(g_0 - g)dg$$

$$= \int_G a(g)\chi(g)f(g_0 - g)dg$$

$$= \int_G a_1(g)\chi(g_0 - g)dg \in A$$

where $a_1(g) = a(g)\chi(g) \in L^1(\mathrm{G})$.

Hence we have that aT is a weighted shift operator.

3.9 In the same space $L^2(G)$ consider another group algebra $\tilde{A}$, namely the algebra generated by operators a_{g_0} : $L^2(G) \rightarrow L^2(G)$ where

$$(a_{g_0} f)(g) = f(g - g_0) \quad .$$

This means that $\tilde{A}$ is the norm closure of the finite sums

$$\tilde{A} = \overline{\left\{ \sum \lambda_{g_i} a_{g_i} \right\}} \quad , \qquad \lambda_{g_i} \in \mathbb{C} \quad .$$

Let T be again an operator of multiplication by a particular character $\chi \in G^*$. Since T is an invertible isometry and

$$T a_{g_0} T^{-1} = \chi(g_0)\, a_{g_0} \in \tilde{A}$$

it follows that $T\tilde{A}T^{-1} = \tilde{A}$ and consequently an operator

$$aT \quad , \qquad a \in \tilde{A}$$

is a weighted shift operator.

3.10 Let $G = \{z: |z| = 1\}$ be the unit circle on the complex plane. Let us denote by S the singular integral operator acting on the space $L^2(G)$ by the formula

$$(Sf)(z) = \frac{1}{\pi i} \int_G \frac{f(\tau)}{z - \tau}\, d\tau$$

where a principal value integral is used on the right. It is known that S is an isometry and $S^2 = I$ (see 2.49). Let A be the algebra in $L(L^2(G))$ generated by operators of multiplication by continuous functions and compact operators.

For any continuous function a we have that $[SaS - a]$ is a compact operator (see 2.49), so $SAS = SAS^{-1} = A$ and aS $(a \in A)$ is a weighted shift operator.

3.11 Let H^2 be a Hardy space (see 3.6), and P be the projection of $L^2(G)$ onto H^2 (here G is a unit circle with normalized Lebesgue measure, and it is known that $H^2 \subset L^2(G)$, see, for example, Hoffman [1]).

For any function $\varphi \in L^\infty(G)$ the operator T_φ: $H^2 \to H^2$ defined by the formula

$$T_\varphi f = P(\varphi \cdot f) \quad , \qquad f \in H^2$$

is called the *Toeplitz operator*.

For any conformal mapping α of the unit disk onto itself consider the operator T: $H^2 \to H^2$ introduced in 3.6. Note that

$$T(T_\varphi)T^{-1} = T_{\varphi\circ\alpha} \quad .$$

Indeed, as $PT = TP$ it follows that

$$T(T_\varphi)T^{-1} = T(P \cdot \varphi)T^{-1} = P(T_\varphi T^{-1}) = P \cdot (\varphi\circ\alpha) = T_{\varphi\circ\alpha} \quad .$$

Thus the operator T generates the automorphism $T \cdot T^{-1}$ at the C^*–algebra generated by Toeplitz operators and $(T_\varphi)T$ is a weighted shift operator.

Remark 3.12 If we do not specify the algebra of weights A then any operator in an arbitrary Banach space can be written in a weighted shift form. For example, set $A = L(E)$. Thus in order to use definition 3.1 effectively it is reasonable to choose A to be essentially narrower than $L(E)$.

In the case of a Hilbert space in fact there exists a way of choosing the minimal suitable algebra A of weights. This algebra is described in Theorem 3.13.

Let b be an arbitrary operator acting on a Hilbert space H. Denote by A_b the C^*–algebra generated by the selfadjoint operators I, bb^*, b^*b, b^2b^{2*}, $b^{2*}b^2$, ...

Theorem 3.13 *Let* b *be an operator acting on a Hilbert space.*

(1) *If* $b = aT$ *is a weighted shift operator with the weight* a *from a* C^**–algebra* A *and with a unitary shift* T *then* $A \supset A_b$.

(2) *If* b *is invertible then* b *can be written as a weighted shift* aT *where* $a \in A_b$ *and* T *is a unitary operator such that* $TA_bT^* = A_b$.

Proof

(1) Let $\hat{T}: a \to TaT^*$ be the automorphism of A generated by T. Then $Ta = \hat{T}(a)T$ and consequently

$$b^n = aT \cdot aT \cdot \ldots \cdot aT = a \cdot \hat{T}(a) \cdot \hat{T}^2(a) \cdot \ldots \cdot \hat{T}^{n-1}(a)T^n \tag{3.1}$$

$$b^{n*} = (T^*a^*)(T^*a^*) \cdot \ldots \cdot (T^*a^*) = T^{*n}\hat{T}^{n-1}(a^*) \cdot \ldots \cdot \hat{T}^2(a^*) \cdot \hat{T}(a^*) \cdot a^* . \tag{3.2}$$

Thus we see that $b^n b^{n*} \in A$ and $b^{n*} b^n \in A$ for all n and it follows 9hat $A_b \subset A$.

(2) Let b be an invertible operator. Let $a = \sqrt{bb^*}$ be the positive square root of bb^* and set $T = a^{-1}b$. Then obviously $b = aT$ and T is unitary. As A_b is a C^*–algebra we have that $a \in A_b$ (see 2.70).

Let us show that $TA_bT^* = A_b$. To do this it suffices to verify that the elements $b^n b^{n*}$ and $b^{n*} b^n$ are mapped into A_b under the mappings $\hat{T}: a' \to Ta'T^*$, $\hat{T}^*: a' \to T^*a'T$.

Consider the elements $a_k = T^k a T^{k*}$, $k = 0, \pm 1, \pm 2, \ldots$. Note that all the operators a_k are positive (as a is positive). Let us show that $a_k \in A_b$.

If $k = 0$ we have that $a_0 = a \in A_b$. Suppose that the assertion is valid for $k = 0, \pm 1, \pm 2, \ldots, \pm(n-1)$. Then it follows that $a_k \in A_b$ for $k = \pm n$. Indeed using (3.1) and (3.2) we have that

$$b^k = a_0 \cdot a_1 \cdot \ldots \cdot a_{k-1} T^k = T^k a_{-k} \cdot a_{-k+1} \cdot \ldots \cdot a_{-1}$$

$$b^{k*} = T^{k*} a_{k-1} \cdot a_{k-2} \cdot \ldots \cdot a_1 \cdot a_0 = a_1 \cdot a_2 \cdot \ldots \cdot a_{-k} T^{k*} .$$

Thus

$$a_0 \cdot a_1 \cdot \ldots \cdot a_{n-1} \cdot a_n \cdot a_n \cdot a_{n-1} \cdot \ldots \cdot a_1 \cdot a_0 = b^{n+1} \cdot b^{(n+1)*} \in A_b$$

and

$$a_{-1} \cdot a_{-2} \cdot \ldots \cdot a_{-n} \cdot a_{-n} \cdot a_{(-n+1)} \cdot \ldots \cdot a_{-1} = b^{n*} \cdot b^n \in A_b . \tag{3.3}$$

By the supposition the elements $a_0, a_1, \ldots, a_{n-1}$, $a_{-1}, a_{-2}, \ldots, a_{-n+1}$ belong to A_b

and are invertible (as b is invertible). Thus, as A_b is a C^*–algebra all the inverses for them also belong to A_b (by Theorem 2.78). Therefore a_n^2 and a_{-n}^2 also belong to A_b and consequently a_n and a_{-n} belong to A_b as well.

For the mappings $\hat{T}$ and $\hat{T}^*$ we have that

$$\hat{T}a_k = a_{k+1} \qquad \text{and} \qquad \hat{T}^* a_k = a_{k-1} \quad .$$

In view of formula (3.3) the elements $b^n b^{n*}$ and $b^{n*} b^n$ can be expressed in terms of a_k. Thus all the elements of the form

$$T(b^n b^{n*})T^* \ , \qquad T^*(b^n b^{n*})T \ , \qquad T(b^{n*} b^n)T^* \ , \qquad T^*(b^{n*} b^n)T$$

belong to A_b.

Since the norms of the mappings $\hat{T}$ and $\hat{T}^*$ do not exceed 1 they can be continued by continuity onto the whole of A_b and by their construction they are mutually inverse.

The proof is complete. □

Remark 3.14 If a C^*–algebra A of weights is commutative then A_b will also be commutative. Thus the case of the commutative weight algebra (for a weighted shift operator $b = aT$) is distinguished by the condition that all the elements $b^k b^{k*}$, $b^{k*} b^k$, $k = \pm 1, \pm 2, \ldots$, mutually commute. An operator b satisfying such conditions is called *central binormal* (see Campbele [1]). Thus central binormal operators (and only them) can be considered as weighted shift operators in a Hilbert space with a commutative weight algebra.

4. The spectral radius of weighted shift operators

We begin the investigation of the conditions for the invertibility of the operators generated by weighted shifts with the consideration of the formula for the spectral radius of a weighted shift. The most effective results will be obtained in the case when A is a uniform algebra.

4.1 Let X be a compact space, and consider a uniform algebra A on X (that is, a uniformly closed subalgebra of $C(X)$ containing constants and separating points of X). Let $\pi: A \to L(E)$ be an isometric representation of A on the space $L(E)$ of bounded linear operators in a Banach space E. Let T be an invertible isometry on E such that $T\pi(A)T^{-1} = \pi(A)$. The automorphism $\hat{T}: \pi(A) \to \pi(A)$ defined by $\hat{T}(b) = TbT^{-1}$, $b \in \pi(A)$, generates the homeomorphism $\alpha: M \to M$ of the maximal ideal space of the algebra A by the formula $\hat{T}(b)(m) = b(\alpha(m))$, $m \in M$.

4.2 The subset V in M is called a *boundary* for A if any function $f \in A$ attains its maximum at some point $m \in V$.

The intersection of all closed boundaries for A is a boundary for A and is called the *Shilov boundary* ∂A of A. We remark that $\partial A \subset X \subset M$ (Gamelin [1]). Since α is generated by an automorphism of A, for any boundary V, $\alpha(V)$ is again a boundary and the same is true for $\alpha^{-1}(V)$. So, ∂A is invariant under α.

4.3 Let (Y, Ω, μ) be a space with a measure ν, $\nu(Y) = 1$, and let $b \in L_\nu^\infty(Y)$. The *geometric mean of b with respect to the measure* ν is

$$S_\nu(b) = \exp \int_Y \ln|b(y)|d\nu \quad , \tag{4.1}$$

where we assume that $S_\nu(a) = 0$ when $\int_Y \ln|b(y)|d\nu = -\infty$.

Recall also that if Y is a compact space then the dual space to $C(Y)$ is the set of all finite complex Borel measures on Y.

Theorem 4.4 *Suppose that* A *is a uniform algebra and let* $\alpha: \partial A \to \partial A$ *be the homeomorphism of* ∂A *generated by the automorphism* $\hat{T}$. *Then the spectral radius* $r(\pi(a)T)$ *of the operator* $\pi(a)T$, $a \in A$, *is given by the formula*

$$r(\pi(a)T) = \max_{\nu \in \Lambda} S_\nu(a) \tag{4.2}$$

where Λ *is the set of measures on* ∂A *which are ergodic* (see **2.18**) *with respect to* α.

Proof Without loss of generality it can be assumed that A is a subalgebra of $C(\partial A)$.

The spectral radius is computed by the formula

$$r(\pi(a)T) = \lim_{n\to\infty} \|[\pi(a)T]^n\|^{\frac{1}{n}}$$

$$= \lim_{n\to\infty} \|\pi(a)\hat{T}(\pi(a))\cdot\hat{T}^2(\pi(a))\cdot\ldots\cdot\hat{T}^{n-1}(\pi(a))T^n\|^{\frac{1}{n}}$$

$$= \lim_{n\to\infty} \max_{x\in\partial A} \left[\prod_{k=0}^{n-1} |a(\alpha^k(x))|\right]^{\frac{1}{n}} \tag{4.3}$$

where we have used the fact that T^n is an isometry and that $\|\pi(a)\| = \max_{x\in\partial A} |a(x)|$, $a \in A$.

Note that if we denote by U the operator on $C(\partial A)$ acting according to the formula $(Uf)(x) = f(\alpha(x))$ we have that $r(aU) = r(\pi(a)\mathrm{T})$ and so it suffices to compute the spectral radius of the operator aU.

Suppose first that $a(x) \neq 0$, $x \in \partial A$. Let ν be any invariant probability measure on ∂A. Then

$$\int_{\partial A} \ln|a(x)|d\nu = \int_{\partial A} \ln|a(\alpha(x))|d\nu \quad .$$

Thus

$$S_\nu(a) = \exp \int_{\partial A} \ln|a(x)|d\nu$$

$$= \exp \frac{1}{n} \sum_{k=0}^{n-1} \int_{\partial A} \ln|a(\alpha^k(x))|d\nu$$

$$\leq \exp \max_{\partial A} \frac{1}{n} \sum_{k=0}^{n-1} \ln|a(\alpha^k(x))|$$

$$= \max_{\partial A} \prod_{k=0}^{n-1} |a(\alpha^k(x))|$$

$$= \|(aU)^n\|^{\frac{1}{n}} \quad .$$

Passing to the limit when $n \to \infty$ we have that

$$S_\nu(a) \le r(aU) \quad ,$$

and so

$$\sup_\nu S_\nu(a) \le r(aU) \quad . \tag{4.4}$$

Now we shall show that there exists an invariant probability measure ν_0 such that $r(aU) = S_{\nu_0}(a)$. For every n let us choose a point $x_n \in \partial A$ such that

$$\max_{\partial A} \prod_{k=0}^{n-1} |a(\alpha^k(x)| = \prod_{k=0}^{n-1} |a(\alpha^k(x_n))|$$

and consider the sequence of bounded linear functions on $C(\partial A)$ given by

$$f_n(\varphi) = \frac{1}{n} \sum_{k=0}^{n-1} \varphi(\alpha^k(x_n)) \quad , \qquad \varphi \in C(\partial A) \quad .$$

Since $\|f_n\| = 1$ we conclude, using the Banach–Alaoglu theorem on weak–* compactness of the unit ball in a dual space, that there exists a limit point f for the sequence $\{f_n\}$ in the weak–* topology. Since for the function $\varphi = \ln|a(x)|$ the sequence $\{f_n(\varphi)\}$ converges we have that

$$\exp f(\ln|a|) = \lim_{n\to\infty} \exp f_n(\ln|a|) = r(aU) \quad . \tag{4.5}$$

Note that f is invariant with respect to α, that is, $f(\varphi) = f(\varphi\circ\alpha)$ for any function $\varphi \in C(\partial A)$. Indeed by the construction of f_n, for any $\varphi \in C(\partial A)$ we have that

$$f_n(\varphi) - f_n(\varphi\circ\alpha) = \frac{1}{n}\left[\varphi(x_n) - \varphi(\alpha^n(x_n)\right] \to 0$$

and as f is a limit point for $\{f_n\}$ we conclude that $f(\varphi) = f(\varphi \circ \alpha)$.

Moreover if $\varphi(x) \geq 0$ then $f(\varphi) \geq 0$ and since $f_n(1) = 1$ we have that $f(1) = 1$. So f defines the invariant probability measure ν_0 on ∂A, that is

$$f(\varphi) = \int_{\partial A} \varphi(x) d\nu_0 \quad .$$

Thus in the case when $a(x) \neq 0$ we obtain the equality

$$r(aU) = S_{\nu_0}(a)$$

and using (4.4) we conclude that

$$r(aU) = \max_{\nu \in \Lambda_0} S_\nu(a) \tag{4.6}$$

where Λ_0 is the set of invariant probability measures with respect to α.

Now let us consider the case when there exists a point $x_0 \in \partial A$ such that $a(x_0) = 0$. In this case $\ln|a(x)|$ is discontinuous and the foregoing reasoning is not valid. Consider the sequence $a_k(x) = |a(x)| + \frac{1}{k}$ and the sequence of operators $b_k = a_k U$ acting on $C(\partial A)$. Since $b_k \to aU$ in the norm of $L(C(\partial A))$ and as the spectral radius possesses the upper semicontinuous property (see 2.12) we conclude that

$$\overline{\lim_{k\to\infty}} \, r(b_k) \leq r(aU) \quad . \tag{4.7}$$

From the inequality $a_k(x) > |a(x)|$ and from the direct computation of $\|(a_k U)^n\|$ and $\|(aU)^n\|$ we obtain that

$$\|(a_k U)^n\| \geq \|(aU)^n\|$$

and so

$$r(a_k U) \geq r(aU) \quad .$$

Passing to the limit when $k \to \infty$ we have that

$$\lim_{k\to\infty} r(b_k) \geq r(aU) \quad . \tag{4.8}$$

From (4.6) and (4.7) it follows that

$$r(aU) = \lim_{k\to\infty} r(b_k) \quad .$$

Let us show that $r(aU)$ is again expressed by the formula (4.6). Suppose that ν is an arbitrary measure from Λ_0. Since a_k is a decreasing sequence and a_k converges to $|a(x)|$ we may (using the Levi theorem) pass to the limit in the equality

$$S_\nu(a_k) = \exp \int_{\partial A} \ln a_k(x) d\nu \quad .$$

Thus we conclude that

$$S_\nu(a) = \lim_{k\to\infty} S_\nu(a_k) \quad .$$

For any k using (4.6) we have that

$$S_\nu(a_k) \leq r(b_k)$$

so passing to the limit we obtain

$$S_\nu(a) \leq r(aU) \quad . \tag{4.9}$$

Let us denote by ν_k an invariant probability measure such that $r(b_k) = S_{\nu_k}(a_k)$.

Let μ be a weak–*limit point of the sequence $\{\nu_k\}$. Then μ belongs to Λ_0 and for the sequence $\{a_k\}$ there exists a subsequence $\left\{\nu_{k_i}\right\}$ such that

$$S_{\nu_{k_i}}(a_k) \to S_\mu(a_k) \quad , \qquad i \to \infty \quad .$$

(The existence of such a subsequence follows as in the foregoing reasoning from the Banach–Alaoglu theorem.)

If $j < i$ we have that

$$r(aU) \le r(b_{k_i}) = S_{\nu_{k_i}}(a_{k_i}) \le S_{\nu_{k_i}}(a_{k_j}) \quad .$$

Passing to the limit when $i \to \infty$ we obtain

$$r(aU) \le S_\mu(a_{k_j}) \quad .$$

Then passing to the limit when $j \to \infty$ (we may do this in view of the Levi theorem) we have that

$$r(aU) \le S_\mu(a) \quad . \tag{4.10}$$

Considering (4.9) and (4.10) we conclude that (4.6) is valid for any $a \in C(\partial A)$.

The ergodic measures, and only them, are the extremal points of the set Λ_0 (Lemma 2.20) so, in the case when $a(x) \neq 0$, we have (using the Choquet–Bishop–de Leeuw theorem (Theorem 2.21) that

$$f(\ln|a|) = \int_\Lambda \nu(\ln|a|)dm(\nu) \tag{4.11}$$

where Λ is the set of ergodic measures ν with respect to α and m is a probability measure. From (4.5), (4.6) and (4.11) we get that there exists an ergodic measure μ_0 such that

$$r(aU) = S_{\mu_0}(\ln|a|)$$

and in this case we have finished.

If there exists a point $x_0 \in \partial A$ such that $a(x_0) = 0$ then, taking a measure $\mu \in \Lambda_0$ such that

$$S_\mu(\ln|a|) = r(aU)$$

we have, by applying the Choquet–Bishop–de Leeuw theorem, that

$$\mu(\ln|a_k|) = \int_\Lambda \nu(\ln|a_k|)dm(\nu)$$

and since $\nu(\ln a_k) \xrightarrow[k\to\infty]{} \nu(\ln|a|)$ monotonically it follows that

$$\mu(\ln|a|) = \lim_{k\to\infty} \mu(\ln a_k) = \int_\Lambda \nu(\ln|a|)dm(\nu)$$

and again we can conclude that there exists an ergodic measure μ_0 such that $\mu(\ln|a|) = \mu_0(\ln|a|)$. Thus

$$S_{\mu_0}(\ln|a|) = r(aU)$$

and the proof is complete. □

Corollary 4.5 *If the function* a *is invertible (in* A) *then the spectrum* σ(π(a)T) *lies in the annulus*

$$\max_\Lambda S_\nu(a) \geq |\lambda| \geq \min_\Lambda S_\nu(a) \quad . \tag{4.12}$$

Proof Computing the spectral radius of $[\pi(a)T]^{-1}$ we have that

$$r([\pi(a)T]^{-1}) = r[T^{-1}\pi(a^{-1})] = r[\pi(a^{-1}\circ\alpha^{-1})T^{-1}]$$

$$= \max_\Lambda \exp \int_{\partial A} \ln|a^{-1}|d\nu = \max_\Lambda \exp\left[-\int_{\partial A} \ln|a|d\nu\right]$$

$$= \max_{\Lambda} S_{\nu}^{-1}(a) \quad .$$

This implies that

$$\min_{\lambda\in\sigma(\pi(a)T)} |\lambda| = \frac{1}{r([\pi(a)T]^{-1})} = \min_{\Lambda} S_{\nu}(a)$$

and thus the spectrum $\sigma(\pi(a)T)$ lies in the annulus

$$\max_{\Lambda} S_{\nu}(a) \geq |\lambda| \geq \min_{\Lambda} S_{\nu}(a) \quad . \quad \square$$

4.6 Looking through the proof of Theorem 4.4 one finds that we actually did not use the invertibility of T. In fact we used only the following properties of A and T:

(1) $\|\pi(a)\| = \max\limits_{x\in\partial A} |a|$;

(2) $\|\pi(a)T^{n}\| = \|\pi(a)\|$;

(3) $T\pi(a) = \pi(a\circ\alpha)T$ where $\alpha\colon \partial A \to \partial A$

is a continuous mapping such that $a \in A$ implies that $(a\circ\alpha) \in A$.

Thus the following result is valid.

Theorem 4.7 *If* A *and* T *satisfy conditions* (1)–(3) *then the spectral radius of* π(a)T *is given by the formula*

$$r(\pi(a)T) = \max_{\nu\in\Lambda} S_{\nu}(a) \quad .$$

4.8 The natural question arises: to what extent can Theorem 4.4 be generalized to cases when a weight algebra A is non–commutative? One of the possible generalizations is the consideration of the determinant in a von Neumann algebra (see Fuglede, Kadison [1] [2]).

Let B be a von Neumann algebra, let b be an invertible element of B and $b = |b|U$ be the polar decomposition of b. Since U is unitary, $|b|$ is invertible and one can define $\ln|b|$ which will be a selfadjoint element of B.

4.9 Suppose that there exists a finite trace φ on B such that $\varphi(I) = 1$. The number $\Delta(b)$ defined by the formula

$$\Delta(b) = \exp[\varphi(\ln|b|)] \qquad (b \text{ is invertible})$$

is called the *determinant* of b.

Remark 4.10 Note that if $b = aT$ is a weighted shift operator with a unitary T then $|b| = |a|$ and consequently

$$\Delta(aT) = \exp[\varphi(\ln|a|)] \quad .$$

Theorem 4.11 (Fuglede, Kadison [2]) *The determinant Δ possesses the following properties*:

(1) $\Delta(\lambda I) = |\lambda|$ *for* $\lambda \neq 0$;

(2) $\Delta(b^*) = \Delta(b) = \Delta(b^* b)^{\frac{1}{2}}$ *for invertible* b ;

(3) $\Delta(b_1 \cdot b_2) = \Delta(b_1) \cdot \Delta(b_2)$ *when* b_1 *and* b_2 *are invertible* ;

(4) $\Delta(\exp b) = |\exp \varphi(b)|$ *for every* $b \in B$;

(5) $\Delta(b) \leq \|b^n\|^{\frac{1}{n}}$ *for invertible* b *and thus* $\Delta(b) \leq r(b)$.

Recalling Remark 4.10 we see that for a weighted shift operator with a unitary shift we always have that

$$r(aT) \geq \Delta(|a|)$$

and naturally the question arises whether there exists a finite trace φ such that

$$r(b) = \Delta_{\varphi}(b) \quad .$$

A slightly weaker question is whether the following formula is valid:

$$r(b) = \sup_{\varphi \in \Phi} \Delta_{\varphi}(b)$$

where Φ is the set of all normalized traces ($\varphi(I) = 1$).

In fact if we are considering only the spectral radius formula we don't need the trace φ to be defined on the whole of B. It should be definitely defined on $|b|, |b^2|, \ldots,$ and so on, and it has to be invariant under the mapping $\hat{T}$: $a' \to Ta'T^*$ and $\hat{T}^*$: $a' \to T^*a'T$. So it has to be defined on the C^*–algebra A_b (see Theorem 3.13) and invariant under the automorphisms $\hat{T}$ and $\hat{T}^*$ of A_b (here $T = |b|^{-1}b$).

So the last question may be weakened even more: for a given invertible operator $b \in B$ is the following formula true?

$$r(b) = \sup_{\varphi \in \Lambda} \Delta_{\varphi}(b)$$

where Λ is the set of all normalized traces on A_b invariant under the automorphism $\hat{T}$, generated by $T = |b|^{-1}b$.

As shown in the following example the answer to the last question (and so the answers to all the previous questions) is negative.

Example 4.12 Let H be the Hilbert space $\mathbb{C}^2$ with the standard scalar product. Consider the operator $b \in L(H)$ defined by the matrix

$$b = \begin{bmatrix} 1 & 1 \\ 0 & 2 \end{bmatrix} .$$

Then we have that

$$bb^* = \begin{bmatrix} 2 & 2 \\ 2 & 4 \end{bmatrix}, \qquad b^*b = \begin{bmatrix} 1 & 1 \\ 1 & 5 \end{bmatrix} .$$

By elementary computation we see that the operators

$$\begin{bmatrix} 1 & 0 \\ 0 & 0 \end{bmatrix}, \qquad \begin{bmatrix} 0 & 0 \\ 0 & 1 \end{bmatrix} \qquad \text{and} \qquad \begin{bmatrix} 0 & 1 \\ 1 & 0 \end{bmatrix}$$

are generated by linear combinations of the operators bb^*, b^*b and I. Thus all selfadjoint operators of $L(H)$ belong to A_b and consequently $A_b = L(H)$.

But there exists a unique normalized trace φ on $L(H)$ namely

$$\varphi(a) = \frac{1}{2}\,\mathrm{Tr}(a) = \frac{a_{11} + a_{22}}{2}$$

where

$$a = \begin{pmatrix} a_{11} & a_{12} \\ a_{21} & a_{22} \end{pmatrix} .$$

Thus

$$\Delta(b) = \exp\left(\frac{\ln 1 + \ln 2}{2}\right) = \sqrt{2} ,$$

and on the other hand $r(b) = 2$. □

Looking through Example 4.12 we see that the main reason for the non–existence of a normalized trace with the property

$$r(b) = \Delta_\varphi(b)$$

is the non–commutativity of the C^*–algebra A_b. On the other when A_b is commutative then the answer to the last question is definitely positive and is stated in the following theorem.

Theorem 4.13 *Suppose that* b *is an invertible operator such that* $\mathrm{A_b}$ *is commutative (that is,* b *is central binormal). Then*

$$\mathrm{r(b)} = \max_{\varphi\in\Lambda} \Delta(\mathrm{b})$$

where Λ *is the set of all normalized traces on* $\mathrm{A_b}$ *which are invariant with respect to the automorphism* $\hat{\mathrm{T}}$: $\mathrm{a} \to \mathrm{TaT}^*$ *where* $\mathrm{T} = |\mathrm{b}|^{-1}\mathrm{b}$.

Proof Follows from Theorems 3.13 and 4.4. □

So to some extent we cannot generalize Theorem 4.4 in this direction too far.

On the other hand there exists another quite natural way of generalizing Theorem 4.4

4.14 Let us consider a weighted shift operator with a matrix coefficient, that is, $A \cong \mathrm{Mat}_N(X)$ (here $\mathrm{Mat}_N(X)$ is the algebra of N–dimensional continuous matrix functions over a compact space X) and let T be an isometry such that $(TaT^{-1})(x) = a(\alpha(x))$, where $\alpha\colon X \to X$ is a homeomorphism.

We confine ourselves here to the case when $\det a(x) \neq 0$.

Let $\tilde{X} = X \times CP^{N-1}$ where CP^{N-1} is an $(N-1)$–dimensional complex projective space (the set of one–dimensional subspaces of $\mathbb{C}^N$). Since $a(x)$ is invertible it maps every one–dimensional subspace onto some one–dimensional subspace, thus $a(x)$ determines a continuous bijective mapping $\bar{a}(x)$ of CP^{N-1} onto itself.

Define the mapping $\beta\colon \tilde{X} \to \tilde{X}$ by the formula

$$\beta(x, \xi) = (\alpha(x), \bar{a}(x)\xi) \quad , \qquad (x, \xi) \in X \times CP^{N-1} \quad .$$

We denote by γ the expansion coefficient of a one–dimensional subspace ξ under $a(x)$:

$$\gamma(x, \xi) = \frac{\|a(x)\eta\|}{\|\eta\|} \quad , \qquad \eta \in \xi \ , \qquad \|\eta\| \neq 0 \quad .$$

Now the spectral radius of aT can be described with the help of the next result.

Theorem 4.15 *Let* a $\in$ A *and* T *be the operators considered in* 4.14 (det a(x) $\neq$ 0). *Then*

$$r(aT) = \max_{\nu \in \Lambda_\beta} S_\nu(\gamma) \tag{4.13}$$

where Λ_β *is the set of ergodic measures on* $\tilde{X}$ *with respect to* β.

Proof Evidently we have

$$\|(aT)^n\| = \max_{\tilde{X}} \prod_{k=0}^{n-1} \gamma(\beta^k(x, \xi))$$

thus

$$r(aT) = \lim_{n\to\infty} \max_{\tilde{X}} \left[\prod_{k=0}^{n-1} \gamma(\beta^k(x, \xi))\right]^{\frac{1}{n}}$$

(cf. (4.3)).

Now the result follows from Theorem 4.4. □

Remark 4.16 We have to point out that in fact the computation of the spectral radius of a weighted shift with a matrix coefficient is far more complicated than the analogous computation in the scalar case. The thing is that in (4.13) the set Λ_β depends not only on α: $X \to X$ but also on the weight a. Thus one has to evaluate this measure for every particular weight a while in the scalar case by formula (4.2) it is enough to obtain the ergodic measures only for α and they will do the job for all the weights a.

Another way to compute the spectral radius of a weighted shift with operator valued weight a (using the Lyapunov exponents) will be presented in Section 25.

Remark 4.17 If instead of one weighted shift operator we consider the sum of such operators, for example

$$b = a_1T_1 + a_2T_2 \;, \qquad a \in A$$

where $T_iAT_i^{-1} = A$, $i = 1,2$, then the problem of the description of the spectrum of b is much more complicated in comparison with the corresponding problem for aT. There is not even an effective formula for the norm $\|a_1T_1 + a_2T_2\|$ in the general case. Thus the computation of the spectral radius in this case needs essentially different ideas. The investigation of these operators will be undertaken in Chapter 3.

5. Computation of the spectral radius. Examples

In this section we apply the results of the previous section to the computation of the spectral radius formulae of certain weighted shift operators.

First of all let us recall some facts from the theory of dynamical systems.

5.1 A point x_0 is called a *non–wandering point* for a homeomorphism $\alpha: X \to X$ if for every neighbourhood $W \ni x_0$ there exists $n \neq 0$ such that $\alpha^n(W) \cap W \neq \emptyset$.

Let $\Omega(\alpha)$ be the set of all non–wandering points for α. It is clear that $\Omega(\alpha)$ is invariant with respect to α and closed.

Lemma 5.2 *The support of any invariant Borel probability measure* ν *under* α *belongs to* $\Omega(\alpha)$.

Proof Let x_0 be any point which does not belong to $\Omega(\alpha)$. Then there exists a neighbourhood $W \ni x_0$ such that $\alpha^k(W) \cap \alpha^j(W) = \emptyset$, $k \neq j$. As ν is invariant we have that

$$\nu\left(\bigcup_{k=0}^{\infty} \alpha^k(W)\right) = \sum_{k=0}^{\infty} \nu(\alpha^k(W)) = \sum_{k=0}^{\infty} \nu(W) \quad .$$

If $\nu(W) > 0$ the series diverges and we have a contradiction. Thus $\nu(W) = 0$ and $x_0 \,\bar{\in}\, \text{supp}\ \nu$. So we conclude that

$$\text{supp}\ \nu \subset \Omega(\alpha) \quad . \quad \square$$

Lemma 5.3 *Let* α: X $\to$ X *be a homeomorphism on the topological space* X *which has a countable base. If a measure* ν *is ergodic with respect to* α *and* supp ν = X *then for almost every point* x $\in$ X *the trajectory of* x *is dense in* X.

Proof (Halmos [1]) Let U_k, $k = \overline{1,\infty}$, be a base for the topology on X. A point x has a non–dense trajectory if and only if there exists an open set U_j from the base such that $x \in \left[X \setminus \bigcup_{n\in\mathbb{Z}} \alpha^n(U_j)\right]$. The set $V_j = \bigcup_{n\in\mathbb{Z}} \alpha^n(U_j)$ is invariant under α, open and non–empty. As supp $\nu = X$ we have that $\nu(V_j) \neq 0$ and as ν is ergodic we conclude that $\nu(X \backslash V_j) = 0$. Now for the set W of all the points with non–dense trajectories we have that

$$W \subset \bigcup_j (X \backslash V_j)$$

and so $\nu(W) = 0$. □

Corollary 5.4 *If ν is an ergodic measure then for almost every point x we have that*

$$\theta_x = \text{supp } \nu$$

where $\theta_x = \overline{\{\alpha^n(x)\}}_{n \in \mathbb{Z}}$ *is the closure of the trajectory of* x.

Corollary 5.5 *If ν is an ergodic measure then there exists a point x_0 such that* $\theta_{x_0} = \text{supp } \nu$.

Now let us recall some facts about the structure of the homeomorphisms of the circle $S^1 = \mathbb{R}^1 (\text{mod } \mathbb{Z})$.

With any orientation–preserving homeomorphism α of the circle we can uniquely associate a number $0 < \tau(\alpha) \leq 1$ called the *rotation number* (Nitecki [1] Section 1.1). The following Theorems 5.6–5.10 are borrowed from Nitecki [1].

Theorem 5.6 *Let α be an orientation–preserving homeomorphism of the circle. Then:*

(1) *The number $\tau(\alpha)$ is rational if and only if $\Omega(\alpha)$ consists only of the periodic points. In this case the smallest period of each point in Ω is equal to* N *where* $\tau(\alpha) = {}^M/N$ *is an irreducible fraction.*

(2) *If $\tau(\alpha)$ is irrational then α can be one of the following types:*

   (a) *a* transitive *homeomorphism, when* $\Omega(\alpha) = S^1$;

   (b) *an* intransitive *homeomorphism, when $\Omega(\alpha)$ is perfect and nowhere dense.*

Theorem 5.7 (Denjoy) *Let* α *be a diffeomorphism of the circle and the derivative* α' *be continuous and has bounded variation. In this case if* $\tau(\alpha)$ *is irrational then* α *is transitive.*

For a number τ, $0 < \tau \leq 1$, we denote by ψ_τ the rotation of the circle by the angle τ which is obtained from the translation $t \to t + \tau$ on the line.

Theorem 5.8 *If* α *is a transitive homeomorphism then* α *is topologically conjugate to the rotation* $\psi_{\tau(\alpha)}$, *that is, there exists a homeomorphism* h: $S^1 \to S^1$ *such that the following diagram is commutative*:

$$\begin{array}{ccc} S^1 & \xrightarrow{\alpha} & S^1 \\ h\downarrow & & \downarrow h \\ S^1 & \xrightarrow{\psi_{\tau(\alpha)}} & S^1 \end{array}$$

Theorem 5.9 *Let* α *be an intransitive homeomorphism. Set* Ω_1 = {the left end points of the intervals of the complement of $\Omega(\alpha)$} *and* $\Omega_2(\alpha) = \Omega(\alpha)\backslash\Omega_1$. *Then there exists a one–to–one continuous mapping* h: $\Omega_2(\alpha) \to S^1$ *(the topology on* $\Omega_2(\alpha)$ *is that induced from* S^1) *such that the following diagram is commutative*:

$$\begin{array}{ccc} \Omega_2(\alpha) & \xrightarrow{\alpha} & \Omega_2(\alpha) \\ h\downarrow & & \downarrow h \\ S^1 & \xrightarrow{\psi_{\tau(\alpha)}} & S^1 \end{array}$$

Theorem 5.10 *Suppose that the homeomorphism* α: $S^1 \to S^1$ *changes the orientation. Then* $\Omega(\alpha)$ *consists of two fixed points and possibly of points of period* 2.

Now we shall compute the spectral radius formulae for the weighted shift operators associated with the homeomorphisms of the circle. Since the classification of the homeomorphisms of the circle is available (Theorems 5.6–5.10) it is possible to obtain an explicit description of all the ergodic measures in this case.

Example 5.11 Consider the operator defined on the space $C(S^1)$ by

$$[(aT)f](x) = a(x)f(\alpha(x)) \tag{5.1}$$

where $a(x) \in C(S^1)$, and α is a homeomorphism of S^1.

If $A = C(S^1)$, then $\partial A = S^1$. The mapping

$$\pi: A \to L(C(S^1)) \qquad [\pi(a)f](x) = a(x)f(x)$$

is an isometric representation of A, and

$$T: C(S^1) \to C(S^1) \qquad [Tf](x) = f(\alpha(x))$$

is an invertible isometry on $C(S^1)$, such that

$$T\pi(a)T^{-1} = \pi(a(\alpha(x))) \quad , \qquad a \in A \quad .$$

Thus we find ourselves satisfying the conditions of Theorem 4.4.

(a) $\tau(\alpha) = {}^{M}/_{N}$ where ${}^{M}/_{N}$ is an irreducible fraction. In this case the smallest period of each point in $\Omega(\alpha)$ is equal to N (Theorem 5.6 (1)). With any point $x \in \Omega(\alpha)$ we can associate the ergodic measure μ_x defined by

$$\mu_x(x) = \mu_x(\alpha(x)) = \dots = \mu_x(\alpha^{N-1}(x)) = \frac{1}{N}$$

and the measure of the remaining part of S^1 is equal to zero. It is evident that any measure ergodic with respect to α is of this form.

Thus if $\tau(\alpha) = {}^{M}/_{N}$, then the spectral radius of the operator aT is computed by the formula

$$r(aT) = \max_{x_0 \in \Omega(\alpha)} \exp \int_{S^1} \ln|a(x)|\,d\mu_{x_0}$$

$$= \max_{x_0 \in \Omega(\alpha)} \left[\prod_{i=0}^{N-1} \left| a(\alpha^i(x_0)) \right| \right]^{\frac{1}{N}} .$$

In the case when all the points are periodic (that is, $\alpha^N(x) = x$) we have that

$$r(aT) = \max_{x \in S^1} \left[\prod_{i=0}^{N-1} |a(\alpha^i(x))| \right]^{\frac{1}{N}} .$$

In particular if $\alpha(x) = x$ $(T \equiv I)$ then the ergodic measures are δ–functions concentrated at points and we return to the spectral radius formula for the operator of multiplication by a function:

$$r(a) = \max_{x \in S^1} |a(x)| .$$

(b) $\tau(\alpha)$ is irrational and α is transitive. Let h be the homeomorphism (mentioned in Theorem 5.8) which establishes the conjugation between α and the rotation $\psi_{\tau(\alpha)}$.

If ν is an invariant measure for α, then the measure μ defined by

$$\mu(\omega) = \nu(h^{-1}(\omega)) , \qquad \omega \subset S^1$$

is invariant for $\psi_{\tau(\alpha)}$. It is evident that in this way we establish a one–to–one correspondence between ergodic measures with respect to α and ergodic measures with respect to $\psi_{\tau(\alpha)}$. But for $\psi_{\tau(\alpha)}$ there exists only one invariant Borel probability measure – the Lebesgue measure. Thus we obtain the spectral radius formula in this case:

$$r(aT) = \exp \int_{S^1} \ln |a(h^{-1}(x))| \mathrm{d}x$$

where we set $\exp \int_{S^1} \ln |a(h^{-1}(x))| \mathrm{d}x = 0$ if the integral diverges.

Remark 5.12

(i) If $a(x) \neq 0$ on S^1 then (4.12) implies that the spectrum of aT lies on the circle of radius

$$\exp \int_{S^1} \ln|a(h^{-1}(x))|\,dx \quad .$$

(ii) If α is a rotation of the circle ($\alpha = \psi_{\tau(\alpha)}$), then $h(x) = x$ and

$$r(aT) = \exp \int_{S^1} \ln|a(x)|\,dx \quad .$$

(c) $\tau(\alpha)$ is irrational and α is intransitive. Let h, $\Omega_1(\alpha)$, $\Omega_2(\alpha)$ be the mapping and the sets mentioned in Theorem 5.9. Note that $\Omega_1(\alpha)$ is a countable set that is invariant under α and consists of non–periodic points (as $\tau(\alpha)$ is irrational α has no periodic points by Theorem 5.6). Thus for any measure ν ergodic with respect to α we have that $\nu(\Omega_1(\alpha)) = 0$. By Lemma 5.2 supp $\nu \subset \Omega(\alpha)$, hence, $\nu(\Omega_2(\alpha)) = 1$. In a way similar to that in (b) one can establish a one–to–one correspondence between measures ergodic with respect to α, and ergodic measures with respect to $\psi_{\tau(\alpha)}$:

$$\mu(\omega) = \nu(h^{-1}(\omega)) \quad , \qquad \omega \subset S^1 \quad .$$

Hence we have that

$$r(aT) = \exp \int_{S^1} \ln|a(x)|\,d\nu \tag{5.2}$$

where supp $\nu = \Omega(\alpha)$ and ν is determined (with the help of h) by the Lebesgue measure on S^1.

We can rewrite formula (5.2) in another way. Let us introduce the function $\tilde{a}$ on S^1 defined by the condition

$$\tilde{a}(x) = a(h^{-1}(x)) \quad , \qquad x \in S^1 \quad .$$

Then

$$r(aT) = \exp \int_{S^1} \ln|\tilde{a}(x)|\,dx \quad .$$

Remark 5.13 If $a(x) \neq 0$ on S^1 then (4.12) implies that the spectrum of aT lies on the circle of radius

$$\exp \int_{S^1} \ln|a(x)|d\nu \quad .$$

(d) Suppose that the homeomorphism α changes the orientation. In this case (see Theorem 5.10) $\Omega(\alpha)$ consists of two fixed points x_1 and x_2 and possibly points of period 2 (we denote the set of these points by $\Omega'(\alpha)$). By reasoning similar to that in case (a) we have that

$$r(aT) = \max \left\{ |a(x_1)|, \ |a(x_2)|, \ \max_{x \in \Omega'(\alpha)} |a(x) \cdot a(\alpha(x))|^{\frac{1}{2}} \right\} \quad .$$

Example 5.14 Let us consider the space $L^p(S^1)$ $1 \leq p \leq \infty$ and an operator

$$[(aT)f](x) = a(x)[\alpha'(x)]^{\frac{1}{p}} f(\alpha(x)) \tag{5.3}$$

where $a(x) \in C(X)$, α is a homeomorphism of S^1 preserving the equivalence class of the normalized Lebesgue measure, $\frac{1}{\infty} = 0$, and $\alpha'(x)$ is the Radon–Nikodym derivative of the measure $\alpha(dx)$ with respect to the Lebesgue measure dx (if α has a derivative at x_0 then it can be assumed that $\alpha'(x_0)$ coincides with the value of the derivative of α at x_0).

Let A be $C(S^1)$. Then $\partial A = S^1$. Take

$$\pi : A \to L(L^p(S^1)) \qquad [\pi(a)f](x) = a(x)f(x) \quad .$$

Then we have that π is an isometric representation of A.

Let T: $L^p(S^1) \to L^p(S^1)$ be defined by

$$[Tf](x) = [\alpha'(x)]^{\frac{1}{p}} f(\alpha(x)) \quad .$$

Then T is an invertible isometry, and

$$T\pi(a)T^{-1} = \pi(a(\alpha(x))) \quad .$$

Since the measures invariant under α are the same as in Example 5.11 we obtain the corresponding spectral radius formulae for the operators of the form aT by using the same argument as in Example 5.11(a)–(d).

Remark 5.15 Note that in fact on the space $C(S^1)$ we can compute the spectral radius for the operator of the form (5.1) even if $\alpha\colon S^1 \to S^1$ is not invertible. In this case we simply use Theorem 4.7 (instead of Theorem 4.4). (We consider α to be invertible simply because in this case we can compute all the ergodic measures explicitly by using Theorems 5.6–5.10.)

If we consider the space $L^p(S^1)$ and an operator of the form (5.3) in the case when α is not invertible then we cannot use Theorem 4.7 because condition (3) of Theorem 4.7 for the operators of the form (5.3) is not valid. The spectral radius formulae for the weighted shift operators associated with non–invertible isometries T will be considered in the second part of the book.

Example 5.16 In the space H^p, $1 \le p \le \infty$, we consider an operator

$$[(aT)f](x) = a(x)[\alpha'(x)]^{\frac{1}{p}} f(\alpha(x)) \tag{5.4}$$

where $a(x)$ is a function analytic in $\mathscr{D}$ and continuous in $\overline{\mathscr{D}}$, $\alpha\colon \overline{\mathscr{D}} \to \overline{\mathscr{D}}$ is a homeomorphism which is conformal on $\mathscr{D}$, $\alpha'(x)$ is the derivative of α in $\mathscr{D}$, and $\frac{1}{\infty} = 0$ (see Example 3.6).

Let A be the algebra of continuous functions in $\overline{\mathscr{D}}$ which are analytic in $\mathscr{D}$. Then A is a uniform algebra and $\partial A = S^1$.

$$\pi\colon A \to L(H^p) \quad , \qquad [\pi(a)f](x) = a(x)f(x)$$

is an isometric representation of A.

$$T\colon H^p \to H^p \quad , \qquad [Tf](x) = (\alpha'(x))^{\frac{1}{p}} f(\alpha(x))$$

is an invertible isometry such that $T\pi(a(x))T^{-1} = \pi[a(\alpha(x))]$.

Thus using Theorem 4.4 and the reasoning of Example 5.11(a), (b), (d) we obtain the corresponding formulae for the spectral radius of the operators of the form (5.4).

Remark 5.17 Since α is analytic and by Denjoy's theorem (Theorem 5.7) it is not possible to realize case (c) of Example 5.11 (an intransitive homeomorphism) in this example.

Example 5.18 Let X be a compact space and $\alpha: X \to X$ be a homeomorphism such that the set $\Omega(\alpha)$ of non–wandering points consists only of periodic points. Consider an operator aT of the form (5.1) (or of the form (5.3) in the case when $X = \text{supp}\ \mu$ for some Borel measure μ and α preserves the equivalence class of μ). In both cases we assume that the operator of multiplication $(af)(x) = a(x)f(x)$ is generated by a continuous function $a \in C(X)$).

Thus we have that $A = C(X)$, $\partial A = X$, and every ergodic measure is concentrated on the trajectory of one of the periodic points $x \in \Omega(\alpha)$.

Using the same arguments as in Example 5.11(a) we have that

$$r(aT) = \max_{x \in \Omega(\alpha)} \left[\prod_{k=0}^{p(x)-1} |a(\alpha^k(x))| \right]^{\frac{1}{p(x)}}$$

where $p(x)$ is the period of $x \in \Omega(\alpha)$.

Example 5.19 Let X be a compact space with a countable base and G be a compact group acting on X by homeomorphisms. This means that we have a continuous mapping $\Phi: G \times X \to X$ such that $e(x) = x$ and $g_1(g_2(x)) = (g_1 \cdot g_2)(x)$, $g_i \in G$, $x \in X$, where $g(x) = \Phi(g,x)$.

Consider an operator $aT_{g_0} : C(X) \to C(X)$:

$$\left[(aT_{g_0})f\right](x) = a(x)f(g_0(x))$$

where $a \in C(X)$, and g_0 is a particular element of G.

In order to compute the spectral radius of the operator aT we have to compute all the ergodic measures (with respect to g_0: $X \to X$).

Let ν be any such measure. By Corollary 5.4 there exists a point x_0 such that $\theta_{x_0} = \text{supp}\ \nu$.

Let G_0 be the closed subgroup in G generated by g_0, and $G_{x_0} \subset G_0$ be the stationary subgroup of the point x_0, that is, $G_{x_0} = \{g \in G_0, g(x_0) = x_0\}$. Now θ_{x_0} is homeomorphic to G_0/G_{x_0} and under this homeomorphism α goes into multiplication by g_0. On G_0/G_{x_0} there exists a unique invariant normalized measure, namely, the normalized Haar measure ν'. Under the homeomorphism $G_0/G_{x_0} \cong \theta_{x_0}$ the Haar measure goes into the ergodic measure concentrated on θ_{x_0} and for any $f \in C(X)$ we have that

$$\int_X f(x)d\nu = \int_{G_0/G_{x_0}} \tilde{f}_{x_0}(g)d\nu'$$

where $\tilde{f}_{x_0}(g)$ is the function on G_0/G_{x_0} generated by the function $f_{x_0}(g)$ on G_0 given by the formula

$$f_{x_0}(g) = f(g(x_0)) \quad , \qquad g \in G_0 \quad .$$

Note that

$$\int_{G_0/G_{x_0}} \tilde{f}_{x_0}(g)d\nu' = \int_{G_0} f_{x_0}(g)dg$$

where dg is the normalized Haar measure on G_0. So finally we have that

$$\rho(aT_{g_0}) = \max_{x \in X} \exp \int_{G_0} \ln|a(g(x))|dg \quad . \tag{5.5}$$

Example 5.20 Let us consider some concrete applications of formula (5.5). Suppose $T^n = \mathbb{R}^n \pmod{\mathbb{Z}^n}$ is the n–dimensional torus. Consider the standard regular shift operator T on the space $C(T^n)$

$$(Tf)(x) = f(x + h) \qquad h \in T^n$$

and a weighted shift operator aT: $C(T^n) \to C(T^n)$

$$(aTf)(x) = a(x)f(x + h)$$

$a \in C(T^n)$.

We can use the argument of Example 5.19. Take $G = T^n$, and let G_0 be the closed subgroup in T^n generated by h. This subgroup is isomorphic to the group $T^l \times \mathbb{Z}_N$ where l is the number of rationally independent numbers among $(h_1, h_2, ..., h_n) = h$, and N is an integer depending on the type of correlations between $h_1, ..., h_n$. On every torus T^l forming the group G_0 the Haar measure of G_0 is equal to $\frac{1}{N}$ dx where dx is the normalized Lebesgue measure on T^l.

Combining Lemma 5.2 and the argument of Example 5.19 we obtain the following method of computation of the ergodic measures (which actually includes all the examples considered).

5.21 Suppose α: $X \to X$ is a homeomorphism of a compact space X, and $Y \subset X$ is a closed invariant subset of X under α. The action of α on Y is said to be included in an action of a topological group G on Y if there exists an element $g_0 \in G$ such that

$$\alpha(x) = g_0(\mathrm{x}) \quad , \qquad x \in Y \quad .$$

Here $g(x)$ is the action of G on Y by homeomorphisms (see Example 5.19).

Theorem 5.22 *Let* α *be a homeomorphism of a compact space* X *which has a countable base of the topology. If the action of* α *on* $\Omega(\alpha)$ *is included in an action of a compact group* G *then a Borel measure* ν *on* X *is ergodic with respect to* α *iff there exists a point* $\mathrm{x}_0 \in \Omega(\alpha)$ *such that*

$$\int_X f(x)d\nu = \int_{G_0} f(g(x_0))dg \quad , \qquad f \in C(X)$$

where G_0 *is the closed subgroup of* G *generated by* α *and* dg *is the normalized Haar measure on* G_0.

6. Discontinuity of the spectral radius. Approximations of the spectrum. It is known that the spectral radius is upper semicontinuous (see 2.12) but in general not continuous, that is, the spectral radius can decrease greatly even when the perturbation of an operator is small. The well–known example which shows the discontinuity of the spectral radius was presented by Kakutani (see Halmos [2]). We begin this section with the observation that the discontinuity of the spectral radius is rather natural for weighted shift operators and obtain a simple example of an operator which possesses this property.

Proposition 6.1 *Consider the space* $S^1 = \mathbb{R} \pmod 1$, *and a function* $a \in C(S^1)$ *such that*

$$a(x) \geq 0 \quad , \qquad a(x_0) = 0 \quad , \qquad \int_{S^1} \ln|a(x)|dx = 0 \quad .$$

Let h *be an irrational number. Then the spectral radius of an operator*

$$[(aT_h)f](x) = a(x)f(x + h) \tag{6.1}$$

in any of the spaces $C(S^1)$ *and* $L^p(S^1)$, $1 \leq p \leq \infty$, *is equal to* 1 *and there exists a sequence of weighted shift operators* b_k *which converge in norm to* aT_h *and all the operators* b_k *are nilpotent. Thus in particular* $r(b_k) = 0$.

Proof By Examples 5.11(b), and 5.14 we have that

$$r(aT_h) = \exp \int_{S^1} \ln|a(x)|dx = 1$$

in any of the spaces $C(S^1)$ and $L^p(S^1)$, $1 \le p \le \infty$.

Let us choose any sequence of functions $a_k \in C(S^1)$ such that

$$a_k \ge 0 \quad , \qquad \max_{S^1} |a_k(x) - a(x)| \to 0$$

and every function a_k vanishes on a certain neighbourhood $V_k \ni x_0$.

Let us define the operators b_k by the formula

$$(b_k f)(x) = [(a_k T_h)f](x) = a_k(x)f(x + h) \quad .$$

Then we have that

$$\|b_k - b\| = \max_{S^1} |a_k(x) - a(x)| \to 0$$

and

$$r(b_k) = \exp \int_{S^1} \ln|a_k(x)|\mathrm{d}x = 0$$

as a_k vanishes on V_k.

Let us show that as in the classical example of Kakutani all the operators b_k are nilpotent.

Since h is irrational the set of points of the form $\{x_0 - jh, j \in \mathbb{Z}\}$ is dense on the circle S^1. Hence when k is fixed the system of sets $V_k - jh$, $j \in \mathbb{Z}$, is an open cover of S^1. By the compactness of S^1 this cover has a finite subcover, that is, there exists a number n_k such that

$$\bigcup_{j=1}^{n_k-1} (V_k - jh) = S^1 \quad .$$

As

$$b_k^n = \left(\prod_{j=0}^{n-1} a_k(x + jh)\right) T_h^n$$

the coefficient $\prod_{j=0}^{n-1} a_k(x + jh)$ vanishes on the set $\bigcup_{j=0}^{n-1} (V_k - jh)$. Thus when $n = n_k$ the coefficient and the operator $b_k^{n_k}$ is equal to zero. □

6.2 The foregoing result shows that if one slightly changes the weight a in the operator aT then the operator itself changes slightly (in norm) while its spectrum may change greatly.

On the other hand if one slightly changes the shift h in (6.1) then under any arbitrary small perturbation the difference between the corresponding weighted shift operators will not be small. For example if T_h is an operator on $L^p(S^1)$ acting by the formulae

$$(T_h f)(x) = f(x + h)$$

then

$$\|T_h - T_\tau\| = 2$$

for any $\tau \neq h$.

So it seems quite natural that in such situations there is no hope for the spectra of aT_h and aT_τ to be close to each other. Nevertheless we shall show now that if h is irrational then

$$\sigma(aT_\tau) \xrightarrow[\tau \to h]{} \sigma(aT_h) \quad .$$

We recall that the *Hausdorff distance* between two closed bounded subsets A and B of a metric space (M, ρ) is the number defined by

$$\rho_H(A, B) = \sup_{\substack{x \in A \\ y \in B}} (\rho(x, B), \quad \rho(y, A))$$

where

$$\rho(x, B) = \inf_{y \in B} \rho(x, y) \quad .$$

Theorem 6.3 *If* aT_h *is an operator on* $L^p(S^1)$, $1 \leq p \leq \infty$, *or* $C(S^1)$ *defined by* (6.1) *and* h *is an irrational number then*

$$\sigma(aT_\tau) \xrightarrow[\tau\to h]{} \sigma(aT_h) \tag{6.2}$$

in the Hausdorff metrics.

Proof If τ is irrational then $\sigma(aT_h) = \sigma(aT_\tau)$ (see **8.5**). Thus we have to consider only the situation when τ is a rational number.

Let us first of all calculate the explicit form of the spectrum in this case.

Lemma 6.4 *If* $\tau = {}^m/_n$ *is an irreducible fraction then*

$$\sigma(aT_\tau) = \left\{\lambda \in \mathbb{C}: \exists\ x \in S^1,\ \lambda^n = \prod_{k=0}^{n-1} a(x + k\tau)\right\} . \tag{6.3}$$

Proof A number λ is a regular number for the operator $b = aT_\tau$ iff the equation

$$\lambda u(x) - a(x)u(x + \tau) = f(x) \tag{6.4}$$

has only one solution (belonging to the space where aT_τ acts).

Implementing the substitution $x \to x + h$ in (6.4) one obtains the system

$$\begin{aligned} &\lambda u(x) - a(x)u(x+\tau) &&= f(x) \\ &\lambda u(x+\tau) - a(x+\tau)u(x+2\tau) &&= f(x+\tau) \\ &\cdots\cdots\cdots\cdots\cdots\cdots\cdots\cdots\cdots\cdots\cdots\cdots \\ &\lambda u(x+(n-1)\tau) - a(x+(n-1)\tau)u(x) &&= f(x+(n-1)\tau) \end{aligned} \tag{6.5}$$

which we shall consider as a system of linear algebraic equations with respect to the numbers $u(x)$, $u(x+h)$, ..., $u(x+(n-1)h)$. The determinant of this system is

$$\Delta(x, \lambda) = \lambda^n - \prod_{k=0}^{n-1} a(x + k\tau) . \tag{6.6}$$

If $\Delta(x, \lambda) \neq 0$ then the system (6.5) has a solution, thus there exists only one function $u(x)$ satisfying (6.4). This function has the form

$$u(x) = \sum_{k=0}^{n-1} c_k(x)f(x + kh)$$

where $c_k(x)$ are continuous (since $a(x)$ is continuous) functions. Thus if $f(x) \in L^p(S^1)$ (or $C(S^1)$) then $u(x) \in L^p(S^1)$ (or $C(S^1)$).

So we conclude

$$\sigma(aT_\tau) \subset \{\lambda\colon \exists x,\ \Delta(x, \lambda) = 0\} = \left\{\lambda\colon \exists x,\ \lambda^n = \prod_{k=0}^{n-1} a(x + k\tau)\right\} .$$

Now let us show that the opposite inclusion is also true.

Let λ be such that $\Delta(x_0, \lambda) = 0$ for some x_0. Suppose that $\lambda I - b$ is invertible. Then there exists a δ–neighbourhood $O(\lambda I - b)$ consisting of invertible operators.

One can easily find a function $\bar{a} \in C(S^1)$ such that

(α) $\bar{a}(x_0 + k\tau) = a(x_0 + k\tau)$, $\qquad k = \overline{0,n-1}$

(β) $\bar{a}(x)$ is constant in some neighbourhood of every point

$\qquad x_0 + k\tau$, $\qquad k = \overline{0,n-1}$

(γ) $|\bar{a}(x) - a(x)| < \delta$, $\qquad x \in C(S^1)$.

The property (γ) means that the operator $d = \lambda I - \bar{a}T_\tau$ belongs to the above–mentioned neighbourhood $O(\lambda I - b)$. On the other hand we shall show now that $\lambda I - \bar{a}T_\tau$ is not invertible and thus arrive at a contradiction.

Let $\lambda \neq 0$. Take any function $u(x)$ which is zero outside of the neighbourhood of x_0 mentioned in (β).

Resolving the homogeneous system (6.5) (substituting $a(x)$ with $\bar{a}(x)$) we find the values of $u(x)$ in the neighbourhoods of the point $x_0 + k\tau$, $k = \overline{1,n-1}$. Thus we obtain the function $u(x)$ such that $du = 0$ and consequently d is not invertible.

If $\lambda = 0$ we have $a(x_0 + k\tau) = 0$ for some $k \in \overline{0,n-1}$. In this case choose $u(x)$ in an arbitrary way in the neighbourhood of this point mentioned in (β). By an argument analogous to the foregoing one we have again that $du = 0$, and the lemma is proved. □

Let us continue the proof of the theorem. By **8.5** we have

$$\sigma(aT_h) = \begin{cases} \lambda \in \mathbb{C} : |\lambda| = S(a) , & \text{if } a(x) \neq 0 \\ \lambda \in \mathbb{C} : |\lambda| \leq S(a) , & \text{if } \exists x_0,\ a(x_0) = 0 \end{cases} \tag{6.7}$$

where $S(a) = \exp \int_0^1 \ln|a(x)|\mathrm{d}x$. Consider first the case $a(x) \neq 0$.

The sum

$$\frac{1}{n} \sum_{k=0}^{n-1} \ln|a(x + k\tau)| = \frac{1}{n} \sum_{k=0}^{n-1} \ln\left|a\left(x + \frac{k}{n}\right)\right|$$

is the integral sum for the Riemann integral

$$\int_0^1 \ln|a(t)|\mathrm{d}t \quad .$$

Thus for any $\varepsilon_1 > 0$ there exists an N_{ε_1} such that for every $n > N_{\varepsilon_1}$ and any x

$$\left|\frac{1}{n} \sum_{k=0}^{n-1} \ln|a(x + k\tau)| - \int_0^1 \ln|a(t)|\mathrm{d}t\right| < \varepsilon_1 \quad .$$

Thus taking a sufficiently small ε_1 we have, for all $\lambda \in \sigma(aT_\tau)$, $\tau = \frac{m}{n}$,

$$S(a) - \varepsilon \leq \exp \left|\int_0^1 \ln|a(t)\mathrm{d}t - \varepsilon_1\right| \leq |\lambda| \leq \exp \left(\int_0^1 \ln|a(t)|\mathrm{d}t + \varepsilon_1\right)$$
$$\leq S(a) + \varepsilon$$

for any $n > N_{\varepsilon_1}$.

Thus for every $\lambda \in \sigma(aT_\tau)$ we have

$$\rho(\lambda, S(a)) = \rho(\lambda, \sigma(aT_h)) < \varepsilon \quad . \tag{6.8}$$

Now observe that $\sigma(aT_\tau)$ is invariant under multiplication by the numbers $e^{i\frac{2\pi}{n}k}$, $k = 0,n-1$ (see 8.5). It follows that if n is sufficiently large then for every $\lambda_0 \in \{\lambda: |\lambda| = S(a)\}$ we have

$$\rho(\lambda_0, \sigma(aT_\tau)) < 2\varepsilon \quad . \tag{6.9}$$

Thus for sufficiently large n and $\tau = {}^m/_n$ we have

$$\rho_H(\sigma(aT_\tau) \ , \quad \sigma(aT_h)) < 2\varepsilon \quad .$$

Obviously since h is irrational we have

$$\tau = \frac{m}{n} \to h \Rightarrow n \to \infty$$

which implies

$$\rho_H(\sigma(aT_\tau) \ , \quad \sigma(aT_h)) \xrightarrow[\tau \to h]{} 0 \quad .$$

Now let there exist an $x_0 \in S^1$ for which $a(x_0) = 0$. Then for any $\lambda \in \sigma(aT_\tau)$ and any $\varepsilon_1 > 0$ one has

$$|\lambda| = \left[\prod_{k=0}^{n-1} \left|a\left(x + \frac{k}{n}\right)\right|\right]^{\frac{1}{n}} \le \left[\prod_{k=0}^{n-1} \left(\left|a\left(x + \frac{k}{n}\right)\right| + \varepsilon_1\right)\right]^{\frac{1}{n}}$$

$$= \exp \frac{1}{n} \sum_{k=0}^{n-1} \ln\left(\left|a\left(x + \frac{k}{n}\right)\right| + \varepsilon_1\right) \quad .$$

The last sum here is the integral sum for the Riemann integral. So, for any $\varepsilon > 0$ taking sufficiently small ε_1 we have for sufficiently large n

$\left(\tau = {}^{m}/_{n}\right)$ that

$$|\lambda| \leq \exp \int_0^1 \ln(|a(x)| + \varepsilon_1)dx + \varepsilon_1 = S(|a| + \varepsilon_1) + \varepsilon_1 < S(a) + \varepsilon .$$

Since by (6.7) we have in this case

$$\sigma(aT_h) = \{\lambda : |\lambda| \leq S(a)\}$$

it follows that

$$\lambda \in \sigma(aT_\tau) \Rightarrow \rho(\lambda, \sigma(aT_h)) < \varepsilon \quad . \tag{6.10}$$

On the other hand

$$[r(aT_\tau)]^n = \max_{S^1} \left[\prod_{k=0}^{n-1} \left|a\left(x + \frac{k}{n}\right)\right|\right] \geq S\left(\prod_{k=0}^{n-1} \left|a\left(x + \frac{k}{n}\right)\right|\right) = [S(a)]^n \quad .$$

So when x is changing, the value $|\lambda|$ is continuously changing from 0 (when $x = x_0$) up to $r(aT_\tau)$ which satisfies

$$S(a) \leq r(aT_\tau) \leq S(a) + \varepsilon \quad .$$

In particular for any $0 \leq t \leq S(a)$ there exists $\lambda_0 \in \sigma(aT_\tau)$ such that $|\lambda_0| = t$, and since $\sigma(aT_\tau)$ is invariant under multiplication by $e^{i\frac{2\pi}{n}k}$ $\left(\tau = {}^{m}/_{n}\right)$ all the numbers of the form $\left\{\lambda_0 \; e^{i\frac{2\pi}{n}k}\right\}$ belong to $\sigma(aT_\tau)$ which again implies that for sufficiently large n

$$\lambda \in \sigma(aT_h) \Rightarrow \rho(\lambda, \sigma(aT_\tau)) < \varepsilon \quad , \tag{6.11}$$

which in view of

$$\tau = \frac{m}{n} \to h \Rightarrow n \to \infty$$

finishes the proof. □

Remarks 6.5

(1) We have proved Theorem 6.3 by obtaining an explicit description of $\sigma(aT_\tau)$ and $\sigma(aT_h)$. If one succeeds in proving this (or an analogous) theorem without using the explicit description of $\sigma(aT_h)$ one obtains the possibility of investigating the spectrum $\sigma(aT_h)$ with the help of approximations by $\sigma(aT_\tau)$ with rational τ. The benefit of such an approach is that if τ is rational then $\left\{T_\tau^k\right\}_{k=\overline{0,n}}$ forms a finite group and the invertibility conditions for $b = \sum_{k=0}^{n-1} a_k T_\tau^k$ can be written explicitly (see Section 15) while the conditions for the invertibility of $\sum_{k=0}^{n-1} a_k T_h^k$ with an irrational h are much more complicated and in general cannot be written explicitly.

In fact the idea of investigating the algebraic properties (and in particular the invertibility conditions) of operators of the form

$$b_h = \sum a_k T_h^k \quad , \qquad h \text{ is irrational}$$

by the approximations

$$b_\tau = \sum a_k T_\tau^k \quad , \qquad \tau \text{ is rational}, \quad \tau \to h$$

has much in common with the procedure of passing from periodic to almost–periodic functions.

(2) It is worth mentioning again that we have found that precisely the spectrum $\sigma(aT_\tau)$ as a *whole set* tends to $\sigma(aT_h)$ while $\tau \to h$. On the other hand if we fix some point λ then it seems to be true (at first

sight) that λ is a regular point for aT_h (h irrational) iff for all the rational τ which are sufficiently near to h, λ is a regular point for aT_τ.

But this statement is false. For example, the spectrum $\sigma(T_h)$ is $\{\lambda: |\lambda| = 1\}$ and the spectrum $\sigma(T_\tau)$ ($\tau = {}^m/_n$) is $\left\{e^{i\frac{2\pi}{n}k}\right\}_{k=\overline{0,n-1}}$. Thus if $\lambda = e^{i2\pi\nu}$ where ν is irrational then ν is a regular point for every T_τ and is a spectral point for T_h.

So what is the property of $\lambda = e^{i2\pi\nu}$ which distinguishes it from the regular points of aT_h?

The additional property which distinguishes the spectral and the regular points of aT_h in terms of aT_τ can be formulated in the following way:

λ is a regular point for aT_h *iff, for all the rational τ which are sufficiently near to* h *the operators* $(\lambda I - aT_\tau)$ *are invertible and their inverses* $(\lambda I - aT_\tau)^{-1}$ *are uniformly bounded.*

The development of this idea, as well as the approximation approach as a whole, will be treated in Section 27.

(3) Observe that $\sigma(aT_{\tau_1}) = \sigma(aT_{\tau_2})$ if $\tau_1 = {}^{m_1}/_n$ and $\tau_2 = {}^{m_2}/_n$ (here ${}^{m_i}/_n$, $i = 1,2$, are irreducible fractions). In particular in this case the trajectories $\{x + k\tau_1\}$ and $\{x + k\tau_2\}$ under the mappings $x \to x + \tau_i$, $i = 1,2$, coincide (as sets). If $h \to \infty$ then these trajectories are disposed on the circle more and more thickly, tending (if $\tau \to h$) to the trajectories of the mapping $x \to x + h$.

On the other hand if τ is rational and $h \to \tau$ then the trajectories of the mapping $x \to x + h$ differ greatly from the trajectories of $x \to x + \tau$.

So this is to some extent an 'explanation' of why

$$\sigma(aT_\tau) \xrightarrow[\tau\to h]{} \sigma(aT_h) \qquad (h \text{ irrational})$$

and on the other hand

$$\sigma(aT_h) \underset{h\to\tau}{\not\longrightarrow} \sigma(aT_\tau) \qquad (\tau \text{ rational}).$$

7. The number of connected components of the spectrum of weighted shift operators.

In this section we find an estimate for the number of connected components of the spectrum of a weighted shift operator in the case when the coefficient is taken from a commutative algebra.

In particular, let $C(M)$ be the space of continuous functions on a compact space M and $A \cong C(M)$ be an isometric representation of $C(M)$ by operators acting on a Banach space E. Let T be an invertible isometry on E such that $TAT^{-1} = A$ and α be the homeomorphism of M generated by the automorphism $\hat{T}$ of A: $\hat{T}(a) = TaT^{-1}$, $a \in A$.

7.1 For a subset $V \subset M$ let us denote by $A(\alpha, V)$ the subalgebra of $C(M)$ consisting of all the functions which are constant on V and invariant with respect to α. By $M(\alpha, V)$ we denote the maximal ideal space of $A(\alpha, V)$. For any element $a \in A$ the symbol $V(a)$ stands for the set $\{m \in M \mid a(m) = 0\}$.

By $|\sigma(aT)|$ we denote the set of numbers of the form $|z|$ where $z \in \sigma(aT)$.

Theorem 7.2 *Suppose that* $1 \bar{\in} |\sigma(\mathrm{aT})|$. *Then the spectral projection*

$$\mathrm{P} = \frac{1}{2\pi \mathrm{i}} \int_{|\lambda|=1} (\lambda - \mathrm{aT})^{-1} d\lambda$$

belongs to $\mathrm{A}(\alpha, \mathrm{V(a)})$.

Proof As $1 \bar{\in} |\sigma(aT)|$ it follows by the definition of $|\sigma(aT)|$ that no point of the circle $|\lambda| = 1$ belongs to the spectrum $\sigma(aT)$.

Consider the projection

$$P = \frac{1}{2\pi \mathrm{i}} \int_{|\lambda|=1} (\lambda - aT)^{-1} d\lambda$$

and subspaces $E_1 = \operatorname{Im} P$ and $E_2 = \operatorname{Ker} P$. We have that

$$E = E_1 \oplus E_2$$

where both subspaces E_i, $i = 1,2$, are invariant with respect to aT and

$$\sigma(aT)|_{E_1} \subset \{z \in \mathbb{C} \mid |z| < 1\} \quad ,$$

$$\sigma(aT)|_{E_2} \subset \{z \in \mathbb{C} \mid |z| > 1\} \quad .$$

Hence there exists a positive integer n_0 and a number ε, $0 < \varepsilon < 1$, such that:

(1) $\|(aT)^n e_1\| \leq (1 - \varepsilon) \|e_1\|$ for any $e_1 \in E_1$ when $n > n_0$;

(2) $\|(aT)^n e_2\| \geq (1 + \varepsilon) \|e_2\|$ for any $e_2 \in E_2$ when $n > n_0$.

The key points of the proof are the following two steps.

Step I *Both subspaces* E_1 *and* E_2 *are invariant for all operators* $a' \in A$.

Proof Let e be any vector of E. Then there exists a unique decomposition of the form $e = e_1 + e_2$, $e_i \in E_i$, $i = 1,2$. Thus

$$\|(aT)^n e\| = \|(aT)^n (e_1 + e_2\| \geq \|(aT)^n e_2\| - \|(aT)^n e_1\| \quad .$$

Conditions (1) and (2) imply that

$$\|(aT)^n e_2\| \xrightarrow[n\to\infty]{} \infty \quad , \qquad \|(aT)^n e_1\| \xrightarrow[n\to\infty]{} 0 \quad .$$

Hence, in the case when $e_2 \neq 0$, we have that

$$\|(aT)^n e\| \xrightarrow[n\to\infty]{} \infty$$

and when $e_2 = 0$ we have that

$$\|(aT)^n e\| = \|(aT)^n e_1\| \xrightarrow[n\to\infty]{} 0 \ .$$

So we conclude that the subspace E_1 is determined by the following condition

$$e \in E_1 \Leftrightarrow (aT)^n e \xrightarrow[n\to\infty]{} 0 \ . \tag{7.1}$$

Condition (7.1) implies that E_1 is an invariant subspace for every operator $a' \in A$.

Indeed for an arbitrary vector $e \in \mathrm{E}_1$ and any operator $a' \in A$ we have that

$$\|(aT)^n \ a'e\| = \|(a' \circ \alpha^n)(aT)^n e\|$$

$$\leq \|a'\| \ \|(aT)^n e\| \xrightarrow[n\to\infty]{} 0 \ .$$

Hence, in view of (7.1) $a'e \in E_1$, that is, $a'E_1 \subset E_1$.

Let us show now that E_2 is also invariant with respect to an arbitrary $a' \in A$.

Note first that the following general result about the spectral decomposition of an operator and its adjoint is valid.

Lemma 7.3 *Suppose that the spectrum of an operator* b *(acting on a Banach space* E*) does not meet the unit circle. Then for the spectral decompositions* $\mathrm{E} = \mathrm{E}_1 \oplus \mathrm{E}_2$ *and* $\mathrm{E}^* = \mathrm{E}_{1*} \oplus \mathrm{E}_{2*}$ *(corresponding to the decompositions of the spectra of the operator* b *and of its adjoint* b^* *by the unit circle) we have that*

$$\mathrm{E}_2 = {}^{\perp}\mathrm{E}_1^* \qquad \textit{and} \qquad \mathrm{E}_1 = {}^{\perp}\mathrm{E}_2^* \ .$$

Proof Since the spectrum $\sigma(b)$ does not meet the circle the same is true for the spectrum $\sigma(b^*)$. Thus we can consider the spectral decompositions

$$E = E_1 \oplus E_2$$
$$E^* = E_{1*} \oplus E_{2*}$$

where

$$\sigma(b)|_{E_1} \subset \{z \in \mathbb{C} | \; |z| < 1\}$$
$$\sigma(b)|_{E_2} \subset \{z \in \mathbb{C} | \; |z| > 1\}$$

and

$$\sigma(b^*)|_{E_{1*}} \subset \{z \in \mathbb{C} | \; |z| < 1\}$$
$$\sigma(b^*)|_{E_{2*}} \subset \{z \in \mathbb{C} | \; |z| > 1\} \quad .$$

Using the explicit form of the spectral projection $P_{E_{1*}}$ onto the subspace E_{1*} (see Riesz, Sz.–Nagy [1], §147) we have that

$$P_{E_{1*}} = \lim_{n\to\infty} (I - b^{*n})^{-1} = \lim_{n\to\infty} (I - b^{n})^{-1^*} = P^*_{E_1} \quad .$$

This implies that ${}^{\perp}E_1^* = E_2$.

Indeed, let $e \in E$ be a vector such that $\langle e, P_{E_{1*}} e^* \rangle = 0$ for any vector $e^* \in E^*$. Then $\langle P_{E_1} e, e^* \rangle = 0$ and consequently $P_{E_1} e = 0$, that is, $e \in E_2$. On the other hand, if $e \in E_2$ then $P_{E_1} e = 0$ and for any vector $e^* \in E^*$ we have that

$$0 = \langle P_{E_1} e, e^* \rangle = \langle e, P^*_{E_1} e^* \rangle = \langle e, P_{E_{1*}} e^* \rangle \quad .$$

Thus

$$e \in E_2 \Leftrightarrow \langle e, e^* \rangle = 0 \qquad \text{for} \qquad e^* \in E_{1*}$$

that is, ${}^{\perp}E_{1*} = E_2$.

The equality ${}^{\perp}E_{2*} = E_1$ can be established by passing to the projection $I - P_{E_{1*}} = I - P^*_{E_1}$, and the proof of the lemma is complete. □

Now we finish Step I.

As the spectrum $\sigma(aT)$ does not meet the unit circle, neither does the spectrum $\sigma((aT)^*)$ and we can consider the spectral decomposition for $(aT)^*$:

$$E^* = E_{1*} \oplus E_{2*}$$

where

$$\sigma((aT)^*)|_{E_{1*}} \subset \{z \in \mathbb{C} | \; |z| < 1\}$$

$$\sigma((aT)^*)|_{E_{2*}} \subset \{z \in \mathbb{C} | \; |z| > 1\} \quad .$$

By reasoning similar to that above (see the proof of the inclusion $a'E_1 \subset E_1$, $a' \in A$) we have that

$$a'^* E_{1*} \subset E_{1*} \qquad \text{for any } a' \in A \quad . \tag{7.2}$$

Now for arbitrary vectors $e_2 \in E_2$ and $e_1^* \in E_{1*}$ by Lemma 7.3 we have that

$$\langle e_2, e_1^* \rangle = 0$$

and using (7.2) we conclude that

$$0 = \langle e_2, a'^* e_1^* \rangle = \langle a'e_2, e_1^* \rangle \quad .$$

Thus $a'e_2 \in E_2$, that is, $a'E_2 \subset E_2$, and the proof of Step I is complete.

Step II *Both subspaces* E_1 *and* E_2 *are invariant with respect to* T.

Proof First we prove that $TE_1 \subset E_1$. If a is invertible this is evident since in this case

$$TE_1 = a^{-1}[(aT)E_1] \subset a^{-1}E_1 \subset E_1$$

where we used the invariance of E_1 with respect to aT and Step I.

Suppose that a is not invertible and fix some positive number $C \geq 2 \max_{m \in M} |a(m)|$. The Tietze theorem (Kuratowski [1], §14.1V) on the extension of a continuous function from a closed set onto the whole of a normal space implies that there exists a function $g_k \in C(M)$, $k = 1,2,$ such that:

(i) $|g_k(m)| \leq C$;

(ii) $g_k(m) = \frac{|a(m)|}{a(m)}$ when $|a(m)| \geq \frac{1}{k}$.

In view of Step I we have that

$$[g_k aT]E_1 \subset g_k E_1 \subset E_1 \quad , \qquad k = 1,2,\ldots \quad ,$$

and from the construction of g_k it follows that

$$g_k a \xrightarrow[k \to \infty]{} |a| \quad .$$

Hence we obtain that

$$[|a|T]E_1 \subset E_1 \quad . \tag{7.3}$$

Let $e \in E_1$. We have to prove that in the representation

$$Te = e_1 + e_2 \qquad e_i \in E_i \quad , \qquad i = 1,2,$$

the vector e_2 is zero. Indeed we have that

$$[|a|T]e = |a|e_1 + |a|e_2 \quad .$$

By Step I we have that $|a|e_i \in E_i$, $i = 1,2$, and in view of (7.3) we have that

$$|a|e_1 + |a|e_2 \in E_1 \quad .$$

It follows then that $|a|e_2 = 0$.

Let us show that this implies $e_2 = 0$. For every $k = 2,3,\ldots$ let φ_k denote a certain function satisfying the conditions

$$0 \le \varphi_k(m) \le 1 \quad , \qquad \varphi_k(m) = \begin{cases} 0 \, , & \text{when} \quad |a(m)| \le \frac{1}{k} \\ 1 \, , & \text{when} \quad |a(m)| \ge \frac{1}{k-1} \, . \end{cases}$$

Such a function exists by Urysohn's Lemma. Note that $\varphi_k e_2 = 0$. Indeed, suppose on the contrary that $\varphi_k e_2 \neq 0$, and denote by $|a|_k$ the operator defined by the condition

$$|a|_k = \begin{cases} |a(m)| \, , & \text{when} \quad |a(m)| > \frac{1}{k} \\ \frac{1}{k} \, , & \text{when} \quad |a(m)| \le \frac{1}{k} \, . \end{cases}$$

It is evident that $|a|_k$ is an invertible operator and $|a|\varphi_k = |a|_k\varphi_k$. Since $\varphi_k e_2 \neq 0$ and $|a|_k$ is invertible we have that

$$|a|_k(\varphi_k e_2) \neq 0 \quad .$$

On the other hand

$$0 = \varphi_k(|a|e_2) = (|a|\varphi_k)e_2 = |a|_k(\varphi_k e_2) \quad .$$

Thus we obtain a contradiction. Hence $\varphi_k e_2 = 0$. As the operator (aT) is an invertible operator on the subspace E_2 there exists a vector $e_3 \in E_2$ such that $(aT)e_3 = e_2$. By the construction of the functions φ_k we have that

$$\varphi_k a \xrightarrow[k \to \infty]{} a \quad .$$

Hence

$$0 = \varphi_k e_2 = \varphi_k a T e_3 \xrightarrow[k \to \infty]{} aTe_3 = e_2 \quad ,$$

so we have proved that

$$TE_1 \subset E_1 \quad .$$

The proof of the inclusion $TE_2 \subset E_2$ can be obtained by an argument similar to that above (see the proof of the inclusion $a'E_2 \subset E_2$ in Step I) by passing to the dual space E^*.

Hence the proof of Step II is complete.

Now we finish the proof of the theorem.

Since T is an invertible operator, $E = E_1 \oplus E_2$ and $TE_i \subset E_i$ it follows that $TE_i = E_i$. So using conditions (1) and (2) we obtain that

(1′) $$\|(aT)^n\|_{E_1} = \left\| \left[\prod_{i=0}^{n-1} (a \circ \alpha^i) \right] T^n \right\|_{E_1} = \left\| \prod_{i=0}^{n-1} (a \circ \alpha^i) \right\|_{E_1} < 1 - \varepsilon$$

when $n > n_0$;

(2′) If $n > n_0$ then in the subspace E_2 there exists an operator

$$\left[\prod_{i=0}^{n-1} (a \circ \alpha^i) |_{E_2} \right]^{-1} \quad \text{and} \quad \left\| \left[\prod_{i=0}^{n-1} (a \circ \alpha^i) |_{E_2} \right]^{-1} \right\| < \frac{1}{1 + \varepsilon} \quad .$$

Condition (2′) follows from the fact that for any vector $e \in E_2$ and $n > n_0$ there exists a vector $e_n \in E_2$ such that $T^{-n}e = e_n$ and hence

$$\begin{aligned} \left\| \prod_{i=0}^{n-1} (a \circ \alpha^i) e \right\| &= \left\| \prod_{i=0}^{n-1} (a \circ \alpha^i) T^n (T^{-n} e) \right\| \\ &= \|(aT)^n (T^{-n} e)\| \\ &> (1 + \varepsilon) \; \|T^{-n} e\| \\ &= (1 + \varepsilon) \; \|e\| \quad . \end{aligned}$$

For brevity let g denote the operator $\prod_{i=0}^{n-1} (a \circ \alpha^i)$. Then Step I shows that both E_1 and E_2 are invariant under g and thus

$$\sigma(g) = \sigma(g)|_{E_1} \cup \sigma(g)|_{E_2}$$

and conditions (1′) and (2′) imply that

$$(1'') \quad \sigma(g)|_{E_1} \subset \{z \in \mathbb{C} \mid |z| < 1\}$$

$$(2'') \quad \sigma(g)|_{E_2} \subset \{z \in \mathbb{C} \mid |z| > 1\} \quad .$$

Consequently there exists a decomposition $E = E_1' \oplus E_2'$ where E_i' $(i = 1,2)$ are invariant with respect to g and

$$(1''') \quad \sigma(g)|_{E_1} = \sigma(g)|_{E_1'}$$

$$(2''') \quad \sigma(g)|_{E_2} = \sigma(g)|_{E_2'} \quad .$$

It follows then that $E_i = E_i'$. Indeed suppose $e = e_1 + e_2$, $e_i \in E_i'$, and $e_2 \neq 0$. Then

$$\|g^n e\| = \left\|g^n(e_1 + e_2)\right\| = \left\|g^n e_1 + g^n e_2\right\| \geq \left|\left\|g^n e_2\right\| - \left\|g^n e_1\right\|\right| \quad .$$

Conditions (1′) and (2′) imply that

$$\left\|g^n e_1\right\| \to 0 \qquad \left\|g^n e_2\right\| \to \infty \qquad \text{as} \quad n \to \infty \quad .$$

Hence the subspace E_1 is distinguished by the condition

$$E_1 = \{e \in E \mid g^n e \to 0\} \quad .$$

But the same condition distinguishes also the subspace E_1' (as $\sigma(g)|_{E_1} = \sigma(g)|_{E_1'}$). Thus $E_1 = E_1'$. Now let us show that $E_2 = E_2'$. Assume the opposite and suppose, for example, that there exists a vector $e_2' \in E_2'$ such that $e_2' \,\bar{\in}\, E_2$. Then $e_2' = e_1 + e_2$ where $e_i \in E_i$ and $e_1 \neq 0$. Since

E_2' is invariant under g and the operator ${}^{g}|E_2'$ is invertible there exists, for any $n > 0$, a vector $e_2'(n) \in E_2'$ such that

$$g^n e_2'(n) = e_2' \quad .$$

Let $e_2'(n) = e_1^n + e_2^n$, where $e_i^n \in E_i$. Note that necessarily $e_1^n \neq 0$ because otherwise we would have that

$$e_2' = g^n(e_2'(n)) = g^n e_2^n \in E_2 \quad .$$

Condition (2‴) implies that

$$e_2'(n) \to 0 \qquad \text{as} \quad n \to \infty \quad .$$

On the other hand

$$e_1 + e_2 = e_2' = g^n e_2'(n) = g^n e_1^n + g^n e_2^n \quad .$$

Since $g^n E_i = E_i$, $i = 1,2$, and the decomposition $e_2' = e_1 + e_2$ is unique we conclude that $g^n e_i^n = e_i$, $i = 1,2$. Therefore, since $e_1 \neq 0$, conditions (1′) and (2′) imply that

$$\left\|e_1^n\right\| \to \infty \qquad \text{and} \qquad \left\|e_2^n\right\| \to 0 \qquad \text{when} \quad n \to \infty \quad .$$

Hence

$$\|e_2'(n)\| = \left\|e_1^n + e_2^n\right\| \geq \left|\left\|e_1^n\right\| - \left\|e_2^n\right\|\right| \to \infty \qquad \text{when} \quad n \to \infty$$

and we have arrived at a contradiction.

In a similar manner one can prove that there does not exist a vector $e_2 \in E_2$ such that $e_2 \bar{\in} E_2'$. Thus $E_2 = E_2'$.

Let P_1 be the projection defined by the characteristic function $\chi_\Delta(m)$, where

$$\Delta = \{m \in M | \ |g(m)| < 1\} \quad .$$

Obviously $\chi_\Delta(m)$ is a continuous function.

By the definition of P_1 the subspace Im P_1 has the properties of E_1 and the subspace Ker P_1 has those of E_2 (see (1″) and (2″)), and consequently by the above reasoning we have that Im $P_1 = E_1$ and Ker $P_1 = E_2$.

Since $TE_i = E_i$ $(i = 1,2)$ the set Δ is invariant under α. Note also that for the function $P_1(m) = \chi_\Delta(m)$ we have that $P_1(V(a)) = 1$. Thus $P_1(m) \in A(\alpha, V(a))$ and we have finished. □

Corollary 7.4 *If* M *has a finite number* K *of connected components then the number of connected components of* $|\sigma(\mathrm{aT})|$ *does not exceed* K.

Proof Assume the opposite. Let U_i, $i = \overline{1,k+1}$, be certain $(k + 1)$ connected components of $|\sigma(aT)|$. Then there exist $(k + 1)$ points r_i, $i = \overline{0,k+1}$, such that:

(1) $r_0 = 0 < r_1 < \dots < r_k < r_{k+1} = r(aT)$;

(2) the points r_i, $i = \overline{1,k}$, do not belong to the set $|\sigma(aT)|$;

(3) $U_i \subset [r_{i-1}, r_i]$.

Consider the projections P_i where

$$P_1 = \frac{1}{2\pi i} \int_{|\gamma|=r_1} (\lambda - aT)^{-1} d\lambda \quad ;$$

P_i, $i = \overline{2,k}$, is the spectral projection corresponding to the part of the spectrum of aT lying in the annulus with radii r_{i-1} and r_i;

$$P_{k+1} = I - \frac{1}{2\pi i} \int_{|\gamma|=r_k} (\lambda - aT)^{-1} d\lambda \quad .$$

By Theorem 7.2 all of the projections P_i, $i = \overline{1,k+1}$, belong to

$A(\alpha, V(a))$. So considering P_i as functions on M we can determine $(K + 1)$ sets

$$M_i = \{m \in M \mid P_i(m) = 1\} \quad .$$

The sets M_i are open and closed and $M_i \cap M_j = \emptyset$ when $i \neq j$. Thus we have arrived at a contradiction and the corollary is proved. □

Remark 7.5 Note that $M(\alpha, V(a))$ may be connected even when M is not connected. For example, consider $A = l^\infty(\mathbb{Z})$ and $\alpha(z) = z - 1$. Then the maximal ideal space M of A is totally disconnected while on the other hand only constant functions are invariant under α. Hence for any $a \in A$ we have that $M(\alpha, V(a))$ consist only of a single point.

Remark 7.6 We can obviously strengthen Corollary 7.4 by considering connected components of $M(\alpha, V(a))$ instead of connected components of M. Theorem 7.2 leads to the following result.

Theorem 7.7 *Let* aT, a ∈ A (A ≅ C(M)) *be a weighted shift operator. Then there exists a mapping* f *(depending only on* a *and* α*) of the set of connected components of the space* M(α, V(a)) *onto the set of connected components of the set* |σ(aT)| *such that for any open and closed set* $\Delta \subset |\sigma(aT)|$ *the corresponding Riesz spectral projection* P_Δ *is the operator of multiplication by a function* χ *in* A(α, V(a)) *such that* supp $\chi = f^{-1}(\Delta)$.

Proof Let U be a particular connected component of $|\sigma(aT)|$. Then U belongs to the closed interval $[0, r(aT)]$ and hence U is either a point or a closed interval, that is,

$$U = [u_1, u_2]$$

where possibly $u_1 = u_2$.

Suppose first that $u_1 \neq 0$. We can obviously choose closed intervals $\left[r_i^1, r_i^2\right]$, $i = \overline{1,\infty}$, such that the points r_i^j, $j = 1,2$, $i = \overline{1,\infty}$, do not belong to the set $|\sigma(aT)|$ and

$$U = [u_1, u_2] = \bigcap_{i=\overline{1,\infty}} \left[r_i^1, r_i^2\right] .$$

For every interval $\left[r_i^1, r_i^2\right]$ the circles of radii r_i^1 and r_i^2 do not meet the spectrum $\sigma(aT)$. Let us consider the Riesz projections P_{0,r_i^1}, $P_{r_i^1,r_i^2}$, $P_{r_i^2,r(aT)}$ corresponding the parts of the spectrum lying:

(a) in the interior of the disk of radius r_i^1 ;

(b) in the annulus with radii r_i^1 and r_i^2 ;

(c) outside the disk of radius r_i^2 .

By Theorem 7.2 these projections belong to $A(\alpha, V(a))$, that is, we can consider them as continuous functions on $M(\alpha, V(a))$. Hence to each interval $[j, k]$ ($[j, k]$ is $\left[0, r_i^1\right]$ or $\left[r_i^1, r_i^2\right]$ or $\left[r_i^2, r(aT)\right]$) there corresponds a set $M_{j,k} \subset M(\alpha, V(a))$ by the rule

$$M_{j,k} = \{m \in M(\alpha, V(a)) \mid P_{j,k} = 1\} .$$

Finally, to the set U there naturally corresponds the set

$$M_U = \bigcap_{i=\overline{1,\infty}} M_{r_i^1,r_i^2} .$$

By construction each $M_{r_i^1,r_i^2}$ consists of entire connected components of $M(\alpha, V(a))$ so the set M_U also consists of entire connected components of $M(\alpha, V(a))$. Obviously M_U does not depend on the choice of the sequence of intervals $\left[r_i^1, r_i^2\right]$, $i = \overline{1,\infty}$.

Suppose now that $U = [0, u_2]$ and $[0, r_i]$, $i = \overline{1,\infty}$, is a sequence of intervals such that $r_i > u_2$, r_i does not belong to the set $|\sigma(aT)|$, and $r_i \to u_2$. Consider the sequence of projections P_{0,r_i} defined above and the corresponding sets M_{0,r_i}. Define M_U by

$$M_U = \bigcap_{i=\overline{1,\infty}} M_{0,r_i} \quad .$$

By construction M_U does not depend on the choice of the sequence r_i and consists of entire connected components of $M(\alpha, V(a))$. Note that if $U' \neq U''$ then there exists a positive number r_0 such that $r_0 \bar{\in} |\sigma(aT)|$ and $U' \subset [0, r_0]$, $U'' \subset [r_0, r(aT)]$. As $r_0 \bar{\in} |\sigma(aT)|$ it follows that $M_{0,r_0} \cap M_{r_0,r(aT)} = \emptyset$ and hence $M_{U'} \cap M_{U''} = \emptyset$.

Consequently we have constructed a decomposition of $M(\alpha, V(a))$ into disjoint sets M_U (depending, by the reasoning of the proof of Theorem 7.2, only on a and α) where each M_U consists of entire connected components of $M(\alpha, V(a))$. Obviously the mapping of the $M(\alpha, V(a))$ into the set of connected components of $|\sigma(aT)|$ defined by the rule

$$f\colon M_U \to U$$

is the one we were looking for. Thus we have finished. □

8. Examples. The symmetry of $\sigma(aT)$. The spectrum and Fredholm spectrum.

8.1 Let aT be a classical bilateral weighted shift operator in $l^p(\mathbb{Z})$, $1 \leq p \leq \infty$, defined in 3.2. The algebra A of weights here is $l^\infty(\mathbb{Z})$ and the only functions which are invariant under the automorphism $\hat{T}\colon l^\infty(\mathbb{Z}) \to l^\infty(\mathbb{Z})$

$$[\hat{T}(a)](k) = a(k - 1) \qquad a \in l^\infty(\mathbb{Z})$$

are precisely the constant functions. Hence Theorem 7.7 implies that $|\sigma(aT)|$ is connected.

In fact in this case the whole of the spectrum $\sigma(aT)$ is connected. Indeed consider the operator $S_h\colon l^p(\mathbb{Z}) \to l^p(\mathbb{Z})$ of multiplication by the function e^{ihn}, $h \in \mathbb{R}$. S_h is an invertible isometry and

$$S_h^{-1}(aT)S_h = e^{-ih}aT \quad .$$

Consequently if a number λ_0 belongs to $\sigma(aT)$ then $e^{ih}\lambda_0$ also belongs to $\sigma(aT)$ and so the whole of the circle $|\lambda| = |\lambda_0|$ belongs to $\sigma(aT)$. Thus, as $|\sigma(aT)|$ is connected $\sigma(aT)$ is connected as well. So the spectrum of aT can be only:

(i) an annulus (when a is invertible (that is its values are separated from 0));

(ii) a disk (when a is not invertible).

In this example the key point was that the only functions in $l^\infty(\mathbb{Z})$ which are invariant with respect to the mapping $z \to z - 1$ are precisely the constant functions. The generalization of this property is the so called indecomposability of measure.

8.2 Let (X, μ) be a space with a measure μ and $\alpha: X \to X$ be a measurable mapping. The measure μ is said to be *indecomposable* with respect to α if for any measurable invariant set ω, under α, we have that

$$\mu(\omega) = 0 \qquad \text{or} \qquad \mu(X\backslash\omega) = 0 \quad .$$

Obviously a measure μ is indecomposable with respect to α iff the only functions $\varphi \in L^\infty_\mu(X)$ which are invariant under the mapping $\varphi \to \varphi\circ\alpha$ are the constant functions.

Lemma 8.3 *Let $\alpha:(X, \mu) \to (X, \mu)$ be an invertible measurable mapping which preserves the equivalence class of measure μ. In the space $L^p_\mu(X)$, $1 \le p \le \infty$, consider an operator*

$$[(aT_\alpha)f](x) = a(x)[\alpha'(x)]^{\frac{1}{p}} f(\alpha(x)) \qquad \left(\frac{1}{p} = \infty\right)$$

where $a \in L^\infty_\mu(X)$ (see 3.3).

Suppose that μ is indecomposable under α. Then $|\sigma(aT_\alpha)|$ is connected.

Proof Obvious (by definition 8.2 and Theorem 7.2).

8.4 Let E be a space $C(\mathbb{R})$ or $L^p(\mathbb{R})$, $1 \leq p \leq \infty$. Consider an operator aT: $E \to E$ defined by the formula

$$[(aT)f](x) = a(x)\, f(x + h)$$

where $a \in C(\mathbb{R})$ and h is a given number. The Lebesgue measure on $\mathbb{R}$ is definitely decomposable under the mapping $x \to x + h$ but the only *continuous projections* $p \in C(\mathbb{R})$ are $p = I$ or $p = 0$. Hence by Theorem 7.2 we have that $|\sigma(aT)|$ is connected.

Actually in this example we can say more. Let S_t: $E \to E$ be an invertible isometry defined by the formula

$$(S_t\, f)(x) = e^{itx} f(x) \qquad t \in \mathbb{R} \quad .$$

Then

$$S_t^{-1}(aT)S_t = e^{ith}(aT) \quad .$$

Since t is an arbitrary number we conclude (as in **8.1**) that $\sigma(aT)$ is invariant under rotations and consequently $\sigma(aT)$ is connected (as $|\sigma(aT)|$ is connected). Therefore $\sigma(aT)$ in this example can be only one of the following forms:

(a) an annulus (when a is invertible, that is, a is separated from zero);
(b) a disk (when a is not invertible).

In particular cases we can compute the radii of an annulus, or a disk.

For example, let a be a function which has finite limits

$$\lim_{x \to +\infty} a(x) = a(+\infty) \qquad \text{and} \qquad \lim_{x \to -\infty} a(x) = a(-\infty) \quad .$$

In this situation we can consider a to be an element of the algebra $A \subset C(\mathbb{R})$ of continuous functions which have finite limits at $+\infty$ and $-\infty$. Obviously the maximal ideal space $M(A)$ is equal to $[\mathbb{R} \cup \{+\infty\} \cup \{-\infty\}]$ and the corresponding homeomorphism α: $M(A) \to M(A)$ has the form

$$\alpha(x) = \begin{cases} x + h & , \quad x \in \mathbb{R} \\ +\infty & , \quad x = +\infty \\ -\infty & , \quad x = -\infty \end{cases} .$$

Hence the only ergodic measures under α are the Dirac point measures $\delta(+\infty)$ and $\delta(-\infty)$. Consequently (by Theorem 4.4)

$$r(aT) = \max\{|a(+\infty)| \;,\; a|(-\infty)|\}$$

and we conclude that $\sigma(aT)$ is

(a) the annulus

$$\min\{|a(+\infty)|, \; |a(-\infty)|\} \le |\lambda| \le \max \{|a(+\infty)|, \; |a(-\infty)|\}$$

when $a(x) \neq 0$, $x \in \mathbb{R}$, $x = +\infty$, $x = -\infty$;

(b) the disk

$$|\lambda| \le \max\{|a(+\infty)|, \; |a(-\infty)|\}$$

when $a(x_0) = 0$ for some $x_0 \in [\mathbb{R} \cup \{+\infty\} \cup \{-\infty\}]$.

8.5 Consider an operator aT: $C(S^1) \to C(S^1)$ defined by

$$[(aT)f](x) = a(x)f(\alpha(x))$$

where $a(x) \in C(S^1)$ and α is a homeomorphism of S^1 (see Example 5.11). Since S^1 is connected we conclude by mapping Corollary 7.4 that $|\sigma(aT)|$ is connected.

If α is a rotation of the circle then the spectrum of aT obtains a certain symmetry property. Indeed let the rotation α be given by the formula

$$\alpha(x) = x + \tau \qquad (S^1 = \mathbb{R}^1(\text{mod } 1)) .$$

Consider the operators S_n $n = \pm 1, \pm 2, \ldots$ of multiplication by the function $e^{i2\pi nx}$. Then

$$S_n^{-1}(aT)S_n = e^{i2\pi n\tau}(aT) \quad .$$

Hence if a number λ_0 belongs to $\sigma(aT)$ then all the numbers $\lambda = e^{i2\pi n\tau}\lambda_0$, $n = \pm 1, \pm 2, \ldots$ also belong to $\sigma(aT)$.

In the case when τ is irrational the numbers $e^{i2\pi n\tau}\lambda_0$, $n = \pm 1, \pm 2, \ldots$, are dense in the circle $|\lambda| = |\lambda_0|$ and consequently (as the spectrum is a closed set) the whole of the circle $|\lambda| = |\lambda_0|$ is contained in $\sigma(aT)$.

Thus when τ is irrational the spectrum $\sigma(aT)$ is connected and invariant under rotations. So (using Example 5.11(b) and Remark 5.12) we conclude that $\sigma(aT)$ has the form:

(a) the circle

$$|\lambda| = \int_{S^1} \ln|a(x)|\,dx$$

when $a(x) \neq 0$, $x \in S^1$;

(b) the disk

$$|\lambda| \leq \int_{S^1} \ln|a(x)|\,dx$$

when $\exists x_0 \in S^1$, $a(x_0) = 0$.

Remark Using Example 5.14 we see that the foregoing argument describes the corresponding spectral properties of the operator aT defined by (5.3) acting on $L^p(S^1)$, $1 \leq p \leq \infty$.

8.6 The reasoning of the subsections **8.1**, 8.4 and **8.5** can be generalized in the following way.

Let G be a locally compact abelian group and μ be the Haar measure on G. Let $C(G)$ be the space of all bounded continuous functions on G with the norm

$$\|f\|_{C(G)} = \sup_G |f(g)| \quad .$$

Let g_0 be a certain element in G and $G_0 \subset G$ be the closed subgroup generated by g_0. Consider an operator aT: $E \to E$ where E is $C(G)$ or $L^p_\mu(G)$, $1 \le p \le \infty$, given by the formula

$$[(aT)f](g) = a(g)\, f(g + g_0)$$

where $a \in C(G)$, and $+$ is the group operation.

Theorem 8.7 *The spectrum* $\sigma(\mathrm{aT})$ *of the operator* aT *possesses the following properties*:

(a) $\sigma(\mathrm{aT})$ *is the invariant under multiplications by the values of characters* $\chi(g_0)$, $\chi \in \hat{G}_0$;

(b) *if* g_0 *has an infinite order then* $\sigma(\mathrm{aT})$ *is invariant under rotations*;

(c) *if* ${}^{G}/G_0$ *is connected then* $|\sigma(\mathrm{aT})|$ *is connected*;

(d) *if* G *is a compact group with a countable base of the topology,* g_0 *is an element of infinite order, and* ${}^{G}/G_0$ *is connected then the spectrum* $\sigma(\mathrm{aT})$ *has the form*:

  (i) *the annulus*

$$\min_{g\in G} \exp \int_{G_0} \ln|a(g+\gamma)|\,d\gamma \le |\lambda| \le \max_{g\in G} \exp \int_{G_0} \ln|a(g+\gamma)|\,d\gamma$$

  when $a(g) \neq 0$, $g \in G$;

  (ii) *the disk*

$$|\lambda| \le \max_{g\in G} \exp \int_{G_0} \ln|a(g+\gamma)|\,d\gamma$$

  when $\exists g' \in G$, $a(g') = 0$.

  Here $d\gamma$ *is the normalized Haar measure on* G.

Proof

(a) Let $\hat{G}_0 \ni \chi$ be any element of the dual group $\hat{G}_0$ of G_0. Then χ can be continued as a character onto the whole of G (see Pontryagin [1] §40, Theorem 55). Consider the operator S_χ of multiplication by the function χ. Obviously S_χ is an invertible isometry on E and we have that

$$S_\chi^{-1}(aT)S_\chi = \chi(g_0)aT$$

and thus $\sigma(aT)$ is invariant under the multiplications by $\chi(g_0)$.

(b) Let g_0 be an element of infinite order. Then the values $\{\chi(g_0)\}$, $\chi \in \hat{G}_0$, are dense in S^1. Indeed this is always true when the set $\{\chi(g_0)\}_{\chi\in\hat{G}_0}$ is infinite (see Pontryagin [1], §36, A). So suppose on the contrary that the set $\{\chi(g_0)\}_{\chi\in\hat{G}_0}$ is finite. In this case we have that $\{\chi(g_0)\}_{\chi\in\hat{G}_0}$ is a finite subgroup of S^1 and thus it is cyclic, that is, isomorphic to $\mathbb{Z}_r$ (see Pontryagin [1] §36, A) so let us consider $\{\chi(g_0)\}_{\chi\in\hat{G}_0}$ to be $\mathbb{Z}_r$. For the element $rg_0 \in G_0$ and any character $\chi \in \hat{G}_0$ we have that

$$\chi(rg_0) = r\chi(g_0) = 0 \pmod{r} \quad .$$

But the characters should distinguish the points of G_0. Thus we arrive at a contradiction and (b) is proved.

(c) Obviously the algebra of continuous bounded functions on G which are invariant under the mapping $a(g) \to a(g + g_0)$ is naturally isomorphic to the algebra of bounded continuous functions on ${}^G/G_0$. Thus if ${}^G/G_0$ is connected then an element $P \in C\left({}^G/G_0\right)$ can be a projection only when $P = I$ or $P = 0$. Therefore by applying Theorem 7.2 we obtain (c).

(d) If g_0 has infinite order and G/G_0 is connected then by (b) and (c) we have that the spectrum $\sigma(aT)$ is invariant under rotations and is connected. Thus $\sigma(aT)$ is:

(i) an annulus (when a is invertible);

(ii) a disk (when a is not invertible).

By applying Corollary 4.5 and Example 5.19 we obtain (d). □

Remark 8.8

(1) If an element g_0 has finite order then in fact the investigation of $\sigma(aT)$ is even simpler than in the case when g_0 has an infinite order. The explicit description of the conditions for invertibility of operators of the form $\sum a_i T^i$ in this case will be presented in a later part of the book (see Sections 15 and 18).

(2) The intrinsic reasons for the spectrum to possess the above–noted symmetry will be considered in Section 13.

8.9 Although the form of the spectrum $\sigma(aT)$ in **8.1–8.5** has some common properties the internal structure of the spectrum in these examples can be different.

Recall that the set $\sigma_F(b)$ of numbers λ such that $(\lambda I - b)$ is not Fredholm is called the *Fredholm spectrum* of an operator b (see **2.9**).

Consider the classical bilateral weighted shift aT (see **8.1**) with a coefficient a such that $a(0) = 0$ and $|a(z)| > \delta > 0$, $z = \pm 1, \pm 2, \ldots$. Let a_1 and p be the functions defined by the formula

$$a_1 = \begin{cases} a(z) & \text{when } z \neq 0 \\ 1 & z = 0 \end{cases}$$

$$p(z) = \delta(z, 0) \quad , \quad \text{the Kronecker symbol.}$$

Then we have that

$$aT = a_1T - pT \quad .$$

Note that a_1T is an invertible operator and pT is a compact operator (as p is a compact operator). Therefore aT is Fredholm. But aT is not invertible (since $a(0) = 0$). Thus $\sigma(aT) \neq \sigma_F(aT)$ in this case.

Now let us consider a weighted shift operator aT: $L^p(S^1) \to L^p(S^1)$, $1 \leq p \leq \infty$ or $C(S^1)$, where $a \in C(S^1)$ and α is an irrational rotation of the circle (see **8.5**). Let us show that in this case $\sigma(aT) = \sigma_F(aT)$ always.

Indeed note first of all that $\sigma_F(aT)$ is invariant under rotations (this can be proved by the same argument as for $\sigma(aT)$ (see **8.5**).

If a is invertible then $\sigma(aT)$ is the circle $|\lambda| = r(aT)$ (see **8.5**) and since the Fredholm spectrum is always non–empty there exists a point λ_0 on this circle such that $\lambda_0 \in \sigma_F(aT)$. By the invariance of $\sigma_F(aT)$ under rotations we have that $\sigma(aT) = \sigma_F(aT)$ in this case.

Suppose now that a is not invertible. If $r(aT) = 0$ then $\sigma(aT) = \sigma_F(aT) = \{0\}$ and we are done. So let $r(aT) \neq 0$. In this case (see **8.5**) we have that $\sigma(aT)$ is the disk

$$|\lambda| \leq r(aT) = \exp \int_{S^1} \ln|a(x)|\,dx \quad .$$

Suppose that $\sigma_F(aT) \neq \sigma(aT)$. This means that there exists a point λ_0, $|\lambda_0| \leq r(aT)$, such that the operator $(\lambda_0 I - aT)$ is Fredholm. Since the property of being Fredholm is stable under small (in norm) perturbations we conclude that there exists an ε–neighbourhood of the operator $\lambda_0 I - aT$ such that all the operators in this neighbourhood are Fredholm; in particular, without loss of generality, we can assume that $\lambda_0 \neq 0$. We shall show that in this neighbourhood there exists a non–Fredholm operator and thus we shall arrive at a contradiction.

Let x_0 be a point on the circle such that $a(x_0) = 0$. Since $a(x)$ is continuous there exists a δ–neighbourhood of x_0 such that $|a(x)| < \varepsilon/2$ on this neighbourhood. Consider the functions $a_t(x)$, $0 \leq t < +\infty$, defined by the formulae $a_t(x) = \varphi_t(x)\cdot\psi(x)$ where

$$\varphi_t(x) = \begin{cases} |a(x)| \ , & |a(x)| \geq \frac{\varepsilon}{2} \\ \frac{\varepsilon}{2} \ , & |a(x)| < \frac{\varepsilon}{2} \ , \ |x-x_0| \geq \delta \\ \frac{\varepsilon}{2}\exp[t(x-x_0+\delta)(x-x_0-\delta)] \ , & |x-x_0| < \delta \end{cases}$$

and $\psi(x)$ is a continuous function satisfying the conditions

$$|\psi(x)| = 1$$

$$\psi(x) = \frac{a(x)}{|a(x)|} \qquad \text{when} \quad |a(x)| \geq \frac{\varepsilon}{2}$$

(such a $\psi(x)$ does exist).

Note that:

(i) a_t is invertible for all t, $0 \leq t < +\infty$;

(ii) $|a_t(x) - a(x)| < \varepsilon$, $x \in S^1$;

(iii) $\int_{S^1} \ln|a_0(x)|dx \geq \int_{S^1} \ln|a(x)|dx$
and thus $r(a_0T) \geq r(aT)$;

(iv) the function $r(a_tT) = \exp \int_{S^1} \ln|a_t(x)|dx$ changes continuously from $r(a_0T)$ to 0 when t changes from 0 to $+\infty$.

Since a_t is invertible then (by the foregoing reasoning) $\sigma_F(a_tT) = \{\lambda: |\lambda| = r(a_tT)\}$. Conditions (iii) and (iv) imply that there exists a t_0 such that $r(a_{t_0}T) = |\lambda_0|$ and consequently $(\lambda_0 I - a_{t_0}T)$ is non–Fredholm. But condition (ii) implies that $\|a_{t_0}T - aT\| < \varepsilon$. Thus we arrive at a contradiction. Therefore $\sigma_F(aT) = \sigma(aT)$. □

9. Weighted shifts with measurable coefficients. Disconnectedness of the spectrum.

In this section we consider some properties of weighted shift operators in $L^p(S^1)$ in the case when a shift operator is generated by a transitive homeomorphism of the circle. We obtain some results that show that although every transitive homeomorphism is topologically conjugate to a rotation of the circle (see Theorem 5.8) the spectral properties of weighted shift operators generated by a transitive homeomorphism may be very different from those of operators generated by rotations.

9.1 Let $\alpha: S^1 \to S^1$ be a transitive homeomorphism. This means that (see Theorem 5.8) there exists a homeomorphism $h: S^1 \to S^1$ and an irrational positive number τ, $0 < \tau < 1$, such that

$$h(\alpha(h^{-1}(x))) = x + \tau \equiv \psi_\tau(x) \quad .$$

We assume that the derivatives α' and $(\alpha^{-1})'$ belong to $L^\infty(S^1)$ and thus the operator $T_\alpha: L^p(S^1) \to L^p(S^1)$, $1 \le p \le \infty$, defined by the formula

$$(T_\alpha f)(x) = |\alpha'(x)|^{\frac{1}{p}} f(\alpha(x)) \qquad \left(\frac{1}{\infty} = 0\right)$$

is an invertible isometry.

Consider a weighted shift operator aT_α where $a: L^p(S^1) \to L^p(S^1)$ is an operator of multiplication by a function $a(x) \in L^\infty(S^1)$.

Lemma 9.2 *The spectrum* $\sigma(aT_\alpha)$ *is invariant under rotations.*

Proof Let S_n, $n = \pm1, \pm2, \ldots$, be the operator of multiplication by the function $e^{2\pi i n h(x)}$. Since $h[\alpha(x)] = h(x) + \tau$ we have that

$$S_n^{-1}(aT_\alpha)S_n = e^{i2\pi n\tau} aT \quad .$$

Thus (as in 8.5) the spectrum $\sigma(aT_\alpha)$ is invariant under rotations. □

Lemma 9.3 *If the Lebesgue measure is indecomposable with respect to* α *then* $\sigma(\mathrm{a}T_\alpha)$ *is connected.*

Proof If the Lebesgue measure is indecomposable with respect to α then $|\sigma(aT)|$ is connected by Lemma 8.3. Now Lemma 9.2 implies that $\sigma(aT)$ is connected. □

For an arbitrary homeomorphism β: $S^1 \to S^1$ with derivative β' belonging to $L^\infty(S^1)$ we define the operator T_β: $L^p(S^1) \to L^p(S^1)$ by the formula

$$(T_\beta f)(x) = [\beta'(x)]^{\frac{1}{p}} f(\beta(x)) \quad .$$

Lemma 9.4 *If* $h(x)$ *and* $h^{-1}(x)$ *are absolutely continuous then the operator* T_h *establishes the conjugation*

$$T_h^{-1}(aT_\alpha)T_h = a_1T_{\psi_\tau} \tag{9.1}$$

(where $a_1(x) = a(h^{-1}(x)))$ *and the spectrum* $(\sigma(aT_\alpha)$ *is connected.*

Proof Since $h(x)$ and $h^{-1}(x)$ are absolutely continuous we have that T_h is an invertible isometry. A routine computation shows that the equality (9.1) is true. As the Lebesgue measure is ergodic with respect to ψ_τ we have, by Lemma 9.3, that $\sigma(a_1T_{\psi_\tau}) = \sigma(aT_\alpha)$ is connected. □

Obviously under the conditions of Lemma 9.4 the operator T_α is conjugate to the operator T_{ψ_τ}. The next theorem shows that the condition of absolute continuity in Lemma 9.4 is essential.

Theorem 9.5 *If* $h(x)$ *is not absolutely continuous then* T_α *is not conjugate to* T_{ψ_τ} *in the space* $L^p(S^1)$, $1 \le p < \infty$.

Proof The numbers $e^{i2\pi n\tau}$ are single eigenvalues for the operator T_{ψ_τ} and the corresponding eigenvectors generate the whole of $L^p(S^1)$, $1 \le p < \infty$.

Under conjugation this property has to be stable. Let us show that if $h^{-1}(x)$ is not absolutely continuous then T_α does not possess this property.

Suppose that $\lambda = 1$ is the eigenvalue of T_α and f_0, $\|f_0\|_{L^p} = 1$ is the corresponding eigenfunction, that is,

$$\alpha'(x)^{\frac{1}{p}} f_0(\alpha(x)) = f_0(x) \quad . \tag{9.2}$$

Obviously $|f_0(x)|$ satisfies the same equation, hence we can assume that $f_0(x) \geq 0$. Since $S^1 = \mathbb{R} \pmod{\mathbb{Z}}$ we can continue $\alpha(x)$ in a natural way, and we can continue $f_0(x)$ by periodicity onto the whole of $\mathbb{R}$, and thus we can assume that (9.2) is valid on the whole of $\mathbb{R}$.

Consider the function

$$\varphi(x) = \int_0^x f_0(\xi)^p d\xi \qquad x \in \mathbb{R} \quad .$$

$\varphi(x)$ determines an absolutely continuous mapping of S^1 onto itself because we have that

$$\varphi(1) - \varphi(0) = \int_0^1 f_0(\xi)^p d\xi = 1 \quad .$$

Note that (9.2) implies that

$$\varphi(\alpha(x)) = \int_0^{\alpha(x)} f_0(\xi)^p dx = \int_{\sigma^{-1}(0)}^{X} \alpha'(\eta) f_0(\eta)^p d\eta$$

$$= \int_{\alpha^{-1}(0)}^{x} f_0(\eta)^p d\eta = \varphi(x) - \varphi(\alpha^{-1}(0)) \quad .$$

Therefore

$$\varphi(\alpha(x)) - \varphi(x) = -\varphi(\alpha^{-1}(0)) = \tau_0 \equiv \text{const.}$$

Suppose that $f_0(x) \neq 0$ almost everywhere. Then $\varphi\colon S^1 \to S^1$ is an absolutely continuous homeomorphism which has inverse φ^{-1} and

$$\varphi(\alpha(\varphi^{-1}(x))) = x + \tau_0 \quad .$$

Thus in this case α is conjugate to the rotation ψ_{τ_0}. Since the rotation number is stable under a conjugation, and a homeomorphism establishing a conjugation is defined uniquely up to a constant we conclude that

$$\tau_0 = \tau \quad ; \qquad h(x) = \varphi(x) + h(0) \quad .$$

This means that $h(x)$ is absolutely continuous and we arrive at a contradiction.

In view of this $f_0(x)$ must be equal to zero on a set which has non–zero measure. Since

$$h(\alpha(x)) = h(x) + \tau$$

the functions $f_k(x) = \mathrm{e}^{\mathrm{i}2\pi n h(x)} f_0(x)$ are eigenfunctions for the operator T_α with the corresponding eigenvalues equal to $\mathrm{e}^{\mathrm{i}2\pi n\tau}$. But these functions do not span the whole of $L^p(S^1)$, and the theorem is proved. □

Consideration of Lemmas 9.3 and 9.4 naturally leads to the question of whether there exists a transitive homeomorphism α such that the spectrum of aT_α is not connected. The example below shows that such a situation can occur. (In view of Lemma 9.4 in this case a homeomorphism establishing the conjugation with the rotation of the circle will necessarily not be absolutely continuous.)

Example 9.6 In order to construct an example of an operator aT_α with disconnected spectrum we have to construct (in view of Lemma 9.3) a homeomorphism $\alpha\colon S^1 \to S^1$ which is transitive but such that the Lebesgue measure is decomposable with respect to α. When such a homeomorphism α is obtained then it is easy to present a coefficient $a \in L^\infty(S^1)$ such that $\sigma(aT_\alpha)$ is disconnected. We simply choose an invariant set $A \subset S^1$ such that $0 < \lambda(A) < 1$ (λ is the Lebesgue measure) and let

$$a(x) = \begin{cases} 1 , & x \in A \\ 2 , & x \bar{\in} A \end{cases} .$$

Then $\sigma(aT_\alpha)$ is a pair of circles $|z| = 1$ and $|z| = 2$.

We shall divide the construction of the desired homeomorphism into a number of steps.

Step I *Construction of a perfect closed set* A *with zero Lebesgue measure which possesses the following property*:

$$x_1, x_2 \in A \Rightarrow x_1 - x_2 \neq k\tau \pmod 1 \qquad k = \pm1, \pm2, \ldots$$

where τ *is any given irrational number.*

The set A is constructed in an analogous way to the construction of the standard Cantor set. The main idea is illustrated in Figs. 4 and 5.

We choose the first set $U_1 = \left[0, \frac{1}{3}\right] \cup \left[\frac{2}{3}, 1\right]$. Then we choose squares $W_1^1, W_1^2, W_1^3, \ldots, W_4^4$ at the angles of the squares W_1, W_2, W_3, W_4 in such a way that all W_i^j, $i = \overline{1,4}$, $j = \overline{1,4}$, have rational coordinates of their angles and no one of the squares intersects with the lines $x - y = \tau \pmod 1$ and $x - y = -\tau \pmod 1$. As τ is irrational we can always do this. Now the set U_2 is the union of the bases of $W_1^1, W_1^2, W_2^1, W_2^2$. By the construction of U_2, it possesses the property

$$x_1, x_2 \in U_2 \Rightarrow |x_1 - x_2| \neq \tau \pmod 1 .$$

Continuing this procedure (see Fig. 5) we can consequently choose $U_3, U_4, \ldots$ which possess the property:

(i) $U_n \neq \emptyset$, $n = 1,2,\ldots$;

(ii) $U_{n+1} \subset U_n$;

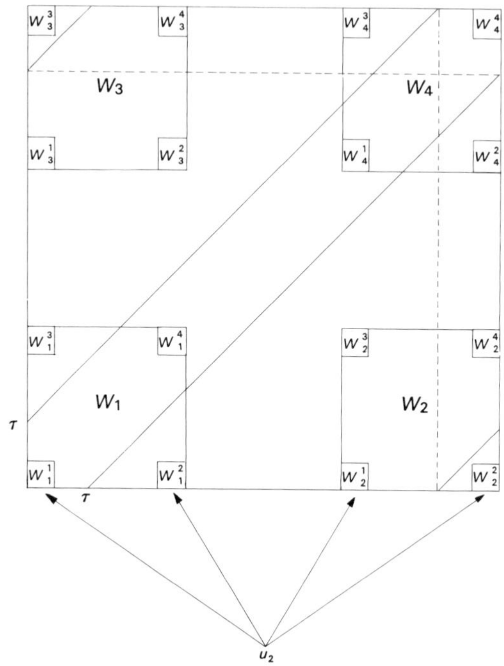

W_3^3
W_3^4
W_4^3
W_4^4
W_3
W_4
W_3^1
W_3^2
W_4^1
W_4^2
W_1^3
W_1^4
W_2^3
W_2^4
W_1
W_2
τ
W_1^1
W_1^2
W_2^1
W_2^2
τ
u_2

Fig. 4

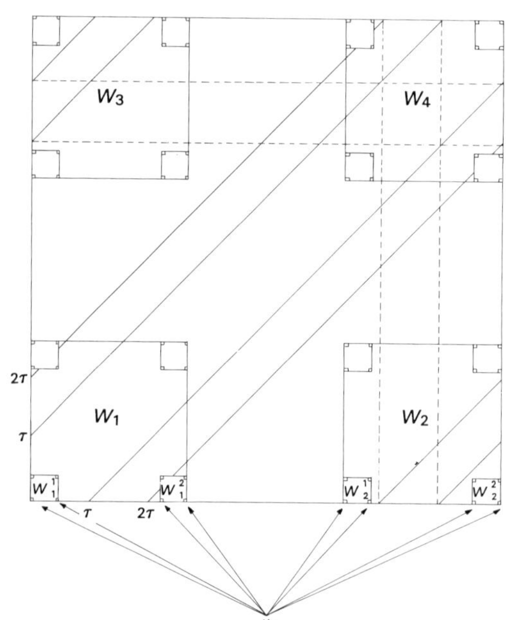

W_3
W_4
2τ
W_1
W_2
τ
W_1^1
W_1^2
W_2^1
W_2^2
τ
2τ
u_3

Fig. 5

(iii) U_n is closed ;

(iv) $x_1, x_2 \in U_n \Rightarrow |x_1 - x_2| \neq k\tau$, $k = 1,2,..,n-1$.

Finally define

$$A = \bigcap_{n=1}^{\infty} U_n$$

and we have finished.

Step II *Construction of a Borel measure* μ *with the properties*:

(i) $\mu(A) \neq 0$;

(ii) $\mu(S^1) = 1$;

(iii) $\frac{1}{2}\,\mu(\omega) \leq \mu(\omega + \tau) \leq 2\mu(\omega)$
for any μ*–measurable set* $\omega \subset S^1$.

Choose any Borel normalized measure μ_0 such that supp $\mu_0 = A$. For any $n = \pm 1, \pm 2, \ldots$ set $A_n = A + n\tau$. By the construction of A we have that $A_n \cap A = \emptyset$. Let μ_n be the measure defined by the formula

$$\mu_n(\omega) = \mu_0(\omega - n\tau) \quad .$$

Set

$$\mu = \frac{1}{3} \sum_{n=-\infty}^{+\infty} \frac{1}{2^{|n|}}\, \mu_n \quad .$$

Obviously μ possesses the properties we are looking for.

Step III *Construction of the desired mapping* α.

Let us define the mapping $h\colon S^1 \to S^1$ by the formula

$$h(x) = \int_0^x d\mu \quad . \tag{9.3}$$

Since τ is an irrational number, and by the construction of the measure μ, we have that:

(i) $h(x)$ is monotone and continuous;

(ii) h maps S^1 onto S^1;

(iii) if $x_1 \neq x_2$ then $h(x_1) \neq h(x_2)$.

Thus h is a homeomorphism of the circle. Note that formula (9.3) implies that

$$\mu(\omega) = \lambda(h(\omega)) \tag{9.4}$$

for any Borel set ω (λ is the Lebesgue measure).

Now set

$$\alpha = h \circ \psi_\tau \circ h^{-1} \quad .$$

In view of property (iii) of the measure μ and (9.4) we have that

$$\frac{1}{2}\lambda(\omega) \leq \lambda(\alpha(\omega)) \leq 2\lambda(\omega)$$

for any λ–measurable set ω. Consequently α preserves the equivalence class of the Lebesgue measure. By construction, α is conjugate to the rotation ψ_τ and hence is transitive. On the other hand the Lebesgue measure is obviously decomposable with respect to α, namely we can choose any subset $B \subset A$ with $0 < \mu(B) < \mu(A)$ and then set

$$B' = \bigcup_{n=-\infty}^{+\infty} \alpha^n(h(B)) \quad .$$

The set B' is invariant under α and

$$0 < \lambda(B') = \mu\left(\bigcup_{n=-\infty}^{+\infty} \psi_\tau(B)\right) < 1 \quad .$$

Thus we have finished. □

10. Examples. Functional differential equations. Solvability.

In this section we present examples of application of the results of the previous sections (concerning the spectral properties of weighted shift operators) to the problem of the solvability of different classes of functional differential equations.

Example I. Functional differential equations of neutral type.

10.1 Let $L^2(\omega)$ be the space of equivalence classes of ω–periodic measurable function (on $\mathbb{R}$) taking complex values for which

$$\|x\| = \left[\int_0^\omega |x(t)|^2 \mathrm{d}t\right]^{\frac{1}{2}} < \infty \quad .$$

We denote by $W_2^m(\omega)$ the space of ω–periodic functions having absolutely continuous derivatives up to order $(m - 1)$ and such that the derivative of order m belongs to $L^2(\omega)$ with the norm

$$\|x\|_m = \left[\sum_{k=0}^{m} \int_0^\omega |x^{(k)}(t)|^2 \mathrm{d}t\right]^{\frac{1}{2}} \quad .$$

Consider a linear functional differential equation with transformed argument of neutral type of the form

$$Bx \equiv a_{m_0}(t)x^{(m)}(t) + a_{m_1}(t)x^{(m)}(t + h)$$

$$+ \sum_{j=1}^{l} \sum_{k=0}^{m-1} a_{k_j}(t)\; x^{(k)}(t + h_j) = f(t) \qquad (10.1)$$

where a_{k_j} ($k \in \overline{0,m}$, $j \in \overline{0,l}$) are ω–periodic functions, $f \in L^2(\omega)$, and $h_j \in R$.

Let us look through the problem of the existence of ω–periodic solutions of (10.1) belonging to $W_2^m(\omega)$.

10.2 Equation (10.1) is said to be *normally solvable* in the space $W_2^m(\omega)$ if:

(1) the homogeneous equation (10.1) (i.e. $f \equiv 0$) has a finite number n_1 of linearly independent solutions in $W_2^m(\omega)$;

(2) there exist a finite number n_2 of linearly independent functions $y_k \in L^2(\omega)$, $k = \overline{1,n_2}$, such that the equation (10.1) has a solution $x \in W_2^m(\omega)$ iff n_2 orthogonality conditions are satisfied:

$$\int_0^\omega f(t)\ \overline{y_k(t)}\ dt = 0 \qquad k = \overline{1,n_2} \ .$$

The number $n_1 - n_2$ is called the *index* of the equation.

If $n_1 = n_2$ then the equation is said to satisfy the *Fredholm alternative.*

From the operator point of view the normal solvability of (10.1) is equivalent to the fact that the operator

$$B\colon W_2^n(\omega) \to L^2(\omega)$$

defined by the left–hand part of the equation is Fredholm. The index of the equation is ind B. The functions y_k, $k = \overline{1,n_2}$, are the solutions of the homogeneous adjoint equation $B^* y \equiv 0$.

Remark For ordinary differential equations and equations resolved with respect to the highest derivative (that is when $a_{m_1} \equiv 0$ in (10.1)) the Fredholm alternative for periodic solutions is always satisfied. In this case, however, it may well occur that $n_1 \neq 0$.

For neutral–type equations and in particular for equations of the form (10.1) the normal solvability is not necessarily the case. For example, for the equation

$$x^{(1)}(t) - x^{(1)}\left(t - \frac{\omega}{2}\right) + x(t) - x\left(t - \frac{\omega}{2}\right) = 0$$

every function belonging to $W_2^1\left(\frac{\omega}{2}\right) \subset W_2^1(\omega)$ is a solution. Thus the homogeneous equation has infinitely many linearly independent solutions. That is why the problem of obtaining conditions for the normal solvability of (10.1) and conditions for the Fredholm alternative to be satisfied has its own interest.

The next statement reduces the investigation of the above–mentioned problems for equation (10.1) to the investigation of an auxiliary functional equation containing a weighted shift.

Along with equation (10.1) consider the operator b in $L^2(\omega)$

$$(bx)(t) = a_{m_0}(t)x(t) + a_{m_1}(t)x(t + h) \quad . \tag{10.2}$$

Proposition 10.3 *Let the coefficients* $\mathrm{a}_{\mathrm{k}_\mathrm{j}}(\mathrm{t})$, $\mathrm{k} = \overline{0,\mathrm{m}-1}$, *in* (10.1) *belong to* $\mathrm{L}^\infty(\omega)$. *Then the operator* B *is Fredholm iff* b *in* (10.2) *is Fredholm and* ind B = ind b.

Proof The expression

$$B_1 x \equiv \sum_{j=1}^{l} \sum_{k=0}^{m-1} a_{k_j} x^{(k)}(t + h_j)$$

defines the bounded operator $\tilde{B}_1$ acting from $W_2^m(\omega)$ into $L^2(\omega)$ and the bounded operator $\approx\!\!\!\!B_1$ acting from $W_2^{m-1}(\omega)$ into $L^2(\omega)$. Therefore $\tilde{B}_1$ can be expressed as

$$\tilde{B}_1 = \overset{\approx}{B}_1 \circ J_m$$

where J_m is the operator of imbedding $W_2^m(\omega)$ into $W_2^{m-1}(\omega)$. Since J_m is compact $\tilde{B}_1$ is compact as well.

As

$$B = b \cdot \frac{d^m}{dt^m} + \tilde{B}_1$$

and in view of the stability of the class of Fredholm operators under compact perturbations it follows that B is Fredholm iff $b \cdot \frac{d^m}{dt^m}$ is Fredholm and their indexes coincide. The operator $\frac{d^m}{dt^m}$ as an operator from $W_2^m(\omega)$ into $L^2(\omega)$ is Fredholm and $\operatorname{ind}\left(\frac{d^m}{dt^m}\right) = 0$. Thus $b \cdot \frac{d^m}{dt^m}$ is Fredholm iff b is Fredholm and $\operatorname{ind} b = \operatorname{ind}\left(b \cdot \frac{d^m}{dt^m}\right) = \operatorname{ind} B$. □

This proposition makes it possible to apply the results obtained in the previous sections to the calculation of the conditions for normal solvability of equation (10.1).

Theorem 10.4 *Let* $a_{m_0}, a_{m_1} \in C(\omega)$, $a_{k_j} \in L^\infty(\omega)$, $k = \overline{0,m-1}$, *and let* $\frac{h}{\omega}$ *be an irrational number. Then equation* (10.1) *is normally solvable in* $W_2^m(\omega)$ *iff one of the following two conditions is satisfied*:

(1) $a_{m_0}(t) \neq 0$ *and*

$$S(a_{m_0}) > S(a_{m_1}) \tag{10.3}$$

or

(2) $a_{m_1}(t) \neq 0$ *and*

$$S(a_{m_1}) > S(a_{m_0}) \tag{10.4}$$

where $S(a) = \exp \frac{1}{\omega} \int_0^\omega \ln|a(t)|dt$, *the geometric mean of* a *over* $[0, \omega]$.

If one of the conditions (1) *or* (2) *is satisfied then equation* (10.1) *has zero index, that is, the obtained conditions are necessary and sufficient for the Fredholm alternative to be satisfied.*

Proof Let $a_{m_0}(t) \neq 0$. Then the operator

$$b(x)(t) = a_{m_0}(t)[I + a(t)\ x(t + h)]$$

where $a(t) = \dfrac{a_{m_1}(t)}{a_{m_0}(t)}$ and we conclude that b is Fredholm iff (-1) does not belong to the Fredholm spectrum of the operator

$$(\overline{b}x)(t) = a(t)\ x(t + h) \quad .$$

As shown in **8.9** the spectrum of $\overline{b}$ in $L^2(\omega)$ coincides with the Fredholm spectrum (one can identify $L^2(\omega)$ and $L^2(S^1)$).

It was shown (see **8.5**) that the spectrum of $\overline{b}$ is:

(i) the circle $|\lambda| = S(a)$ if $a(t) \neq 0$;

(ii) the disk $|\lambda| \leq S(a)$ if there exists t_0 such that $a(t_0) = 0$.

Thus we conclude that if $a_{m_1}(t_0) = 0$ then b is Fredholm iff $S(a) < 1$, which is equivalent to (10.3).

If $a_{m_1}(t) \neq 0$ then b is Fredholm iff $S(a) \neq 1$ which means that (10.3) or (10.4) is satisfied.

If the condition $a_{m_0}(t) \neq 0$ is not satisfied but $a_{m_1}(t) \neq 0$ then the analogous argument shows that (10.4) is the necessary and sufficient condition for b to be Fredholm in this case.

It remains to show that if there exist t_0, t_1 such that $a_{m_0}(t_0) = 0$, $a_{m_1}(t_1) = 0$, then b is not Fredholm.

In order to do this we need the following auxiliary result.

Lemma 10.5 *Let* $p(t) \in C(\omega)$, $p(t_0) = 0$, *and*

$$p_\varepsilon(t) = \begin{cases} |p(t)|\ , & |p(t)| \geq \varepsilon \\ \varepsilon\ , & |p(t)| < \varepsilon\ . \end{cases}$$

Then for any $\varepsilon > 0$ *and any number* C, $0 < C < S(p_{\frac{\varepsilon}{2}})$ *there exists a function* $q(t) \in C(\omega)$ *such that:*

(i) $|p(t) - q(t)| < \varepsilon$;

(ii) $S(q) = C$.

Proof Observe first that for any ε there exists an ω–periodic function $q_0(t) \in C(\omega)$ such that:

(1) $|q_0(t) - p(t)| < \varepsilon$;

(2) $q_0(t) \neq 0$;

(3) $S(q_0) = S(p_{\frac{\varepsilon}{2}})$.

Such a function can be constructed, for example, in the following way: The set $\left\{t = |p(t)| < \frac{\varepsilon}{2}\right\}$ is open and thus consists of a countable unity of intervals (α_k, β_k) where $|p(\alpha_k)| = \frac{\varepsilon}{2}$ and $|p(\beta_k)| = \frac{\varepsilon}{2}$. Let us set

$$q_0(t) = \begin{cases} p(t)\ , & |p(t)| \geq \frac{\varepsilon}{2} \\ \frac{\varepsilon}{2} \exp\left[\arg p(\alpha_k) \dfrac{\beta_k - t}{\beta_k - \alpha_k} + \arg p(\beta_k) \dfrac{t-\alpha_k}{\beta_k - \alpha_k}\right] i, & t \in (\alpha_k, \beta_k)\ . \end{cases}$$

Let us take $\delta > 0$ in such a way that we have $|p(t)| < \frac{\varepsilon}{2}$ when $|t - t_0| < \delta$.

Set

$$\varphi(t) = \begin{cases} 1 , & |t - t_0| > \delta \\ \exp[(t - t_0 - \delta)(t - t_0 + \delta)] , & |t - t_0| \le \delta \end{cases}$$

over $[0, \omega]$ and let us continue this function over the whole of $\mathbb{R}$ in a periodic way.

Consider the set of functions

$$q_l(t) = q_0(t)[\varphi(t)]^l , \qquad l \ge 0 .$$

For every $l \ge 0$ we have

$$|q_l(t) - p(t)| \le \varepsilon \tag{10.5}$$

and

$$S(q_l) = S(q_0) \cdot [S(\varphi)]^l . \tag{10.6}$$

Since $S(\varphi) < 1$ (10.6) implies that when l changes from 0 to $+\infty$ $S(q_l)$ changes continuously from $S(q_0)$ to 0. Thus for any C, $0 < C < S(q_0) = S(p_{\frac{\varepsilon}{2}})$, there exists l such that $S(q_l) = C$. The proof of the lemma is complete. □

Let us now finish the proof of the theorem. Assume on the contrary that $a_{m_0}(t_0) = 0$ and $a_{m_1}(t_1) = 0$ but b is Fredholm. In this case there exists an ε–neighbourhood of b consisting of Fredholm operators. By Lemma 10.5 there exists functions $q_0(t)$ and $q_1(t)$ such that

$$q_i(t) \ne 0 \qquad i = 1,2 \tag{10.7}$$

$$\left|a_{m_i}(t) - q_i(t)\right| < \frac{\varepsilon}{2} \tag{10.8}$$

and

$$S(q_i) = \frac{\varepsilon}{4} \tag{10.9}$$

(the last property follows from the lemma and the observation that $S(p_\varepsilon) \ge \varepsilon$).

In view of (10.8) the operator

$$(qx)(t) = q_0(t)\ x(t) + q_1(t)\ x(t + h)$$

belongs to the ε–neighbourhood of b and in (10.7) and (10.9) q is not Fredholm (as (–1) belongs to the Fredholm spectrum of the operator

$$(\bar{q}x)(t) = \frac{q_1(t)}{q_0(t)}\ x(t + h) \quad).$$

Thus we arrive at a contradiction which means that b is not Fredholm. □

Example II. Singular integral operator with a shift.

10.6 Consider the space $L^2(\Gamma)$ where Γ is a smooth closed curve on a complex plane. Recall (see 2.52) that a singular integral operator in $L^2(\Gamma)$ is an operator of the form

$$A = a(t)I + b(t)S + K \tag{10.10}$$

where I is the identity operator, S is the Cauchy singular integral operator

$$(Sx)(t) = \frac{1}{\pi i} \int_\Gamma \frac{x(\tau)}{\tau - t}\, d\tau$$

$a(t)$ and $b(t)$ are continuous functions, and K is a compact operator.

The operator S is bounded in $L^2(\Gamma)$ and possesses the properties (see **2.49**)

$$S^2 = I \tag{10.11}$$

$$aS - Sa \in \mathscr{K} \tag{10.12}$$

for every $a \in C(\Gamma)$ (here $\mathscr{K}$ is the set of compact operators in $L^2(\Gamma)$).

From Theorem 2.54, the operator A in (10.10) is Fredholm iff

$$a(t) \pm b(t) \neq 0 \qquad t \in \Gamma$$

and

$$\text{ind } A = \text{ind}\left[\frac{a(t) - b(t)}{a(t) + b(t)}\right] . \tag{10.13}$$

Let Γ be the unit circle $|t| = 1$. Consider a singular integral functional operator B obtained by attaching to A the weighted shift operator T_h

$$(T_h x)(t) = x(e^{ih}t) \quad .$$

The operator B has the form

$$B = aI + bS + cT_h + dST_h + K \tag{10.14}$$

where $a, b, c, d \in C(\Gamma)$ and K is compact.

Let us show that the problem of the Fredholmity of the operator B can be reduced to the investigation of some functional operators.

Let

$$\begin{aligned} u_1(t) &= a(t) + b(t) \\ u_2(t) &= a(t) - b(t) \\ v_1(t) &= c(t) + d(t) \\ v_2(t) &= c(t) - d(t) \quad . \end{aligned}$$

Consider two functional operators in $L^2(\Gamma)$:

$$\begin{aligned} b_1 &= u_1 I + v_1 T_h \\ b_2 &= u_2 I + v_2 T_h \quad . \end{aligned} \tag{10.15}$$

Theorem 10.7 *The operator* B (10.14) *is Fredholm in* $L^2(\Gamma)$ *iff both* b_1 *and* b_2 *in* (10.15) *are invertible in* $L^2(\Gamma)$.

Proof By (10.11) the operators

$$P = \frac{I + S}{2} \qquad \text{and} \qquad Q = \frac{I - S}{2} = I - P$$

are projections and

$$PQ = QP = 0 \quad . \tag{10.16}$$

One can verify (see 2.55) that P is in fact the projection onto the 'analytical' part of the Fourier expansion of the function.

For our purposes it is convenient to rewrite B in the form

$$B = b_1 P + b_2 Q + K \quad . \tag{10.17}$$

B is Fredholm iff it has a regularizer (see **2.8**), that is, an operator R such that

$$BR = I + K_1 \quad , \qquad RB = I + K_2$$

where K_1, K_2 are compact operators.

Let b_i, $i = 1,2$, be invertible. Let us show that in this case the operator

$$R = b_1^{-1} P + b_2^{-1} Q \tag{10.18}$$

is a regularizer of B and thus B is Fredholm.

Hereinafter we denote by K (in calculations) different compact operators.

Observe that

$$T_h S = S T_h \quad .$$

Thus (10.12) implies

$$b_i S = S b_i + K \quad .$$

By multiplying this equality by b^{-1} from the left and from the right we obtain

$$S = b_i^{-1} S b_i + K = b_i S b_i^{-1} + K$$

which implies

$$b_i^{-1} P = P b_i^{-1} + K$$

$$b_i^{-1} Q = Q b_i^{-1} + K \quad .$$

Using (10.16), (10.17) and (10.18) and these relations we have

$$BR = (b_1P + b_2Q + K)(b_1^{-1}P + b_2^{-1}Q)$$

$$= b_1Pb_1^{-1}P + b_1Pb_2^{-1}Q + b_2Qb_1^{-1}P + b_2Qb_2^{-1}Q + K$$

$$= P^2 + b_1b_2^{-1}PQ + b_2b_1^{-1}QP + Q^2 + K$$

$$= P + Q + K = I + K \quad .$$

An analogous computation shows that

$$RB = I + K \quad .$$

Thus we have proved that the invertibility of b_i, $i = 1,2$, implies the Fredholmity of B. Let us now verify that the converse is also true.

Suppose that, say, b_1 is not invertible. Let us prove that this implies that B is not Fredholm.

(α) Consider first the case when either Ker $b_1 \neq 0$ or Im b_1 is not closed. In this situation there exists a sequence $x_k \in L^2(\Gamma)$, $\|x_k\| = 1$, such that

$$b_1x_k \to 0 \quad . \tag{10.19}$$

Without loss of generality one can assume that every x_k has a finite Fourier expansion

$$x_k(t) = \sum_{l=-p_k}^{p_k} c_{kl}t^l \quad . \tag{10.20}$$

For any fixed h there exists a sequence n_k such that

$$e^{in_k h} \to 1$$

and (10.21)

$$n_k > p_k + n_{k-1} + p_{k-1} \quad .$$

Consider the sequence

$$y_k(t) = t^{n_k} x_k(t) \quad .$$

By the choice of n_k, y_k is an orthonormal sequence (and thus is not compact) and weakly tends to zero.

As

$$b_1 y_k(t) = u_1(t) t^{n_k} x_k(t) + v_1(t) e^{in_k h} t^{n_k} T_h(x_k(t))$$

$$= t^{n_k} b_1 x_k(t) + t^{n_k}\left(e^{in_k h} - 1\right) v_1(t) T_h x_k(t)$$

we conclude by (10.19) and (10.21) that

$$b_1 y_k \to 0 \quad .$$

By the choice of n_k all the coefficients in the Fourier expansion of y_k having negative numbers are zero, thus

$$Py_k = y_k \qquad \text{and} \qquad Qy_k = 0$$

and it follows that

$$By_k = b_1 Py_k + b_2 Qy_k = b_1 y_k \to 0 \quad .$$

It is known (see for example Glazman [1]) that if for an operator L there exists a non–compact sequence ξ_k weakly tending to zero such that $L\xi_k \to 0$ then L is not Fredholm.

Thus the foregoing argument shows that B is not Fredholm.

(β) If Ker $b_1 = 0$ and Im b_1 is closed then since b_1 is not invertible one has Ker $b_1^* \neq 0$ and the argument goes analogously by passing to B^*. The case when b_2 is not invertible can be considered in the same way (taking $y_k(t) = t^{-n_k} x_k(t)$). Thus the theorem is proved. □

Remark It is worth mentioning that the sufficiency in the theorem was proved by using only the property

$$b_i S = S b_i + K \tag{10.22}$$

while the proof of necessity used the very special form of b_i. This is not incidental. In fact for arbitrary b_i in (10.17) (even satisfying (10.22)) B may be Fredholm while b_i are not.

Now this theorem and the results obtained while considering Example I make it possible to calculate the explicit conditions for B to be Fredholm.

Theorem 10.8 *Let* $^{h}/2\pi$ *be an irrational number. A singular integral functional operator* B *of the form* (10.14) *is Fredholm in* $L^2(\Gamma)$ *iff for every* i = 1,2 *one of the following conditions is satisfied:*

(α) $u_i(t) \neq 0 \qquad S(u_i) > S(v_i)$

(β) $v_i(t) \neq 0 \qquad S(u_i) < S(v_i)$.

The index of B *is expressed by the formulae*

$$\operatorname{ind} B = \begin{cases} \operatorname{ind}\left[\dfrac{u_2(t)}{u_1(t)}\right] & \text{if } \begin{array}{l} S(u_1) > S(v_1) \\ S(u_2) > S(v_2) \end{array} \\ \operatorname{ind}\left[\dfrac{v_2(t)}{u_1(t)}\right] & \text{if } \begin{array}{l} S(u_1) > S(v_1) \\ S(u_2) < S(v_2) \end{array} \\ \operatorname{ind}\left[\dfrac{u_2(t)}{v_1(t)}\right] & \text{if } \begin{array}{l} S(u_1) < S(v_1) \\ S(u_2) > S(v_2) \end{array} \\ \operatorname{ind}\left[\dfrac{v_2(t)}{v_1(t)}\right] & \text{if } \begin{array}{l} S(u_1) < S(v_1) \\ S(u_2) < S(v_2) \end{array} \end{cases}$$

Proof By Theorem 10.7 B is Fredholm iff both b_i, $i = 1,2$, are invertible. The conditions (α) and (β) are necessary and sufficient for b_i to be invertible (see the proof of Theorem 10.4). Thus we only have to obtain the index formula.

Let for example

$$u_1(t) \neq 0 \qquad S(u_1) > S(v_1)$$
$$u_2(t) \neq 0 \qquad S(u_2) > S(v_2) \quad .$$

Consider the set of operators

$$B_\tau = (u_1 + \tau v_1 T_h)P + (u_2 + \tau v_2 T_h)Q \qquad \tau \in [0,1] \quad .$$

By the part of the theorem already proved all B_τ, $\tau \in [0,1]$, are Fredholm. The set B_τ depends continuously on τ, so in view of the homotopic stability of the index we have

$$\operatorname{ind} B = \operatorname{ind} B_1 = \operatorname{ind} B_\tau = \operatorname{ind} B_0 = \operatorname{ind}(u_1 P + u_2 Q) = \operatorname{ind}\left[\frac{u_2}{u_1}\right]$$

where in the last equality we used (10.13).

The remaining cases can be considered in an analogous way. □

Example III. Pseudodifferential difference operator.

10.9 Let $X = \mathrm{T}^m$ be an m–dimensional torus realized as $\mathbb{R}^m/\mathbb{Z}^m$. By T_h, $h \in \mathbb{R}^m$, we denote the shift operator in a functional space on T^m

$$(T_h u)(x) = u(x + h) \quad . \tag{10.23}$$

Consider the pseudodifferential different operator of the form

$$B = A_0(x, D) + A_1(x, D)T_h \tag{10.24}$$

where A_i, $i = 0{,}1$, are pseudodifferential operators of order r acting from $H^s(\mathrm{T}^m)$ into $H^{s-r}(\mathrm{T}^m)$ (see **2.42**).

The spectral radius formula obtained in Example 5.19 will make it possible to find the conditions on the symbols of A_i under which B is Fredholm as an operator acting from $H^s(\mathrm{T}^m)$ into $H^{s-r}(\mathrm{T}^m)$.

Consider the action of the torus T^m on the manifold $\mathrm{T}^m \times S^{m-1} = S^*X$ by the formula

$$g_0(x, \xi) = (x + g_0, \xi) \quad , \qquad g_0 \in \mathrm{T}^m \quad .$$

Let H be the minimal closed subgroup in T^m containing h. For any continuous function $d(x, \xi)$ we denote by $S_{(x_0,\xi_0)}(d)$ the number

$$S_{(x_0,\xi_0)}(d) = \exp \int_H \ln|d(x_0 + g,\xi_0)|\,dg$$

where dg is the normalized Haar measure on H.

Theorem 10.10 *The satisfaction of one of the following conditions* (α) *or* (β) *is a sufficient condition for the operator* B (10.24) *to be Fredholm as an operator from* $\mathrm{H}^{\mathrm{s}}(\mathrm{T}^{\mathrm{m}})$ *into* $\mathrm{H}^{\mathrm{s-r}}(\mathrm{T}^{\mathrm{m}})$:

(α) $\sigma(A_0(x,\xi)) \neq 0$ *on* S^*X *and*

$$S_{(x,\xi)}(\sigma(A_0)) > S_{(x,\xi)}(\sigma(A_1)) \quad , \quad (x,\xi) \in S^*X \ ;$$

(β) $\sigma(A_1(x,\xi)) \neq 0$ *on* S^*X *and*

$$S_{(x,\xi)}(\sigma(A_1)) > S_{(x,\xi)}(\sigma(A_0)) \quad , \quad (x,\xi) \in S^*X \ ;$$

where $\sigma(A_i)$ *are the symbols of* $A_i(x, D)$, $i = 0,1$. *If one of these conditions is satisfied then* ind $B = 0$.

Proof Observe first that without loss of generality we can reduce the consideration to the case of operators of order 0 in $L^2(T^m)$.

Indeed in view of **2.41** Remark (2) and **2.42** there exists a pseudodifferential operator Λ of order 1 with the symbol $\sigma(\Lambda) = \|\xi\|$ which for every $s \in \mathbb{R}$ is the bijection of $H^s(T^m)$ onto $H^{s-1}(T^m)$. Then for any operator A acting from $H^s(T^m)$ into $H^{s-r}(T^m)$ the operator $\tilde{A} = \Lambda^{s-r} A\Lambda^{-s}$ acts in $L^2(T^m)$, is Fredholm iff A is Fredholm and ind A = ind $\tilde{A}$. In view of Theorem 2.48 we have

$$\tilde{B} = \Lambda^{s-r} B\Lambda^{-s} = \tilde{A}_0(x, D) + \tilde{A}_1(x, D)T_h + K$$

where K is compact.

Since $\tilde{B}$ acts in $L^2(T^m)$, ind B = ind $\tilde{B}$, $\tilde{A}_i(x, D)$ are pseudodifferential operators of order 0, and $\sigma(A_i)|_{S^*X} = \sigma(\tilde{A}_i)|_{S^*X}$, in all the following considerations we can assume that $s = 0$ and $r = 0$.

Now we prove the sufficiency of the conditions (α), (β). Let (α) be satisfied. Then a pseudodifferential operator R with the symbol $[\sigma(A_0)]^{-1}$ is a regularizer for A_0, that is,

$$RA_0 = I + K \qquad \text{and} \qquad A_0R = I + K$$

(recall again that by K we denote different compact operators). Thus we can rewrite B in the form

$$B = A_0(I - CT_h) + K$$

where C is a pseudodifferential operator of order 0 with the symbol $\sigma(C) = -\dfrac{\sigma(A_1)}{\sigma(A_0)}$.

Since A_0 is Fredholm and ind $A_0 = 0$ (see Agranovich [1]) the operator B is Fredholm iff $B_1 = I - CT_h$ is Fredholm and ind B = ind B_1.

Now we present two different arguments for the verification of the Fredholmity of B_1:

(i) the 'classical' pseudodifferential argument;
(ii) the algebraic argument.

I. The pseudodifferential argument. Let us construct a regularizer for B_1. Consider the series

$$I + CT_h + (CT_h)^2 + \dots \quad . \tag{10.25}$$

If it were a convergent series its sum would be the inverse of B_1. Unfortunately (10.25) need not converge under condition (α) of the theorem.

Let us show that there exist compact operators K_j, $j = \overline{1,\infty}$, such that the series

$$I + (CT_h + K_1) + \left((CT_h)^2 + K_2\right) + \dots + \left((CT_h)^n + K_n\right) + \dots \tag{10.26}$$

converges.

We have

$$(CT_h)^n = CT_h \cdot CT_h \cdot \dots \cdot CT_h = C\left(T_h C T_h^{-1}\right)\left(T_h^2 C T_h^{-2}\right) \cdot \dots \cdot \left(T_h^{n-1} C T_h^{1-n}\right) T_h^n$$

and by Theorem 2.48 $\sigma\left(T_h^j C T_h^{-j}\right)(x, \xi) = \sigma(C)(x + jh, \xi)$. Thus

$$(CT_h)^n = C_n T_h^n$$

where C_n is a pseudodifferential operator with the symbol

$$\sigma(C_n)(x, \xi) = \prod_{j=0}^{n-1} \sigma(C)(x + jh, \xi) \quad .$$

In the space $L^2(S^*X)$ consider the operator

$$(Vu)(x, \xi) = \sigma(C)(x, \xi)\ u(x + h, \xi) \quad .$$

By Example 5.19 the inequality in the condition (α) is equivalent to the condition

$$r(V) < 1 \quad .$$

Take some q, $r(V) < q < 1$; then there exists a constant L such that

$$\|V^n\| = \max_{(x,\xi)} |\sigma(C_n)(x, \xi)| < Lq^n \quad .$$

In view of Theorem 2.46 there exists a compact operator K'_n such that

$$\|C_n + K'_n\| \le Lq^n \quad .$$

Since T_h is unitary it follows that

$$\|(CT_h)^n + K_n\| = \|C_nT_h^n + K_n\| = \|(C_n + K'_n)T_h^{-n}\| \le Lq^n$$

(here $K_n = K'_nT_h^n$) which implies the convergence of the series (10.26).

Let R be the sum of (10.26). Then

$$RB_1 = R(I - CT_h) = \lim_{n\to\infty} \left(I + \sum_{j=1}^{n} (CT_h)^j + K_j \right)(I - CT_h)$$
$$= I + \lim_{n\to\infty} \left[-(CT_h)^{n+1} - K_{n+1} + \sum_{j=1}^{n} K_j(I - CT_h) + K_{n+1} \right] . \tag{10.27}$$

Since $\|(CT_h)^{n+1} + K_{n+1}\| \to 0$ and the limit in (10.27) exists it follows that the sequence $\left\{\sum_{j=1}^{n_1} K_j(I - CT_h) - K_{n+1}\right\}$ converges to the considered limit which is a compact operator.

Thus the constructed operator R is a left–hand regularization for B_1. In an analogous way one can verify that R is a right–hand regularizer. So B_1 is Fredholm.

II. The algebraic argument. Let $^L/K$ be the factor algebra of the algebra of linear bounded operators in $L^2(\mathrm{T}^m)$ with respect to the ideal of compact operators. As is known (see 2.61) an operator B_1 is Fredholm iff its factor class $[B_1]$ is invertible in $^L/K$. But we have already noticed that

$$\|[(CT_h)^n]\| = \|C_n\| = \|V^n\|$$

which implies

$$r([CT_h]) = r(V) < 1$$

and in particular $[B_1] = [I - CT_h]$ is invertible in $^L/K$. Thus B_1 is Fredholm.

Now let us find the index of B_1 under the condition (α).

In this case the part of the theorem already proved implies that the set B_t, $t \in [0,1]$, of operators

$$B_t = I - tCT_h$$

consists of Fredholm operators.

Since B_t realizes the homotopy between $B_1 = I - CT_h$ and $B_0 = I$ it follows that

$$\operatorname{ind} B_1 = \operatorname{ind} I = 0 \quad .$$

The sufficiency of the condition (β) for the Fredholmity of B and the corresponding argument for the index in this case can be easily carried out in an analogous way. □

Remark In fact, the satisfaction of one of the conditions (α) or (β) is not only sufficient but also necessary for B to be Fredholm but we shall not prove this yet.

The reason is that in comparison with the one–dimensional case considered in the previous example the direct pseudodifferential proof of the necessity of these conditions is rather bulky and needs the use of some specific properties of pseudodifferential operators and the particular form of the considered operator B (cf. the pseudodifferential variant of the proof of sufficiency).

On the other hand the algebraic variant (cf. *II*) of the proof of necessity needs the development of the corresponding C^* tool (which is not yet ready). So we leave the proof of the necessity (as well as the consideration of many other related problems) until the corresponding general theory is built in Chapter 5.

Example IV. Boundary–value problems with the conditions set on the whole of the boundary

10.11 Consider the problem: given the string oscillation equation

$$\frac{\partial^2 u}{\partial t^2} = \frac{\partial^2 u}{\partial x^2} \tag{10.28}$$

find the solution $u(x, t)$ of (10.28) on $\mathbb{R}^2$ satisfying the periodicity condition

$$u(x + 1, t) = u(x, t) \tag{10.29}$$

and the conditions

$$u\big|_{t=0} = g_1(x) \tag{10.30}$$

$$a(x)\,\frac{\partial u}{\partial x} + b(x)\,\frac{\partial u}{\partial t}\bigg|_{t=h} = g_2(x) \tag{10.31}$$

where a, b, g_1, g_2 are given periodic functions (with period 1) and h is a fixed number.

The problem (10.28)–(10.31) is, in fact, a particular case of the problem (1.7), (1.9) and (1.10) from the Introduction.

The additional condition here is the periodicity condition. The physical meaning of this problem was discussed in the Introduction. The periodicity condition means that we are considering a closed string (a ring).

Observe that the periodicity condition does not imply the periodicity of the functions φ_1 and φ_2 by means of which one expresses the solution $u(x, t)$ of (10.28):

$$u(x, t) = \varphi_1(x - t) + \varphi_2(x + t) \quad . \tag{10.32}$$

For example, if $\varphi_1(x) = x$, $\varphi_2(x) = -x$ then

$$u(x, t) = x - t - (x + t) = -2t$$

is periodic with respect to x.

On the other hand the derivatives φ_1' and φ_2' have to be periodic.

In fact, substituting $u(x, t)$ in the form (10.32) into (10.29) and setting $t = -x$ we obtain (by taking the derivative) that

$$\varphi_1'(2x) = \varphi_1'(2x + 1) \quad .$$

Thus φ_1' is periodic. The periodicity of φ_2' can be verified similarly.

Observe also that φ_1 and φ_2 are not determined uniquely, but if we set $\varphi_2(0) = 0$ then the expression of $u(x, t)$ in terms of φ_1 and φ_2 will be unique.

Note that φ_2 can be determined uniquely if φ_2' is given and the condition $\varphi_2(0) = 0$ is fixed.

Now expressing φ_1 by means of φ_2 from (10.30) and substituting this expression into (10.31) one obtains the equation

$$(\bar{b}\varphi)(x) \equiv a_1(x)\varphi(x) + a_2(x)\varphi(x + 2h) = g(x) \tag{10.33}$$

where

$$\varphi(x) = \varphi_2'(x - h) \quad , \qquad a_1(x) = -a(x) + b(x) \quad ,$$
$$a_2(x) = a(x) + b(x) \quad , \qquad g(x) = g_2(x) - [a(x) - b(x)]g_1'(x - h)$$

(cf. the calculations in the Introduction).

Let $C_1^l(\mathbb{R})$ be the space of l times continuously differentiable functions on $\mathbb{R}$ which are periodic with period 1. Our foregoing considerations imply the following result.

Lemma 10.12 *Let* a, b $\in C_1^1(\mathbb{R})$. *The problem* (10.28)–(10.31) *has a unique solution belonging to* C^2 *for any*

$$g_1 \in C^2(\mathbb{R}) \quad , \qquad g_2 \in C_1^1(\mathbb{R}) \tag{10.34}$$

iff the operator $\mathfrak{b}$ *defined by the left–hand side of equation* (10.33) *is invertible as an operator from* $C_1^1(\mathbb{R})$ *into* $C_1^1(\mathbb{R})$.

Lemma 10.13 *The operator* $\bar{\mathfrak{b}}$ *is invertible in* $C_1^1(\mathbb{R})$ *iff the operator* $\bar{\bar{\mathfrak{b}}}$ *is invertible, where* $\bar{\bar{\mathfrak{b}}}$ *is defined by the same formula* (10.33) *in* $C_1(\mathbb{R})$.

Proof The operator

$$D\colon \psi \to \psi' + \psi$$

establishes an isomorphism between $C_1^1(\mathbb{R})$ and $C_1(\mathbb{R})$. The inverse D is given by the formula

$$[D^{-1}v](x) = e^{-x}\left(\int_0^x e^s v(s)ds + \frac{1}{1 - e^{-1}}\int_0^1 e^s v(s)ds\right) .$$

The operator $D\bar{b}D^{-1}$ deals in $C_1(\mathbb{R})$ and this space is isomorphic to the space of continuous functions over the unit circle S^1. $D\bar{b}D^{-1}$ is similar to $\bar{b}$ and thus they are invertible (Fredholm) simultaneously. Via a straightforward calculation one could verify that

$$D\bar{b}D^{-1} = \bar{\bar{b}} + K \tag{10.35}$$

where K is a compact operator. Thus $\bar{b}$ is Fredholm iff $\bar{\bar{b}}$ is Fredholm.

By an argument similar to that used in the proof of Theorem 10.4 (see also 8.9 and Part II of the book) one can show that $\bar{\bar{b}}$ is Fredholm iff it is invertible.

Thus if $\bar{b}$ is invertible then $\bar{\bar{b}}$ is also invertible. On the other hand, if $\bar{\bar{b}}$ is invertible then in view of (10.35) $\bar{b}$ is Fredholm and ind $\bar{b}$ = 0. Suppose that $\bar{b}$ is not invertible. Then there exists $f(t) \in C_1^1(\mathbb{R})$ such that

$$\bar{b}f = 0 \quad ,$$

but in this case $f \in C_1(\mathbb{R})$ and since $\bar{\bar{b}}$ acts by the same formula (10.33) we have

$$\bar{\bar{b}}f = 0$$

thus arriving at a contradiction. □

Remark 10.14 While proving Lemma 10.13 we used in an essential way the explicit form of the transformation $\alpha(x) = x + h$. If $\alpha(x)$ has a different form then the analogous statement is not true. The consideration of such cases will be presented in Part II of the book.

Using Lemmas 10.12 and 10.13 and the results on the invertibility conditions of the operator $\bar{\bar{b}}$ (10.33) in $C(S^1)$ (the corresponding consideration for $L^2(S^1)$ was done in the proof of Theorem 10.4, see also Part II of the book) one can obtain the next result.

Theorem 10.15 *Let* a, b $\in C_1^1(\mathbb{R})$. *The problem* (10.28)–(10.31) *has a unique solution belonging to* C^2 *for any* g_1, g_2 *satisfying the conditions* (10.34) *iff the following conditions (depending on* h) *are satisfied:*

(1) *if* $h = 0$ *then* $b(x) \neq 0$;

(2) *if* h *is a rational number and* $2h = {}^m/_n$ *is an irreducible fraction then the condition has the form*

$$\prod_{k=0}^{n-1} a_1(x + 2kh) \neq \prod_{k=0}^{n-1} a_2(x + 2kh) ;$$

(3) *if* h *is irrational the necessary and sufficient condition is the satisfaction of one of the following conditions* (α) *or* (β):

(α) $a_1(x) \neq 0$ *and* $S(a_1) > S(a_2)$;

(β) $a_2(x) \neq 0$ *and* $S(a_2) > S(a_1)$.

Notes and Remarks

Section 3 By now the number of articles in which different properties of weighted composition operators (that is, operators of the form $(Uf)(x) = a(x)f(\alpha(x))$ where f belongs to some functions space $F(X)$ on a set X, $\alpha: X \to X$ and $a(x)$ is some function) are treated is vast. Thus we have no possibility (or intention) of presenting here the whole bibliography concerning this subject. The reader will immediately feel the richness of new structures and the depth and variety of problems if he looks through say Antonevich, Bykadorov [1], Cowen [1], Nikolskij [1], Nordgren [2], Shields [1], though a lot of ideas are not treated there at all. We shall try to make some useful comments when going through the later material. Theorem 3.13 is due to Muhly [1].

Section 4 It seems that the first to examine intensively the spectral properties of weighted composition operators was Parrot [1]. Theorem 4.4 was obtained as a generalization and reinterpretation of the results of Antonevich [4, 8], Antonevich, Ryvkin [1], Antonevich, Vol′pert [1], Karapetyants [1, 2], Karapetyants, Samko [1], Karlovich, Kravchenko [1], Kravchenko [1], Lebedev [3, 5], Mukhamadiev, Sadovskij [1], Parrot [1], and Petersen [1].

We want to emphasize that the spectral radius formulae obtained by Karlovich, Kravchenko [1], Kravchenko [1], and Lebedev [3] also have another form.

The idea of obtaining some sort of general formula like (4.2) had arisen naturally after the appearance of the article by Antonevich [4].

Theorem 4.4 was proved by Lebedev [1, 2] and independently and almost simultaneously by Kitover [1] (the last article concerns the spectrum of a weighted composition operator in $C(X)$, but it is enough to consider only this case when calculating the spectral radius (see the proof of Theorem 4.4)).

In connection with the Fuglede–Kadison determinant we would like to mention the paper by Harpe, Scandalis [1], where its K–theoretical meaning is brought to light, and Lebedev [14] who established the interrelation between this and the Szego–Kolmogorov–Krein theorem.

Theorem 4.15 is due to Antonevich [15]. We would like to mention here the work by Helson [1, 2] and Gromov [2–4] where a very deep and thorough analysis of the spectral structure of unitary weighted shifts generated by rotations of the circle and ergodic automorphisms is provided.

Section 5 In this section we presented a unified approach to the computation of the spectral radius of a weighted shift based on Theorem 4.4. A number of examples considered here were investigated (using other methods) by different authors. A weighted shift operator in $C(S^1)$ in the case when a homeomorphism α: $S^1 \to S^1$ has a finite number of periodic points (particular cases of Example 5.11(a) and (d)) was investigated by Kravchenko [1].

The conditions of invertibility for operators of the form $\sum_{i=1}^{n} a_i T^i$ in the case when $\alpha^n = I$ for some n (that is, α is a periodic homeomorphism) have been rediscovered by a number of authors. The first of them seems to be Babbage [1]. For $L^p(S^1)$, $1 \leq p \leq \infty$, these conditions for example are presented in Litvinchuk [2] (see also Antonevich [14]). We shall investigate the general situation in Section 15.

The spectra of weighted shift operators generated by irrational rotations of the circle (a particular case of Example 5.11(b)) was investigated by Karpetyants [1], and Antonevich, Ryvkin [1].

In connection with this it is worth mentioning the paper by Mukhamadiev, Sadovskij [1] where an estimate for the spectral radius of an irrational rotation weighted shift operator was obtained, and this estimate appeared to be faithful.

Examples 5.11(c) and 5.16 are due to Lebedev [7, 8]. Particular cases of Example 5.18 (weighted shifts generated by diagonal shifts on the torus and Morse–Smale diffeomorphisms) were considered by Antonevich, Vol′pert [1] and Lebedev [3].

The results of **5.19–5.21** are due to Antonevich [4, 8, 14].

Section 6 Proposition 6.1 and Theorem 6.3 are due to Antonevich [5]. In fact these results stimulated the investigation which has brought to light the approximation approach (see Section 27).

Section 7 A number of authors in various situations observed that spectra of particular weighted shift operators are connected. We have to repeat in this connection Parrot [1] and the authors mentioned while discussing Section 5. The results presented here are due to Lebedev [1, 2, 7]. The assertion on connectedness of the spectrum for the space $C(X)$ in the case when X is connected (a particular case of Corollary 7.4) was obtained also by Kitover [1].

Section 8 In this section we presented a number of examples demonstrating the symmetry property of the spectrum of weighted shifts. Such symmetry was noted by the authors already mentioned (see the preceding paragraph) and by others; see, for example, Nordgren [1], Ridge [2], and Singh, Veluchamy [1] who examined this property for the spectrum of composition operators.

We want to emphasize that this was established in different situations with the help of different methods, which has stimulated the search for a unified approach (see Section 13).

The interrelation between the conditions for invertibility and Fredholmity of different classes of operators associated with weighted shifts has been investigated by many authors. For example, the L^p–space case was examined (apart from the above–mentioned authors) by Ridge [1], Singh [1], Azbelev, Rahmatullina [1], Singh, A. Kumar [1], A. Kumar [1], Gupta, Komal [1], Singh, R. Kumar [1, 2], Singh, Dharmdhikari [1], Singh, R. Kumar, David Chandra [1], Singh, Sharma [1], Singh, Veluchamy [2] and others. The problem was treated by a number of methods (for example, Lebedev [5, 6] used a decomposition of weighted shifts into a direct integral). The C^* picture of this problem is drawn in Section 14. The non–Hilbert aspects of this and other problems concerning weighted shifts will be considered in the second part of the book.

Section 9 The results of 9.1–9.5 are due to Antonevich, Ryvkin [3]. Example 9.6 belongs to Ryvkin (unpublished).

Section 10 This section in fact serves as a presentation of the interrelation between the theory of invertibility of operators generated by weighted shifts and the theory of the solvability of a certain sort of functional differential equation.

Proposition 10.3 is a particular case of the result of V.R. Nosov (see Kolmanovskij, Nosov [1]). The statement of this theorem, as bringing to light the role of functional operators while investigating neutral–type operators, is noted by Hale [1]. The proof presented here of this proposition follows the scheme of Antonevich, Ryvkin [2] which contains Theorem 10.4 as well.

Theorems 10.7 and 10.8 are due to Antonevich, Ryvkin [1] and Theorem 10.10 is due to Antonevich [4]. The results of **10.11–10.15** are borrowed from the work by Antonevich [10, 13].

Further treatment of this theme will be presented in the second part of the book.

Chapter 2: The isomorphism theorem

'The density theorem is Kaplansky's great gift to mankind. It can be used every day, and twice on Sundays' G.K. Pedersen *C^*–algebras and their automorphism groups*

'Amenable = unmeanable' Eye–dialect

'А она – подошла к столу и выпила, залпом, еще сто пятьдесят, ибо она была совершенна, а совершенству нет предела...'

Венедикт Ерофеев 'Москва – Петушки'*

In this Chapter we shall obtain one of the most important C^* results of this book – the isomorphism theorem (Corollary 12.17) which will enable us later to solve numerous problems in the theory of functional differential equations.

In Sections 13 and 14 we present the first corollaries of this theorem (linked with the questions raised in Chapter 1), which bring to light the intrinsic reasons for the spectral symmetry (noted in Section 8), give us the possibility of investigating the ideals in the considered algebras, and enable us to find the conditions under which the examined objects are invertible iff they are Fredholm (cf. Section 8).

The isomorphism theorem describes the structure of the resulting C^*–algebras under some condition on the induced group action (named the topologically free action (definition 12.13)) which means that the set of 'periodic' points is 'small' (in Baire's sense). This condition, though rather natural in many ways, does not cover the whole range of reasonable group actions. So in the last section of this chapter (Section 16) we treat the general situation and obtain some versions of the isomorphism theorem in this case.

We want to emphasize here that the main applications of the isomorphism theorem will be given in subsequent chapters. In Chapter 3 we shall use it extensively to establish various invertibility conditions for the considered operators and, in the second part of the book, with its help we shall obtain a symbolic calculus for numerous classes of functional differential equations and the conditions for their solvability.

---

* 'And she came up to the table and drank another glass of vodka in a gulp – since she was perfect and perfection has no limit...'

Venedikt Yerofeyev *Moscow – Petushki*

11. Cross–products. Amenable groups.

This is a preliminary section to the main one (Section 12) of this chapter. Here we introduce some notions which will be used later. All the results of this section are borrowed from Greenleaf [1], Pedersen [1], and Paterson [1].

11.1 Any triple (A, G, α) consisting of a C^*–algebra A, a group G and a homeomorphism $\alpha: G \to \mathrm{Aut}(A)$ is called a *C^*–dynamical system.*

For any C^*–dynamical system (A, G, α) there exists a canonically constructed C^*–algebra $A \times_\alpha G$ called the *cross-product* of the C^*–dynamical system (A, G, α). In the case when G is discrete the algebra $A \times_\alpha G$ is constructed in the following way. We define operations of involution and convolution on the set $K(G, A)$ of formal finite sums

$$b = \sum a_g \otimes g \quad , \qquad \text{where} \quad a_g \in A \quad , \quad g \in G$$

by the equations

$$(a_g \otimes g)^* = \alpha_{g^{-1}}(a_g^*) \otimes g^{-1} \qquad \text{where} \quad \alpha_g = \alpha(g) \tag{11.1}$$

$$(a_g \otimes g)\cdot(a_h \otimes h) = a_g \cdot \alpha_g(a_h) \otimes g\cdot h \quad . \tag{11.2}$$

Straightforward computations show that $K(G, A)$ becomes a *–algebra with convolution as the product. For each $b \in K(G, A)$ define the norm

$$\|b\|_1 = \sum \|a_g\| \quad .$$

Then $K(G, A)$ becomes a normed algebra. The cross–product $A \times_\alpha G$ is the enveloping algebra for $K(G, A)$; this means that $A \times_\alpha G$ is the completion of $K(G, A)$ in the norm

$$\|b\| = \sup_\pi \|\pi(b)\|$$

where the supremum is taken over all the representations of $K(G, A)$ which are continuous with respect to the norm $\|b\|_1$.

Remark 11.2 For the general definition of the cross–product for any arbitrary locally compact group G see for example Pedersen [1]. In this book we shall not use non–discrete groups. (To be more precise even when dealing with non–discrete topological groups (for example in Sections 26, 27) we shall construct the corresponding cross–product as for discrete groups.)

11.3 Though the notion of the cross–product may be introduced for any locally compact group G in the case when G is 'too large' and behaves 'too badly' the cross–product $A \times_\alpha G$ becomes rather 'ugly' and very difficult to work with. The most general groups for which the cross–product has 'reasonable' properties are the so–called amenable groups. Here we introduce this notion and some facts about these groups (recall that we are considering only discrete groups).

11.4 A state m on $l^\infty(G)$ is called *a left–invariant mean* if

$$m(f(s^{-1}g)) = m(f(g))$$

for every $f \in l^\infty(G)$, $s \in G$.

A group G is called *amenable* if there exists a left–invariant mean on $l^\infty(G)$.

There are a number of different criteria for a group G to be amenable. Theorem 11.5 shows some of them.

Theorem 11.5 *The following conditions on* G *are equivalent*:

(i) G *is amenable*;

(ii) *for each finite set* K *and* $\varepsilon > 0$ *there exists a positive function* f *in the unit sphere of* $l^1(G)$ *such that* $\|f(s^{-1}g) - f(g)\|_{l^1(G)} < \varepsilon$ *for every* s *in* G;

(iii) *for every finite set* K *and* $\varepsilon > 0$ *there exists a non–empty finite set* U *such that*

$$\frac{|s \cdot U \,\Delta\, U|}{|U|} < \varepsilon \qquad \textit{for every} \quad s \in K$$

where $|U|$ *is the number of elements on* U.

The following Theorems 11.6–11.10 show that many frequently used groups are amenable.

Theorem 11.6 *Every abelian group is amenable.*

Theorem 11.7 *Every finite group is amenable.*

Theorem 11.8 *Every subgroup of an amenable group is amenable.*

Theorem 11.9 *If* H *is a normal subgroup of* G *such that* H *and* $^{G}/_{H}$ *are amenable then* G *is amenable.*

Theorem 11.10 *If* G *is the union of an increasing net of amenable groups then* G *is amenable.*

Theorems 11.6 and 11.9 show in particular that every solvable group is amenable.

11.11 The cross–product is a universal object and as a rule not very convenient to work with. That is why a lot of investigations have been directed towards obtaining faithful but more accessible representations of the cross–product. When G is amenable we can find many such representations. The most popular among them is the so–called regular representation. Here we recall its construction.

Let (A, G, α) be a C^*–dynamical system and (π, H) be a representation of A in a Hilbert space H. Define a covariant representation $\tilde{\pi}$ of $A \underset{\alpha}{\times} G$ in the space $l^2(G, H)$ by the formulae

$$\tilde{\pi}[(a \otimes e)\xi](g) = \pi(\alpha_g^{-1}(a))\xi(g) \tag{11.3}$$

$$\tilde{\pi}[(I \otimes h)\xi](g) = \xi(h^{-1}g) \tag{11.4}$$

where $\xi \in l^2(G, H)$, $h \in G$.

The representation $\tilde{\pi}$ is called the *regular representation* of $A \underset{\alpha}{\times} G$ induced by the representation π of A.

Theorem 11.12 *If* (A, G, α) *is a* C^**–dynamical system with amenable group* G *and* π *is a faithful representation of* A *then* $\tilde{\pi}$ $(A \underset{\alpha}{\times} G)$ *is a faithful representation of* $A \underset{\alpha}{\times} G$.

Remarks (α) If G is not amenable then the statement of the foregoing theorem is not true (see Pedersen [1], 7.3.9).

(β) Along with the representation (11.3), (11.4) which is the *left* regular representation one can use the *right* regular representation (see Remark 15.3) and the corresponding analogue of Theorem 11.12.

11.13 The key reason why Theorem 11.12 is valid is a particular structure of functionals on $A \underset{\alpha}{\times} G$ in the case when G is amenable. Namely, in this case the dual space of $A \underset{\alpha}{\times} G$ is generated by functionals which have finite supports. More precisely we mean the following. With each functional ϕ on $A \underset{\alpha}{\times} G$ we associate the function Φ: $G \to A^*$ by

$$\Phi(g)(a) = \phi(a \otimes g) \quad , \qquad g \in G \quad , \qquad a \in A \quad .$$

We shall say that a *functional* ϕ *has a finite support* if the corresponding function Φ has a finite support.

Theorem 11.14 *If* G *is amenable then the set of positive functionals with finite supports is weak** *dense in the set of all positive functionals in* $(A \underset{\alpha}{\times} G)^*$.

11.15 Now we take notice of one simple but rather useful observation.

Let $G_1 \oplus G_2$ be the direct sum of two discrete amenable groups. Along with the C^*–dynamical system $(A, G_1 \oplus G_2, \tau)$ we shall consider the following objects:

(1) the C^*–dynamical system (A, G_1, τ_1) where

$$\tau_1(g_1)(a) = \tau(g_1 \oplus e)(a) \quad , \qquad g_1 \in G_1 \quad , \qquad a \in A \quad ;$$

(2) the C^*–dynamical system $(A \underset{\tau_1}{\times} G_1, G_2, \tau_2)$ where the automorphism $\tau_2(g_2)$ is induced by the mapping

$$\tau_2(g_2)(a \otimes g_1) = [\tau(e \oplus g_2)(a)] \otimes g_1, \quad a \in A,\ g_2 \in G_2,\ g_1 \in G_1.$$

Having in mind these notations we arrive at the following result.

Lemma 11.16

$$A \underset{\tau}{\times} [G_1 \oplus G_2] \cong [A \underset{\tau_1}{\times} G_1] \underset{\tau_2}{\times} G_2 \quad . \tag{11.5}$$

Proof Just note that

$$K(G_1 \oplus G_2, A) \cong K(G_2, K(G_1, A))$$

where the algebraic operations are defined in 11.1. Thus the result follows from the definition of the cross–product. □

11.17 Our final point in this section is to link the C^*–tensor product and cross–products with the trivial group action.

Let us recall the definition of the C^*–tensor product of C^*–algebras.

Let A_i, $i = 1,2$, be two C^*–algebras and π_i, $i = 1,2$, be their faithful representations in H_{π_i}.

Consider the Hilbert space $H_{\pi_1} \otimes H_{\pi_2}$ where the inner product is defined by

$$\langle \xi_1 \otimes \xi_2, \eta_1 \otimes \eta_2 \rangle = \langle \xi_1, \eta_1 \rangle \cdot \langle \xi_2, \eta_2 \rangle \quad .$$

Consider also the *–algebra $A_1 \odot A_2$ (the vector tensor product of A_1 and A_2) with the algebraic operations

$$(a_1 \odot a_2)(b_1 \odot b_2) = a_1 b_1 \odot a_2 b_2$$

$$(a_1 \odot a_2)^* = a_1^* \odot a_2^* \quad .$$

Define the representation π: $A_1 \odot A_2 \to L(H_{\pi_1} \otimes H_{\pi_2})$ by

$$\pi(a_1 \odot a_2)(\xi_1 \otimes \xi_2) = \pi_1(a_1)\xi_1 \otimes \pi_2(a_2)\xi_2$$

and consider the norm in $A_1 \odot A_2$

$$\left\|\sum_i a_1^i \odot a_2^i\right\| = \left\|\sum \pi(a_1^i \odot a_2^i)\right\| \tag{11.6}$$

(this norm does not depend on the particular choice of faithful representations π_i of A_i).

The completion of $A_1 \odot A_2$ with respect to this norm is called the C^**–tensor product* of A_1 and A_2 and is denoted by $A_1 \otimes A_2$.

Lemma 11.18 *If* (A, G, α) *is a dynamical system with the trivial group action* α(g) = I, g ∈ G, *then*

$$A \underset{I}{\times} G = A \otimes C^*(G) \quad .$$

Proof Consider any faithful representation π_1 of A in H_{π_1}. Then

$$H_{\pi_1} \otimes l^2(G) \cong l^2(G, H_{\pi_1}) \quad . \tag{11.7}$$

Let π_2: $C^*(G) \to L(l^2(G))$ be the representation induced by the left regular representation of G:

$$(\pi_2(g_0)f)(g) = f(g_0^{-1}g) \quad .$$

Since the mapping

$$\pi(a_{g_0} \otimes g_0)(h \otimes g) = \pi_1(a_{g_0})h \otimes g_0^{-1}g$$

induces on the one hand the regular representation of $A \underset{I}{\times} G$ (see **11.11**), and on the other hand is the one used while defining the norm (11.6) it follows in view of Theorem 11.12 that

$$A \underset{I}{\times} G = A \otimes C^*(G) \quad . \qquad \square$$

12. The isomorphism theorem

In this section we consider one technical result (the isomorphism theorem (Corollary 12.17)) which will give us an opportunity of obtaining numerous applications of algebras generated by weighted shift operators to the investigation of different problems of functional differential equations.

The main object considered here is the following. Let A be a C^*–algebra which is a subalgebra of the algebra $L(H)$ of linear continuous operators acting on a Hilbert space H and let $g \to T_g$, $g \in G$, be a unitary representation of a discrete group G in $L(H)$ such that $T_g A T_{g^{-1}} = A$ for each $g \in G$, so that the operators T_g induce automorphisms $\hat{T}_g$: $A \to A$ by the equations $\hat{T}_g(a) = T_g a T_{g^{-1}}$, $a \in A$.

Consider the C^*–algebra $C^*(A, T_g)$ generated by the algebra A and the operators $\{T_g\}_{g\in G}$. With the algebra $C^*(A, T_g)$ we can associate the C^*–dynamical system $(A, G, \hat{T})$ where the homomorphism $\hat{T}$: $G \to \mathrm{Aut}(G)$ is defined by the formula $\hat{T}(g) = \hat{T}_g$.

Our goal is to find the conditions under which $C^*(A, T_g)$ is uniquely determined (up to *–isomorphism) by its C^*–dynamical system $(A, G, \hat{T})$.

Remark 12.1 In general there may exist non–isomorphic algebras of type $C^*(A, T_g)$ which induce the same C^*–dynamical system.

Example Let A be any C^*–subalgebra (with identity) of $L(H)$ such that the commutant A' contains a unitary operator U which does not belong to A.

Consider two representations of the group $\mathbb{Z}$:

(α) $z \to U^z = T_z$;

(β) $z \to I = T_z^1$, the trivial representation.

Since $U \in A'$ we have that in both cases (α) and (β) the corresponding automorphisms $\hat{T}_z$ and $\hat{T}_z^1$ are trivial:

$$\hat{T}_z(a) = \hat{T}_z^1(a) = a \quad , \qquad a \in A \quad .$$

Thus $C^*(A, \mathbb{Z}, \hat{T}) = C^*(A, \mathbb{Z}, \hat{T}^1)$. On the other hand $C^*(A, T_z^1) = A$ and $C^*(A, T_z) \neq A$. □

12.2 We shall say that an algebra $C^*(A, T_g)$ possesses the *property* (*) if for any finite sum $b = \sum a_g T_g$ we have an inequality

$$\|b\| = \left\|\sum a_g T_g\right\| \geq \|a_e\|$$

where e is the identity of the group G.

Henceforth we suppose that A is a C^*–algebra with a unit.

If an algebra $C^*(A, T_g)$ possesses the property (*) then for any finite sum $\sum a_g T_g$ the coefficient a_e is uniquely determined and therefore (passing to the limit) we can define the mapping

$$N\colon C^*(A, T_g) \to A$$

by $N(b) = a_e$. By applying the property (*) to an element $bT_{h^{-1}}$ we find that the coefficient $a_h = N(bT_{h^{-1}})$ is also uniquely determined.

12.3 Let $A \underset{\alpha}{\times} G$ be the cross–product of a discrete group G and a C^*–algebra A.

We can identify the algebra A with the set of elements of the form

$$a \otimes e \qquad a \in A$$

where e is the unit of G.

Let us define by T_g the element

$$I \otimes g$$

where I is the unit of A. With this identification $A \underset{\alpha}{\times} G$ possesses the property (*). Indeed consider a regular representation $\tilde{\pi}(A \underset{\alpha}{\times} G)$ where π is a faithful representation of A in a Hilbert space H (see **11.11**). Note that

$$\left\|\tilde{\pi}\left(\sum a_g \otimes g\right)\right\| \quad \|a_e\|$$

for every finite sum $\sum a_g \otimes g$.

Indeed let $\eta \in H$, $\|\eta\| = 1$, be any vector such that $\|\pi(a_e)\eta\| \geq \|a_e\| - \varepsilon$ (such a vector exists as π is faithful). Consider the vector $\xi \in l^2(G, H)$ defined by

$$\xi(g) = \delta(e, g)\eta$$

where $\delta(e, g)$ is the Kronecker symbol. From the definition of regular representation we have that if $g_1 \neq g_2$ then

$$\tilde{\pi}(a_{g_1} \otimes g_1)\xi \perp \tilde{\pi}(a_{g_2} \otimes g_2)\xi \quad .$$

Hence

$$\left\|\tilde{\pi}\left(\sum a_g \otimes g\right)\xi\right\|^2 = \|\tilde{\pi}(a_e \otimes e)\xi\|^2 + \left\|\tilde{\pi}\left[\sum_{g \neq e} (a_g \otimes g)\right]\xi\right\|^2 \quad .$$

Consequently since $\|\tilde{\pi}(a_e \otimes e)\xi\| = \|\pi(a_e)\eta\| \geq \|a_e\| - \varepsilon$ we conclude that

$$\left\|\tilde{\pi}\left(\sum a_g \otimes g\right)\right\| \geq \|a_e\| - \varepsilon$$

and as ε is arbitrary we obtain

$$\left\|\tilde{\pi}\left(\sum a_g \otimes g\right)\right\| \geq \|a_e\| \quad .$$

Therefore since $\tilde{\pi}$ is a representation of $A \underset{\alpha}{\times} G$ we have that

$$\left\|\sum a_g \otimes g\right\| \geq \|a_e \otimes e\|$$

and this means that $A \underset{\alpha}{\times} G$ possesses the property (*). □

A useful property of $A \times_\alpha G$ with an amenable group G is that its elements are uniquely determined by their 'coefficients'.

By **12.2** and the foregoing argument, to any element $b \in A \times_\alpha G$ there corresponds the set of 'coefficients'

$$b \to \{N(b\cdot(I \otimes g^{-1}))\}_{g\in G} \quad .$$

Theorem 12.4 *Let* G *be an amenable group. Then an element* b $\in$ A $\times_\alpha$ G *is zero iff all its 'coefficients' are zero.*

Proof Since G is amenable $A \times_\alpha G$ is isomorphic to its regular representation (by Theorem 11.12) and $\tilde{\pi}(A \otimes e) \cong A$. By the construction of this isomorphism (see **11.11**) to each element b there corresponds an operator $\tilde{b}$ on $l^2(G, H)$ which is defined by the operator matrix

$$[\tilde{b}]_{h,g} = \alpha_h[N(b\cdot(I \otimes g^{-1}h))] \qquad h,g \in G \quad .$$

But an operator is zero iff its matrix is zero. □

12.5 Suppose that $C^*(A, T_g)$ possesses the property (*). For each functional ψ on $C^*(A, T_g)$ let us define the function Ψ: $G \to A^*$ by

$$\Psi(g)(a) = \psi(aT_g) \qquad a \in A \quad , \quad g \in G \quad .$$

The definition of this function is correct in view of the property (*). We shall say that a functional $\psi \in [C^*(A, T_g)]^*$ has a finite support if the function Ψ has a finite support.

12.6 For any C^*–algebra $C^*(A, T_g)$ one can define the *standard representation*

$$\pi\colon A \underset{\hat{T}}{\times} G \to C^*(A, T_g)$$

by the formulae

$$\pi(a \otimes e) = a \tag{12.1}$$
$$\pi(I \otimes g) = T_g \qquad a \in A \ , \quad g \in G \ . \tag{12.2}$$

Lemma 12.7 *Suppose that* $C^*(A, T_g)$ *possesses the property* (*). *If* ϕ *is a state on* $A \underset{T}{\hat{\times}} G$ *with a finite support then there exists a state* ϕ' *on* $C^*(A, T_g)$ *such that*

$$\phi = \phi' \circ \pi$$

where π *is the standard representation* (12.1), (12.2).

Proof Let Φ be the function corresponding to ϕ and let $M \subset G$ be the finite support of Φ. Consider the functional f on A defined by

$$f = \Phi(e) \quad .$$

It is evident that $\|f\| = \|\Phi(e)\| = \|\phi(I \otimes e)\| = 1$ where I is the unit of A. For any $a \in A$ we have by the Cauchy–Bunyakowskii–Schwartz inequality

$$|\Phi(g)\ \alpha_g(a)|^2 = |\phi(\alpha_g(a)) \otimes g|^2$$

$$= |\phi[(I \otimes g)(a \otimes e)]|^2 \leq \phi[(I \otimes g)(I \otimes g^{-1})]\cdot\phi[(a^* \otimes e)(a \otimes e)]$$

$$= \phi(a^* a \otimes e) = \Phi(e)(a^* a) = f(a^* a) \quad .$$

Consequently we have

$$|\Phi(g)(\alpha_g(a))|^2 \leq \|a^* a\| = \|a\|^2 \quad ,$$

that is

$$|\Phi(g)(\alpha_g(a))| \leq \|a\| \quad .$$

This means that

$$|\Phi(g)(a)| = \left|\Phi(g)(\alpha_g(\alpha_{g^{-1}}(a)))\right|$$

$$\leq \left\|\alpha_{g^{-1}}(a)\right\| = \|a\| \quad .$$

Now let us consider the functional ϕ' on $C^*(A, T_g)$ defined by

$$\phi'(b) = \sum_{g \in M} \Phi(g)[N(bT_{-g})] \quad .$$

We have

$$|\phi'(b)| \leq |M| \max_{g \in M} \|N(bT_{-g})\| \leq |M| \, \|b\|$$

where the second inequality is a consequence of the property (*) (here $|M|$ is the number of elements in M).

This means that ϕ' belongs to $[C^*(A, T_g)]^*$ and in view of its construction ϕ' is a state and $\phi = \phi' \circ \pi$.

The proof of the lemma is complete. □

This lemma leads to Theorem 12.8 which shows that the property (*) is the main property to be considered when constructing faithful representations of $A \times_\alpha G$.

Theorem 12.8 *Let* (A, G, α) *be a* C^**–dynamical system with* G *a discrete amenable group, let* π *be a faithful representation of* A *and let* $\{T_g\}_{g \in G}$ *be a unitary group in the representation space* H *of* π *such that*

$$\pi(\alpha_g(a)) = \hat{T}_g \, \pi(a) \qquad \textit{for any} \quad a \in A \quad .$$

Then the standard representation $A \times_\alpha G \xrightarrow{\pi} C^*(\pi(A), T_g)$ *defined by the formulae* (12.1), (12.2) *is the* **–isomorphism between* $A \times_\alpha G$ *and* $C^*(\pi(A), T_g)$ *iff the algebra* $C^*(\pi(A), T_g)$ *possesses the property* (*).

Proof Obviously possession of the property (*) is a necessary condition for $C^*(\pi(A), T_g)$ to be isomorphic to $A \times_\alpha G$ in view of 12.3, so we shall prove sufficiency. Consider any positive functional $\phi \in (A \times_\alpha G)^*$ with a finite support. By Lemma 12.7 there exists a positive functional $\phi' \in [C^*(A, T_g)]^*$ such that $\phi = \phi' \circ \pi$. In view of Theorem 11.14 we conclude that the set of functionals of the form $\phi' \circ \pi$ where ϕ' is a positive functional on $C^*(A, T_g)$ is weak* dense in the set of positive functionals on $A \times_\alpha G$. Thus the representation π has zero kernel. □

Corollary 12.9 *Let* $C^*(A_1, T_g)$ *and* $C^*(A_2, V_g)$ *be two* C^**–algebras and let* φ: $A_1 \to A_2$ *be an isomorphism such that*

$$\varphi(T_g\, aT_{g^{-1}}) = V_g\, \varphi(a)V_{g^{-1}} \qquad \forall_g \in G \quad .$$

If G *is amenable and both the algebras* $C^*(A_1, T_g)$ *and* $C^*(A_2, V_g)$ *possess the property* (*) *then* $C^*(A_1, T_g)$ *is* **–isomorphic to* $C^*(A_2, V_g)$ *and the isomorphism is given by*

$$\sum a_g T_g \to \sum \varphi(a_g)\, V_g \quad , \qquad a_g \in A_1 \quad .$$

Remark 12.10 One may observe that Theorem 11.12 follows from **12.3** and Theorem 12.8.

Remark 12.11 In general the property (*) of $C^*(A, T_g)$ is not correlated with the action of $\hat{T}_g$ on A.

Example Consider the space $l^2(\mathbb{Z})$. Let $A \cong \mathbb{C}$ be the algebra of operators of multiplication by constant functions. Consider two unitary representations of the group $\mathbb{Z}$:

(i) $(T_z^1\, \xi)(k) = \xi(k - z)$, bilateral shifts;

(ii) $(T^2_z\ \xi)(k) = \xi(k)$, the trivial representation (where $\xi \in l^2(\mathbb{Z})$).

In both cases we have that for any $a \in A$

$$T^1_z\ aT^1_{-z} = T^2_z\ aT^2_{-z} = a \quad .$$

Thus $\hat{T}^1(z) = \hat{T}^2(z) = I$, the trivial automorphism.

The algebra $C^*(A,\ T^1_z)$ possesses the property (*) since $C^*(A,\ T^1_z)$ is simply the regular representation for $\mathbb{C} \underset{I}{\times} \mathbb{Z}$. On the other hand $C^*(A,\ T^2_z) = A$, and the algebra A of multiplication by constants does not possess the property (*).

12.12 The goal of this section is to obtain the condition on $\hat{T}_g$ under which the algebra $C^*(A,\ T_g)$ will automatically possess the property (*).

Let Prim A be the primitive ideal space of A equipped with the Jacobson topology (see 2.63). An automorphism $\hat{T}_g$ of A induces the homeomorphism t_g of Prim A which can be defined in the following way:

for each $\pi \in \hat{A}$ we set

$$\tau_g(\pi)[a] = \pi[\hat{T}_g^{-1}(a)]$$

and if x = Ker π then we set

$$t_g(x) = \text{Ker}\ \tau_g(\pi) \quad .$$

One can easily verify that

$$t_g \cdot t_h = t_{gh} \quad .$$

We denote by X_g, $g \in G$, the set

$$X_g = \{x \in \text{Prim}\ A\colon\ t_g(x) = x\} \quad .$$

12.13 We say that the group G acts *topologically freely* on A by automorphisms $\hat{T}_g$ if for any finite set $\{g_1,\dots,g_n) \subset G$ $(g_i \neq e)$ the set $\left[\bigcup_{i=1}^{n} X_{g_i}\right]$ has an empty interior.

One can reformulate this definition in a slightly different way as follows.

12.13′ The action of G is said to be topologically free if for any finite set $\{g_1,\dots,g_k\} \subset G$ and an open set $U \subset \operatorname{Prim} A$ there exists a point $x \in U$ such that all the points $t_{g_i}(x)$, $i = \overline{1,k}$, are distinct.

Evidently **12.13′** ⇒ **12.13**.

On the other hand if **12.13** is true then along with any set $\{g_1,\dots,g_k\} \subset G$ one can consider the set

$$\{h_1,\dots,h_n\} = \left\{g_j^{-1} g_i\right\} \qquad i,\ j \in \overline{1,k}\ , \quad i \neq j\ .$$

Since $j \neq i$ we have $g_j^{-1} g_i \neq e$. So by **12.13** the set $W = \bigcup_{s=1}^{n} X_{g_s}$ has an empty exterior.

Thus for any open set U there exists $x \in U \backslash W$. For this point we have

$$t_{g_j^{-1} g_i}(x) \neq x \qquad i \neq j$$

which is equivalent to

$$t_{g_i}(x) \neq t_{g_j}(x) \qquad i \neq j\ .$$

Theorem 12.14 *If* G *acts topologically freely on* A *then* $C^*(A, T_g)$ *possesses the property* (*).

Proof Let F be any finite set in G. Consider an operator $\sum_{g \in F} a_g T_g$. Let π_0 be an irreducible representation of A such that $\|\pi_0(a_e)\| = \|a_e\|$ (such a π_0 exists by Lemma 2.88). Set $x_0 = \operatorname{Ker} \pi_0$. Since the function $\breve{a}_e(x)$ is lower semicontinuous (Theorem 2.87) and $\breve{a}_e(x_0) = \|a_e\|$ there exists an open set $U \subset \text{Prim } A$ such that $\breve{a}_e(x) \geq \|a_e\| - \varepsilon$ for every $x \in U$.

As G acts topologically freely there exists a point $x' \in U$ such that $t_g(x')$, $g \in F$, are distinct. Let π' be an irreducible representation of A such that $\text{Ker } \pi' = x'$ and let ν be any extension of π' up to an irreducible representation of $C^*(A, T_g)$ (see Theorem 2.81). For this representation ν we have

$$H_{\pi'} \subset H_\nu$$

where $H_{\pi'}$ is the representation space for π', and H_ν is that for ν.

Let ξ be a vector in $H_{\pi'}$ such that $\|\xi\| = 1$ and $\|\pi'(a_e)\xi\| \geq \|a_e\| - 2\varepsilon$. First of all we observe that

$$\langle \pi'(a_1)\xi, \nu(a_2T_g)\xi \rangle = 0 \qquad a_1, a_2 \in A \ , \quad g \in F \ .$$

To prove this we need the following lemma.

Lemma 12.15 *Let* A *be a* C^**–subalgebra of* L(H). *If* P_1, $\mathrm{P}_2 \in \mathrm{A}'$ *are two orthogonal projections such that the restrictions*

$$\mathrm{A}|_{\mathrm{H}_{\mathrm{P}_1}} \quad \textit{and} \quad \mathrm{A}|_{\mathrm{H}_{\mathrm{P}_2}}$$

(where $\mathrm{H}_{\mathrm{P}_1} = \mathrm{P}_1(\mathrm{H})$, $\mathrm{H}_{\mathrm{P}_2} = \mathrm{P}_2(\mathrm{H})$*) are both irreducible and these restrictions are distinct representations then*

$$\mathrm{H}_{\mathrm{P}_1} \perp \mathrm{H}_{\mathrm{P}_2} \ .$$

Proof The condition that $A|_{H_{P_1}}$ and $A|_{H_{P_2}}$ are distinct representation means that there exists an operator $a \in A$ such that $P_1a = 0$ and $P_2a \neq 0$ (or $P_1a \neq 0$ and $P_2a = 0$). Consider the case when $P_1a = 0$ and $P_2a \neq 0$ (the second case can be considered in an analogous way). By applying the condition that $P_2 \in A'$ we conclude that there exists a vector $\xi \in H_{P_2}$ such that $P_1 a\xi = 0$ and $P_2 a\xi \neq 0$.

This means that the vector $a\xi \in H_{P_2}$ is orthogonal to H_{P_1}. So we have that

$$\forall h \in H_{P_1} \qquad \langle h, a\xi \rangle = 0 \quad .$$

This means (as H_{P_1} is invariant under A) that for any $b \in A$ we have that

$$0 = \langle b^* h, a\xi \rangle = \langle h, ba\xi \rangle \quad .$$

As $A|_{H_{P_2}}$ is irreducible the vector $(a\xi)$ is cyclic for $A|_{H_{P_2}}$ and this means that $\overline{\{ba\xi\}}_{b\in A} = H_{P_2}$. So we conclude that $H_{P_2} \perp H_{P_1}$ and the proof of the lemma is complete. □

Now in order to prove the equality

$$\langle \pi'(a_1)\xi, \nu(a_2\, T_g)\xi \rangle = 0 \qquad a_1, a_2 \in A \ , \quad g \in F$$

we consider an orthogonal projection P_1 onto the subspace $H_{\pi'}$, and the projection $P_2 = \nu(T_g) P_1 \nu(T_g^*)$. It is clear that $P_1, P_2 \in [\nu(A)]'$ and $\nu(A)|_{H_{P_2}}$ is irreducible. From the definition of P_2 we have that $P_2 \nu(T_g) = \nu(T_g) P_1$. From the construction of the projections P_1 and P_2 we have that

$$P_1 \nu(a) = 0 \Leftrightarrow \pi'(a) = 0 \Leftrightarrow \breve{a}(x') = 0$$

and

$$P_2 \nu(a) = 0 \Leftrightarrow \nu(T_g)\ P_1 \nu(T_g^*)\ \nu(a) = 0$$

$$\Leftrightarrow \nu(T_g)\ P_1\ \nu(\hat{T}_{g^{-1}}(a))\ \nu(T_g^*) = 0$$

$$\Leftrightarrow P_1\ \nu(\hat{T}_{g^{-1}}(a)) = 0 \Leftrightarrow \left[\widebreve{\hat{T}_{g^{-1}}(a)}\right](x') = 0$$

$$\Leftrightarrow \breve{a}(t_g(x')) = 0 \quad .$$

So (since the points x' and $t_g(x')$ are distinct) we conclude that the representations $A|_{H_{P_1}}$ and $A|_{H_{P_2}}$ are distinct.

Applying Lemma 12.15 we find that $P_2 \cdot P_1 = 0$. Now for any vector $\xi \in H_{P_1} = H_{\pi'}$, $a_1, a_2 \in A$ and $g \in F$, we have that

$$\langle \pi'(a_1)\xi,\ \nu(a_2T_g)\xi \rangle = \langle P_1\nu(a_1)\xi,\ \nu(a_2T_g)P_1\xi \rangle$$

$$= \langle P_1\nu(a_1)\xi,\ P_2\nu(a_2T_g)\xi \rangle = \langle P_2 \cdot P_1\nu(a_1)\xi,\ \nu(a_2T_g)\xi \rangle = 0 \quad .$$

To finish the proof of the theorem we observe that

$$\left\| \sum_{g \in F} a_g T_g \right\|^2 \geq \left\| \nu\left(\sum_{g \in F} a_g T_g \right) \xi \right\|^2$$

$$= \|\nu(a_e)\xi\|^2 + \left\| \nu\left(\sum_{g \neq e} a_g T_g \right) \xi \right\|^2$$

$$\geq \|\nu(a_e)\xi\|^2 = \|\pi'(a_e)\xi\|^2 \geq (\|a_e\| - 2\varepsilon)^2$$

and as ε is arbitrary the proof is complete. □

Corollary 12.16 *Let* $C^*(A, T_g)$ *be an algebra such that* G *is amenable and acts topologically freely on* A *by automorphisms* $\hat{T}_g$. *Then* $C^*(A, T_g)$ *is* **–isomorphic to* $A \times_{\hat{T}} G$.

Corollary 12.17 (The isomorphism theorem) *Let* $C^*(A, T_g)$ *and* $C^*(A_1, V_g)$ *be* C^**–algebras and let* $\varphi: A \to A_1$ *be an isomorphism such that*

$$\varphi(T_g a T_{g^{-1}}) = V_g\, \varphi(a) V_{g^{-1}} \qquad g \in G \quad .$$

If G *is amenable and acts topologically freely on* A *by automorphisms* $\hat{T}_g$ *then* $C^*(A, T_g)$ *and* $C^*(A_1, V_g)$ *are* **–isomorphic.*

Proof Observe that in this situation the action of G on A_1 by automorphisms $\hat{V}_g$ is also topologically free. □

Remark 12.18 Note that if the action $\hat{T}_g$ on A is not topologically free then $C^*(A, T_g)$ in general does not possess the property (*) (see the example in Remark 12.11). Thus the condition that G acts topologically freely is the essential condition under which $C^*(A, T_g)$ is uniquely determined by its C^*–dynamical system $(A, G, \hat{T})$.

13. Corollaries of the isomorphism theorem (invertibility, spectral symmetry)

Here we shall obtain some immediate corollaries of the isomorphism theorem showing the algebraic reasons for the spectrum to possess the symmetry noticed in Section 8 and also describing a symmetry in the conditions for the invertibility of elements in $C^*(A, T_g)$.

Proposition 13.1 *Let* $\mathrm{C}^*(\mathrm{A}, \mathrm{T}_g)$ *be a* C^**–algebra which possesses the property* (*). *Then for any character* χ *of the group* G *the elements* $\sum a_g T_g$ *and* $\sum a_g \chi(g) T_g$ *either are simultaneously invertible or simultaneously not invertible.*

Proof Since $[\chi(g)T_g]\, a \left[\chi(g^{-1})T_{g^{-1}}\right] = T_g a T_{g^{-1}}$ for any $g \in G$, the elements T_g and $\chi(g)T_g$ induce the same automorphism of A. By Corollary 12.9 the mapping

$$\sum a_g T_g \to \sum a_g [\chi(g) T_g]$$

induces an automorphism of the algebra $C^*(A, T_g)$ and consequently corresponding elements are simultaneously invertible or are simultaneously not invertible. □

Proposition 13.1 can be used to establish the invariance of the so–called multiparameter spectrum.

13.2 Let $\lambda = (\lambda_1,\dots,\lambda_n) \in \mathbb{C}^n$ and $f(\lambda)$ be a function which has values in the algebra B. The *multiparameter spectrum* of f is the set $\sigma(f)$ of those λ for which $f(\lambda)$ is not invertible.

Proposition 13.3 *Let* $G \cong \mathbb{Z}^n$ *and* $g_1,\dots,g_n$ *be generators of* G. *Let* $C^*(A, T_g)$ *be a* C^**–algebra which possesses the property* (*) *and let the function* $f: \mathbb{C}^n \to C^*(A, T_g)$ *be defined by*

$$f(\lambda) = \sum \lambda^z a_z T^z \tag{13.1}$$

where $\mathbb{Z}^n \ni z = (z_1,\dots,z_n)$ *is a multiindex,* $\lambda^z = \lambda_1^{z_1}\cdot\ldots\cdot\lambda_n^{z_n}$, $T^z = T_{g_1}^{z_1}\cdot\ldots\cdot T_{g_n}^{z_n}$ *and the sum* (13.1) *is finite or is a converging series. Then the multiparameter spectrum* $\sigma(f)$ *is invariant under the transformation*

$$(\lambda_1,\dots,\lambda_n) \to (\delta_1\lambda_1,\dots,\delta_n\lambda_n)$$

where $|\delta_1| = \dots = |\delta_n| = 1$ *i.e. the spectrum* $\sigma(f)$ *is invariant under the action of the torus* T^n, *the dual group for* $\mathbb{Z}^n$.

Proof For any set of complex numbers $\delta = (\delta_1,\dots,\delta_n)$, $|\delta_i| = 1$, the formula $\chi_\delta(z) = \delta_1^{z_1}\cdot\ldots\cdot\delta_n^{z_n}$ defines the character of $\mathbb{Z}^n$ and so the proposition follows from Proposition 13.1. □

13.4 In the case $G = \mathbb{Z}$, instead of writing $C^*(A, T_z)$, we shall write $C^*(A, T)$ (because the group $\{T_z\}_{z\in\mathbb{Z}}$ is generated by a single element $T = T_1$).

Proposition 13.5 *Let* $C^*(A, T)$ *be a* C^**–algebra which possesses the property* (*). *Then for any unitary element* u *from the centre of* A *the map*

$$Q: \sum a_z T^z \to \sum a_z (uT)^z$$

defines an automorphism of $C^*(A, T)$ *and elements* b *and* Q(b) *are simultaneously invertible or are simultaneously not invertible.*

Proof Since $(uT)\ a(uT)^{-1} = TaT^{-1}$ the proposition follows from Corollary 12.9. □

Corollary 13.6 *Let* $C^*(A, T)$ *be a* C^**–algebra which possesses the property* (*). *Then for any complex number* λ *such that* $|\lambda| = 1$ *the elements* $a_0 + a_1T$ *and* $a_0 + \lambda a_1 T$ *are simultaneously invertible or are simultaneously not invertible. In particular, the spectrum of* aT *is invariant under rotations about the point* 0.

In fact we can say even more.

Proposition 13.7 *Let the* C^**–algebra* $C^*(A, T)$ *be such that* $\{\hat{T}^z\}$ *acts topologically freely on* A. *Then for any unitary elements* $u_0, u_1 \in A'$ (*the commutant of* A) *the elements* $a_0 + a_1T$ *and* $u_0a_0 + u_1a_1T$ *are simultaneously invertible or are simultaneously not invertible.*

Proof Since $(u_0^{-1}u_1T)\ a(u_0^{-1}u_1T)^{-1} = TaT^{-1}$ the elements T and $u_0^{-1}u_1T$ induce the same automorphism of A. By Corollary 12.17 the algebras $C^*(A, T)$ and $C^*(A, u_0^{-1}u_1T)$ are isomorphic and the element $a_0 + a_1T$ corresponds to the element $a_0 + a_1u_0^{-1}u_1T$ which is simultaneously invertible or not simultaneously invertible with $u_0a_0 + u_1a_1T$. □

In the case of a commutative algebra A, Proposition 13.7 leads to the following.

Proposition 13.8 *If* A *is a commutative* C^**–algebra and* $\{\hat{T}^z\}$ *acts topologically freely on* A *then elements* $a_0 + a_1T$ *and* $|a_0| + |a_1|T$ *are either simultaneously invertible or simultaneously not invertible.*

Proof Let $a_0 = |a_0|u_0$ and $a_1 = |a_1|u_0$ be the polar decomposition of a_0 and a_1. Then u_0 and u_1 belong to $A'' \subset A'$ and we have

$|a_0| + |a_1|T = a_0u_0^{-1} + a_1u_1^{-1}T$. Applying Proposition 13.7, we have finished. □

13.9 Using the results of Sections 4 and 7 we can now describe the spectrum $\sigma(aT)$. Let M be the maximal ideal space of a commutative C^*–algebra A and $M(\hat{T})$ be the maximal ideal space of the C^*–subalgebra of all $\hat{T}$–invariant elements in A.

Theorem 13.10 *Let* $M(\hat{T})$ *be a connected space and* $C^*(A, T)$ *possesses the property* (*). *Then* $\sigma(aT)$ *has the following form:*

(1) *if* a *is invertible then* $\sigma(aT)$ *is the annulus*

$$\max_{\nu\in\Lambda} S_\nu(a) \geq |\lambda| \geq \min_{\nu\in\Lambda} S_\nu(a) \quad ;$$

(2) *if* a *is not invertible then* $\sigma(aT)$ *is the disk*

$$|\lambda| \leq \max_{\nu\in\Lambda} S_\nu(a)$$

where Λ *is the set of ergodic measures with respect to* t (*where* t *is a homeomorphism of* M *induced by* $\hat{T}$) *and*

$$S_\nu(a) = \exp \int_M \ln|a(x)|d\nu \quad .$$

Proof This follows immediately from Theorem 4.4, Corollary 4.5, Theorem 7.7 and Corollary 13.6. Since (by Corollary 13.6) $\sigma(aT)$ is invariant with respect to rotations about zero, so the set $|\sigma(aT)|$ in the conditions of Theorem 7.7 can be replaced by $\sigma(aT)$. □

13.11 Coincidence of spectra and simultaneous invertibility of operators in different spaces

A rather common situation is when the same formula defines operators in different spaces. The isomorphism theorem gives us a method of establishing relations between certain properties of such operators. Here we shall look through some examples.

Let us consider four Hilbert spaces of N–dimensional vector functions on $\mathbb{R}$:

(1) $H_1 = L_2^N(\mathbb{R})$;

(2) H_2 = periodic functions of period 1 which locally belong to $L_2^N(\mathbb{R})$;

(3) $H_3 = [B^2(\mathbb{R})]^N$, the Besicovitch space of almost periodic functions;

(4) $H_4 = l_2^N(\mathbb{R})$, the space of vector functions each of which does not vanish on a set which has not more than countably many points and such that the series $\sum_{f(x)\neq 0} \|f(x)\|^2$ converges.

The spaces H_1, H_2 and H_4 do not intersect with each other, the spaces H_1 and H_2 are separable, and H_3 and H_4 are non–separable. So these spaces are essentially different.

Let A be a C^*–algebra of continuous periodic matrix functions on $\mathbb{R}$ which have period 1. Consider the formula

$$(bf)(x) = \sum a_k(x)\ f(x + kh) \tag{13.2}$$

where $a_k \in A$ and h is a certain number. The formula (13.2) defines the operators b_i acting in the spaces H_i, $i = \overline{1,4}$. In view of the differences in the spaces H_i some properties of b_i are essentially different. For example the formula

$$(Tf)(x) = f(x + h)$$

defines the shift operators T_i, $i = \overline{1,4}$. The operators T_1 and T_4 have no eigenvalues, T_2 has eigenvalues $\lambda_k = e^{i2\pi kh}$, $k \in \mathbb{Z}$, and for T_3 all the numbers λ, such that $|\lambda| = 1$, are eigenvalues. Nevertheless the isomorphism theorem leads to the following result.

Proposition 13.12 *If* h *is an irrational number then the spectra of the operators* b_i, $i = \overline{1,4}$, *coincide. If* h *is rational then the spectra of* b_1, b_3, b_4 *coincide and the spectrum of* b_2 *may have a different form.*

Proof Consider the algebra B_i, $i = \overline{1,4}$, generated by operators of the form (13.2) in the space H_i, $i = \overline{1,4}$. B_i is a C^*–algebra of the form $C^*(A_i, T_i)$, where A_i is the algebra of operators of multiplication by matrix functions from A on H_i. All A_i, $i = \overline{1,4}$, are isomorphic to A and Prim $A = S^1$ $= \mathbb{R}(\text{mod } 1)$. All the operators T_i, $i = \overline{1,4}$, induce the same automorphism $\hat{T}$ of A and the corresponding homeomorphism of Prim A, $\alpha(x) = x + h$, is a rotation of the circle.

If h is irrational then $\{\hat{T}^k\}$ acts topologically freely and in view of the isomorphism theorem all B_i, $i = \overline{1,4}$, are isomorphic and b_i are mapped into each other under these isomorphisms. So their spectra coincide.

Now let $h = {}^p/_m$ be a rational number. Then the spectrum $\sigma(T_2)$ consists of m points and $\sigma(T_3)$ is the unit circle so $\sigma(T_2) \neq \sigma(T_3)$.

On the other hand $\sigma(b_1) = \sigma(b_3) = \sigma(b_4)$. Indeed one can easily verify that the property (*) is valid for the algebras B_1, B_3, B_4 and so by Corollary 12.9 these algebras are isomorphic. We have finished. □

14. Corollaries of the isomorphism theorem (ideals in $C^*(A, T_g)$, Fredholmity, compactness)

In Section 8 we noticed that in some situations the spectrum and the Fredholm spectrum of a weighted shift operator coincide. Here we shall find the intrinsic reasons for this, linked with the structure of ideals in $C^*(A, T_g)$ and the isomorphism theorem.

In the case of a topologically free action of a group G we can obtain a rather efficient description of ideals in $C^*(A, T_g)$.

Theorem 14.1 (see also Theorem 16.23) *If* $J \neq \{0\}$ *is a closed* **–ideal in* $C^*(A, T_g)$ *and* G *acts topologically freely then* $J \cap A \neq \{0\}$.

Proof Let $A \cap J = \{0\}$. Consider the canonical mapping $\pi: C^*(A, T_g) \to C^*(A, T_g)/_J$. It is evident that $\pi(C^*(A, T_g)) = C^*(\pi(A), \pi(T_g))$ and $\pi(T_g a T_{g^{-1}}) = \pi(T_g)\ \pi(a)\ \pi(T_{g^{-1}})$. As $A \cap J = \{0\}$ we have (applying Corollary 2.77)

$$\pi(A) = {}^{A+J}/_J \cong {}^{A}/_{A\cap J} \cong A \quad .$$

Consequently in view of the isomorphism theorem

$$C^*(A, T_g) \cong C^*(\pi(A), \pi(T_g)) \cong C^*(A, T_g)/_J \quad .$$

This means, in view of Corollary 2.76, that $J = \{0\}$. □

Remarks 14.2

(1) For any ideal $I \subset A$ let us denote by supp I the set

$$\text{supp } I = \{x \in \text{Prim } A,\ x \supset I\} \quad .$$

If J is an ideal in $C^*(A, T_g)$ then $J \cap A$ is evidently invariant under $\hat{T}_g$ and consequently supp$[J \cap A]$ is invariant under t_g for all $g \in G$. On the other hand for any ideal $I \subset A$ we have (by Dixmier [1] 2.9.7) that

$$I = \bigcap_{x \in \text{supp} I} x$$

and if supp I is invariant under all t_g, $g \in G$, then I is invariant under $\hat{T}_g$ and $C^*(I, \mathrm{T}_g)$ is an ideal in $C^*(A, T_g)$.

(2) For any $x \in$ Prim A let $\overline{G(x)}$ be the closure of the orbit of x, that is,

$$\overline{G(x)} = \overline{\{t_g(x)\}}_{g \in G} \quad .$$

By definition $\overline{G(x)}$ is closed and invariant under all t_g, $g \in G$.

If U is any closed set in Prim A which is invariant under all t_g, $g \in G$, then obviously $\overline{G(x)} \subset U$ for all $x \in U$.

Now Theorem 14.1 and the above remarks imply

Proposition 14.3 *If* G *acts topologically freely and* $\overline{\mathrm{G(x)}}$ = Prim A *for all* x $\in$ Prim A *then* $\mathrm{C}^*(\mathrm{A}, \mathrm{T}_g)$ *is simple, that is,* $\mathrm{C}^*(\mathrm{A}, \mathrm{T}_g)$ *does not contain any non–zero* **–ideals.*

Example Consider the algebra A of operators of multiplication by continuous functions on the space $L^2(S^1)$ and the shift operator T induced by an irrational rotation of the circle. Then in view of Proposition 14.3 $C^*(A, T)$ is simple.

As immediate corollaries of Theorem 14.1 we also obtain the following results (hereinafter, up to 14.12, $C^*(A, T)$ is supposed to be a C^*–subalgebra in $L(H)$).

Proposition 14.4 *Consider an algebra* $\mathrm{C}^*(\mathrm{A}, \mathrm{T}_g)$ *where* G *is amenable and acts topologically freely on* A. *If* A *does not contain any compact operator other than the zero operator then* $\mathrm{C}^*(\mathrm{A}, \mathrm{T}_g)$ *does not contain any compact operator other than zero.*

Proposition 14.5 *If* G *is amenable and acts topologically freely on* A *and* A *does not contain any compact operator other than zero then every Fredholm operator in* $\mathrm{C}^*(\mathrm{A}, \mathrm{T}_g)$ *is invertible.*

Proof Let K be the ideal of compact operators in the algebra $L(H)$ of linear bounded operators on the Hilbert space H. By the same argument as in Theorem 14.1 we have

$$C^*(A, T_g)/_K \cong C^*(A, T_g) \quad .$$

As Fredholm operators are simply operators which are invertible modulo the compact operators. We have finished. □

To use Proposition 14.5 we have to have some criteria for A not to contain any compact operator. Here we present a number of such criteria.

Proposition 14.6 *If in each non–empty open set in* Prim A *there exist two non–empty disjoint open subsets then the algebra* A *does not contain compact operators other than zero.*

Proof Let a be any element in A and K be any compact operator. We shall prove that $\|a + K\| \geq \frac{1}{\sqrt{2}} \|a\|$ and this will be enough.

Let $x_0 \in \text{Prim } A$ be a point such that $\breve{a}(x_0) = \|a\|$. By the lower semicontinuity of the function $\breve{a}(x)$ there exists an open set $U \subset \text{Prim } A$ such that $\breve{a}(x) \geq (1 - \varepsilon)\|a\|$ for all $x \in U$. By the condition of the proposition one can find mutually non–intersecting open sets $V_i \subset U$, $i = 1,2,\ldots$. Let I_i be the ideal of all elements in A vanishing on $[\text{Prim } A \backslash V_i]$ and let P_i be the projection on the essential subspace H_i of I_i (that is, a closed subspace generated by vectors $\{a\xi\}$, $\xi \in H$, $a \in I_i$). Since V_i is open, H_i is non–zero.

By Lemma 2.82 $P_i \in \overline{A} \cap \overline{A}'$, where $\overline{A}$ is the strong closure of A and, as $V_i \cap V_j = \emptyset$, $i \neq j$, we have that $P_iP_j = 0$, that is, $H_i \perp H_j$. From $V_i \subset U$ it follows that

$$\|aP_i\| \geq (1 - \varepsilon)\|a\| \quad .$$

Let $\xi_i \in H_i$ be vectors such that $\|\xi_i\| = 1$ and

$$\|aP_i\xi_i\| \geq (1 - 2\varepsilon)\|a\| \quad .$$

Since K is a compact operator we can assume, choosing a suitable subsequence of $\{\xi_i\}$, that

$$\|K(\xi_i - \xi_j)\| \leq \frac{1}{i} \qquad j > i \quad . \tag{14.1}$$

As $\xi_i \perp \xi_j$ and $\|\xi_i\| = \|\xi_j\| = 1$ we have that

$$\|a(\xi_i - \xi_j)\| \geq \|P_i a(\xi_i - \xi_j)\| = \|aP_i(\xi_i - \xi_j)\|$$

$$= \|aP_i\xi_i\| \geq (1 - 2\varepsilon)\|a\|$$

$$= \frac{(1 - 2\varepsilon)}{\sqrt{2}} \|a\| \|\xi_i - \xi_j\|$$

and using this fact and (14.1) we conclude that

$$\|(a + K)(\xi_i - \xi_j\| \geq | \|a(\xi_i - \xi_j)\| - \|K(\xi_i - \xi_j)\| |$$

$$\geq \frac{(1 - 2\varepsilon)}{\sqrt{2}} \|a\| \|\xi_i - \xi_j\| - \frac{1}{i} \frac{\|\xi_i - \xi_j\|}{\sqrt{2}}$$

$$= \frac{1}{\sqrt{2}} \left[(1 - 2\varepsilon)\|a\| - \frac{1}{i}\right] \|\xi_i - \xi_j\| .$$

Hence as ε and i are arbitrary this proves

$$\|a + K\| \geq \frac{1}{\sqrt{2}} \|a\|$$

as required. □

14.7 For any open set $U \subset \text{Prim } A$ let I_U denote the ideal of all elements in A vanishing on $[\text{Prim } A \backslash U]$. We can reformulate the condition of Proposition 14.6 in terms of such ideals.

Proposition 14.8 *Let* A *be a separable* C^**–algebra. If, for every open set* $\text{U} \subset \text{Prim A}$, I_U *does not contain* {0} *as its primitive ideal then* A *does not contain compact operators other than zero.*

Proof By Theorem 2.84, $\text{Prim } I_U \approx U$, and as {0} does not belong to $\text{Prim } I_U$ there exist, by Dixmier [1] 3.91, two non–empty disjoint open subsets in $\text{Prim } I_U$. Now apply Proposition 14.6. □

14.9 Let X be a topological space. A point x_0 in X is called *Hausdorff* if for any x_1 in X such that x_1 does not belong to the closure of x_0 the points x_0 and x_1 have non–intersecting neighbourhoods.

The following result links the 'isolated' Hausdorff points with compact operators.

Proposition 14.10 *Let* A *be a separable* C^**–algebra. If none of the open sets in* Prim A *coincide with the closure of any Hausdorff point then* A *does not contain compact operators other than zero.*

Proof As A is separable the set of Hausdorff points is dense in Prim A (see Dixmier [1] 3.9.4). Take any open $U \subset$ Prim A and any Hausdorff point $x_0 \in U$. By the condition $\overline{\{x_0\}} \neq U$, there exists a point $x_1 \in U$ such that $x_1 \overline{\in} \overline{\{x_0\}}$, and via the Hausdorffness of x_0 these points have disjoint neighbourhoods. Now the result follows from Proposition 14.6. □

14.11 In each of the Propositions 14.6, 14.8 and 14.10 we considered some sufficient conditions for A not to contain any non–zero compact operator. Are these conditions necessary? Let us look through some examples.

(1) Let $X = [0, 1] \cup \{2\}$ and let the measure μ on X be equal to the Lebesgue measure on $[0, 1]$ and $\mu(\{2\}) = 1$. Consider two C^*–subalgebras of the algebra of linear operators on $L^2(X, \mu)$:

(a) A_1, the algebra of operators of multiplication by continuous functions $a \in C(X)$;

(b) A_2, the algebra of operators of multiplication by continuous functions such that $a(1) = a(2)$.

Prim $A_1 = X$ and there are non–zero compact operators in A_1; for example, multiplication by the characteristic function $\chi_{\{2\}}(x)$.

Prim $A_2 = [0, 1]$ and by Proposition 14.6 there are no non–zero compact operators.

For A_1, the open set $\{2\}$ in Prim A_1 does not contain two disjoint open subsets and $\{2\}$ is the closure of a Hausdorff point, and $\{0\}$ is a primitive ideal for $I_{\{2\}}$. So we see that the conditions of Propositions 14.6, 14.8 and 14.10 are essential.

(2) Let $X = [0, 1] \cup [2, 3]$ with the Lebesgue measure μ. A_3 is the algebra of operators of multiplication by functions continuous on [0, 1] and constants on [2, 3] on $L^2(X, \mu)$. Algebras A_1 and A_3 are isomorphic as C^*–algebras. A_1 contains a non–zero compact operator and we can easily see that there are no non–zero compact operator in A_3.

So the conditions of Propositions 14.6, 14.8 and 14.10 are not necessary. Notice in particular that the structure of Prim A does not always give an answer to whether A contains non–zero compact operators or not.

14.12 The objects which are really compatible with the algebraic structure of C^*–algebra are the so–called *compact elements* in A introduced by K. Vala [2].

14.13 An element u of a normed algebra A is called *compact*, if the mapping $a \to uau$ is a precompact operator on A.

Theorem 14.14 (Vala [1]) *Let* A = L(E) *be the algebra of all linear bounded operators on a Banach space* E. *Then* u *is a compact element of* A *iff it is a compact operator on* E.

This theorem shows on the one hand that definition 14.11 is rather natural and on the other hand it follows that if an algebra A is a subalgebra of $L(E)$ then compact operators which are contained in A belong to the set of compact elements of A.

14.15 An element $u \in A$ is called *relatively finite dimensional* if the range of the mapping $a \to uau$ is finite dimensional.

Theorem 14.16 (Ylinen [1]) *A* C^**–algebra* A *contains a non–zero compact element iff* A *contains a relatively one–dimensional projection (that is, the projection* P *such that* dim(PAP) = 1).

With the help of this theorem we can find another criterion for a C^*–algebra not to contain compact elements.

Theorem 14.17 *Let* A *be a separable* C^**–algebra.* A *contains a non–zero compact element iff* Prim A *contains a Hausdorff point* x_0 *with the open closure such that* $I_{\overline{\{x_0\}}}$ *contains a relatively one–dimensional projection.*

Proof

(1) Let P be a relatively one–dimensional projection in $I_{\overline{\{x_0\}}}$. Then it is also a relatively one–dimensional projection in A.

(2) Let P be a relatively one–dimensional projection in A. Consider the set $U = \{x \in \text{Prim } A\text{: } \breve{P}(x) \neq 0\}$. Since P is a projection the values of $\breve{P}(x)$ can be only 0 or 1, so by the lower semicontinuity of $\breve{P}(x)$ we have that U is an open set. As A is separable the set of Hausdorff points is dense in Prim A (Dixmier [1] 3.9.4) and thus U contains a Hausdorff point. U cannot contain more than one Hausdorff point, because if there exist two Hausdorff points $x_1, x_2 \in U$ then by the Dauns–Hofmann theorem (Theorem 2.90) one can find two elements a_1, a_2 in the centre of A such that

$$\breve{a}_1(x_1) = 1 \quad , \qquad \breve{a}_1(x_2) = 0$$

$$\breve{a}_2(x_1) = 0 \quad , \qquad \breve{a}_2(x_1) = 1$$

and

$$(P\breve{a}_i P)(x) = \breve{P}(x)\ \breve{a}_i(x) \quad , \qquad i = \overline{1,2} \quad .$$

Thus in this case $\dim(PAP) \geq 2$. It follows then that U is the closure of some Hausdorff point x_0, and as $\breve{P}(x) = 0$ when $x \,\bar{\in}\, \overline{\{x_0\}}$ we have that $P \in I_{\overline{\{x_0\}}}$. So P is the relatively one–dimensional projection in $I_{\overline{\{x_0\}}}$ as required. □

Remark 14.18 Looking through the proof of Theorem 14.17 we see that the separability of A is needed only to establish the density of the Hausdorff points in Prim A. So the same result is valid for all A which have a dense set of Hausdorff points in Prim A.

In particular we have the following.

Theorem 14.19 *Let* A *be a commutative* C^**–algebra.* A *contains a non–zero compact element iff the maximal ideal space of* A *contains an isolated point.*

Proof As A is a commutative C^*–algebra its maximal ideal space (Prim A) is Hausdorff. Now apply Theorem 14.17 and Remark 14.18. □

In the examples in 14.11 both A_1 and A_3 contain compact elements, though A_3 does not contain compact operators. So these examples show, at the same time, that compact operators and compact elements may be different objects in a C^*–algebra.

Theorem 14.20 *Let* A *be a separable* C^**–algebra and* U *be the union of all open sets which are the closures of the Hausdorff points in* Prim A. *If* a *is a compact element in* A *then* $a \in I_U$.

Proof Let a be a compact element in A. Then as the compact elements from an ideal in A (Ylinen [1], Theorem 1.1) aa^* is also a compact element. By Theorems 3.8 and 4.2 (Ylinen [1]) aa^* has the form

$$aa^* = \sum \lambda_k P_k$$

where P_k are pairwise orthogonal relatively one–dimensional projections of A and the series converges on norm. Looking through the proof of Theorem 14.17 we find that all $P_k \in I_U$. Thus $aa^* \in I_U$. As $\breve{a}\breve{a}^*(x) = [\breve{a}(x)]^2$ for all $x \in$ Prim A we conclude that $a \in I_U$. □

Theorem 14.21 *Let* A *be a separable* C^**–algebra of operators acting on the Hilbert space* H. *Then* $a \in A$ *is a compact operator iff* $a \in I_U$ *and*

$a|_{H_U}$ *is a compact operator, where* H_U *is the essential space for* I_U (I_U *was defined in the previous theorem).*

Proof Let a be a compact operator. Then a is a compact element in A. Thus by Theorem 14.20 $a \in I_U$ so H_U is an invariant space for a and $a|_{H_U}$ is a compact operator. Conversely if $a \in I_U$ and $a|_{H_U}$ is a compact operator then as $a(H \ominus H_U) = 0$ a is a compact operator on H. □

Corollary 14.22 *If* $A|_{H_{\overline{\{x\}}}}$ *does not contain non–zero compact operators for every Hausdorff point* x *which has an open closure then* A *does not contain non–zero compact operators.*

Proof $H_U = \oplus H_{\overline{\{x\}}}$ and all $H_{\overline{\{x\}}}$ are invariant with respect to A. Now apply Theorem 14.21. □

Theorem 14.17 leads to a link between the compact elements in a C^*–algebra and its irreducible representations. The following theorem is a well–known fact.

Theorem 14.23 (Dixmier [1] 4.1.10) *Let* π *be an irreducible representation of a* C^**–algebra* A *on the Hilbert space* H. *If* $\pi(A)$ *contains any non–zero compact operator then* $\pi(A)$ *contains all the compact operators on* H.

For the compact elements we also have the following result.

Theorem 14.24 (Ylinen [1]) *Let* π *be an irreducible representation of a* C^**–algebra* A *on the Hilbert space* H_π. *If* u *is a compact (resp. relatively finite dimensional) element of* A *then* $\pi(u)$ *is a compact operator (resp. an operator with finite dimensional range) on* H_π.

Now Theorems 14.17, 14.23 and 14.24 imply the following.

Theorem 14.25 *Let* A *be a separable* C^**–algebra.* A *contains a non–zero compact element iff* Prim A *contains a Hausdorff point* x_0 *with an open* *closure such that the irreducible representation* π *with* $\operatorname{Ker} \pi = x_0$ *contains all the compact operators on* H_π.

Using Proposition 14.5 and the results of **14.6–14.25** we can immediately obtain a number of results concerning the coincidence of Fredholm and invertible elements in a C^*–algebra. For example:

Theorem 14.26 *Let* A *be a separable* C^**–algebra such that the closure of any Hausdorff point in* Prim A *is not open. If* G *is amenable and acts topologically freely on* A *then every Fredholm operator in* $C^*(A, T_g)$ *is invertible.*

Proof Apply Propositions 14.5 and 14.10. □

Example 14.27 The following example shows that the condition on G to act topologically freely is essential in Theorem 14.1.

Let A_2 be the C^*–algebra considered in **14.11**. It does not contain any non–zero compact operators. Let T be the operator of multiplication by the function

$$t(x) = \begin{cases} -1 , & x \in [0, 1] \\ 1 , & x = 2 . \end{cases}$$

Then $TaT^{-1} = a$ for any $a \in A_2$. So $\mathbb{Z}$ does not act topologically freely by the automorphisms $T^z \cdot T^{-z}$. If $a(2) \neq 0$ then the operator $a + aT$ is a non–zero compact operator belonging to $C^*(A_2, T)$.

14.28 Consideration of Theorem 14.1 leads to an obvious question: whether an ideal $J \subset C^*(A, T_g)$ is determined by its intersection $A \cap J$ with the algebra A?

The following example shows that in general the answer is negative.

Example Consider the C^*–algebra B generated by bilateral weighted shifts on $l^2(\mathbb{Z})$ with the coefficients having limits at infinity, that is, the algebra generated by operators of the form

$$ae_n = a(n)e_n \, , \qquad \text{where} \quad \lim_{n\to\infty} a(n) = a(\infty)$$

and

$$Te_n = e_{n-1} \; .$$

Here e_n, $n \in \mathbb{Z}$, is an orthonormal basis in $l^2(\mathbb{Z})$.

B is an algebra of the form $C^*(A, T)$ where A is the algebra of operators of multiplication by functions having limits at infinity.

Prim $A = \mathbb{Z} \cup \{\infty\}$ and the base of the neighbourhoods of $\{\infty\}$ is

$$U_k = \{n \in \mathbb{Z}: |n| > k\} \qquad k = 1,2,\ldots \; .$$

As $(TaT^{-1})(n) = a(n+1)$ we see that $\{\hat{T}^z\}$ acts topologically freely and by the isomorphism theorem $B \cong A \times_{\hat{T}} \mathbb{Z}$.

Consider the representation π of B into the algebra of operators on $L^2(S^1)$ defined by

$$[\pi(a)f](z) = a(\infty)f(z) \qquad a \in A$$
$$[\pi(T)f](z) = z\cdot f(z) \; .$$

As $B \cong A \times_{\hat{T}} \mathbb{Z}$ and $\{\infty\}$ is a fixed point for the homeomorphism of Prim A corresponding to $\hat{T}$, this is really a representation of B.

Obviously $\pi(B) \cong C(S^1)$. Let J_1 be the ideal in B defined by:

$$J_1 = \{b \in B: \pi(b)(1) = 0\} \; .$$

Consider any $a \in A$ which belongs to J_1. By the construction of J_1 we have $a(\infty) = 0$ and evidently all a satisfying this condition belong to J_1. So

$$A \cap J_1 = \{a \in A: a(\infty) = 0\} \; .$$

Consider now another ideal J_2 of B generated by all the elements $a \in A$ such that $a(\infty) = 0$. Let a be any element in A. Then by Theorem 12.14

$\|a - J_2\| \geq |a(\infty)|$. This means that

$$a \in J_2 \Leftrightarrow a(\infty) = 0$$

and consequently $A \cap J_2 = A \cap J_1$. But $J_2 \neq J_1$. Indeed the element $b = aT - aT^*$ belongs to J_1 for any $a \in A$ and it belongs to J_2 only when $a(\infty) = 0$.

So the example shows that in general we cannot have all the information about the ideal J just by considering its intersection with A. Theorem 14.30 below describes a situation in which $A \cap J$ gives us complete information about J.

14.29 Recall that for any ideal $I \subset A$ we denote by supp I the set

$$\text{supp } I = \{x \in \text{Prim } A: x \supset I\} \quad .$$

Evidently supp I is closed relative to the Jacobson topology.

Let J be any ideal in $C^*(A, T_g)$ and consider the ideal $I = J \cap A$. Then I and consequently supp I are invariant under $\hat{T}_g$. Thus we can consider the homeomorphism τ_g which is the restriction of the homeomorphism t_g (induced by $\hat{T}_g$) to supp I.

If an algebra $C^*(A, T_g)$ possesses the property (*) then to each element $b \in C^*(A, T_g)$ we can associate a unique family of elements $\{\beta_g\}_{g \in G}$ in A where $\beta_g = N(bT_{-g})$ (see 12.2).

Theorem 14.30 *Let* $J \neq \{0\}$ *be a* **–ideal in* $C^*(A, T_g)$ *and let* G *act topologically freely. If* $\{T_g\}$ *acts topologically freely on* $\text{supp}(J \cap A)$ *then an element* $b \in C^*(A, T_g)$ *belongs to* J *if and only iff* β_g *belongs to* $J \cap A$ *for all* $g \in G$.

Proof Consider the canonical mapping

$$C^*(A, T_g) \overset{\pi}{\to} C^*(A, T_g)/_J \cong C^*(\pi(A), \pi(T_g)) \quad .$$

Applying Corollary 2.77 we have that

$$\pi(A) = A + J/_J \cong A/_I$$

where $I = A \cap J \neq \{0\}$ by Theorem 14.1.

The unitary operators $\pi(T_g)$ induce the automorphisms $\widehat{\pi(T_g)}$ of $\pi(A)$. Observe that Prim $\left(A/_I\right)$ = supp I in view of **2.85** and $\widehat{\pi(T_g)}$ induces the homeomorphisms τ_g, so $\{\widehat{\pi(T_g)}\}$ acts topologically freely on $\pi(A) \cong A/_I$. This means (applying Corollary 12.16) that

$$C^*(\pi(A), \pi(T_g)) \cong \left(A/_I\right) \underset{\widehat{\pi(T_g)}}{\times} G \quad .$$

To an element $b \in C^*(A, T_g)$ there corresponds a family $\{\beta_g\}_{g\in G} = \left\{N(bT_{g^{-1}})\right\}_{g\in G}$ and therefore to the element $\pi(b)$ there corresponds the family $\{\pi(\beta_g)\}_{g\in G} = \left\{N(\pi(bT_{g^{-1}}))\right\}_{g\in G}$.

As $C^*(\pi(A), \pi(T_g)) \cong \left(A/_I\right) \underset{\widehat{\pi(T_g)}}{\times} G$ we have (by Theorem 12.4) that

$$\pi(b) = 0 \Leftrightarrow \forall_g \in G \qquad \pi(\beta_g) = 0 \quad .$$

Let $b \in J$. Then $\pi(b) = 0$ and consequently $\pi(\beta_g) = 0$ for all $g \in G$. This means that $\beta_g \in I$ for all $g \in G$ and we have finished. □

15. The isomorphism theorem for a finite group of automorphisms. Invertibility

In the case when the group of automorphisms is finite the isomorphism theorem can be formulated in a rather special form.

15.1 Let A be a C^*–subalgebra (with identity) of the algebra of operators acting on a Hilbert space H and $\{T_g\}_{g \in G}$ be a finite unitary group $|G| = N$ such that $T_g A T_g^{-1} = A$ for every $g \in G$.

Evidently G acts topologically freely on A by the automorphisms $\hat{T}_g$ iff for any open set $U \subset \text{Prim } A$ there exists a point $x \in U$ such that the points $t_g(x)$, $g \in G$, are distinct (cf. **12.13′**).

As G is finite any operator $b \in C^*(A, T_g)$ has the form

$$b = \sum_{g \in G} a_g \, T_g \qquad a_g \in A \quad . \tag{15.1}$$

We shall denote by $\tilde{H}$ the Hilbert space $H^N = \bigoplus_{g \in G} H_g$ $(H_g \approx H)$. With the operator b of the form (15.1) let us associate the operator $\tilde{b}$: $\tilde{H} \to \tilde{H}$ defined by the matrix $\{b_{h,g}\}_{h,g \in G}$ where

$$b_{h,g} = \hat{T}_h\left[a_{h^{-1}g}\right] \quad . \tag{15.2}$$

Consider the mappings π: $A \to L(\tilde{H})$ and V: $G \to L(\tilde{H})$ defined by the formulae

$$\pi(a)(g \otimes \xi) = g \otimes [\hat{T}_g(a)](\xi) \tag{15.3}$$

where $a \in A$, $g \in G$, $\xi \in H$, and

$$V(g_0)(g \otimes \xi) = g \cdot g_0^{-1} \otimes \xi \qquad g_0 \in G \quad . \tag{15.4}$$

As the elements $g \otimes \xi$, $g \in G$, $\xi \in H$, generate $\tilde{H}$ the mappings π and V are well defined.

Theorem 15.2 $C^*(\pi(A), V(g)) \cong A \times_{\hat{T}} G$.

Proof Routine calculations show that

$$\pi(T_{g_0} \, aT_{g_0^{-1}}) = V(g_0) \, \pi(a) \, V(g_0^{-1}) \quad . \tag{15.5}$$

Let us verify that $C^*(\pi(A), V(g))$ possesses the property (*). Indeed

$$\sum_{h\in G} \pi(a_h) \, V(h) \, (g \otimes \xi) = \sum_{h\in G} \pi(a_h) \, (g\cdot h^{-1} \otimes \xi)$$

$$= \sum_{h\in G} g\cdot h^{-1} \otimes \hat{T}_{g\cdot h^{-1}}(a_h)\xi \quad . \tag{15.6}$$

Since g is fixed all the summands in the last sum in (15.6) are pairwise orthogonal. So we have

$$\left\| \sum_{h\in G} \pi(a_h)V(h)(g \otimes \xi)\right\|^2 = \sum_{h\in G} \left\|\hat{T}_{g\cdot h^{-1}}(a_h)\xi\right\|^2 \geq \|\hat{T}_g(a_e)\xi\|^2 \quad .$$

As ξ is arbitrary we conclude

$$\left\| \sum_{h\in G} \pi(a_h)V(h)\right\| \geq \|a_e\| \quad .$$

Now the result follows from Theorem 12.8. □

Remark 15.3 One can easily see that $C^*(\pi(A), V(g))$ is in fact the right regular representation for $A \underset{\hat{T}}{\times} G$ (using the notations of **11.11** (15.3) and (15.4) can be rewritten as

$$[\pi(a)\xi](g) = [\hat{T}_g(a)]\xi(g)$$

$$[V_{g_0}\xi](g) = \xi(g\cdot g_0)$$

cf. (11.3) and (11.4)).

So we can construct an analogous object using the **left** regular representation defined by (11.3), (11.4).

Corollary 15.4 *If* $C^*(A, T_g)$ *possesses the property* (*) *then the mapping* $b \to \bar{b}$ *is an isomorphism of* $C^*(A, T_g)$ *onto* $C^*(\pi(A), V(g))$.

Proof By Theorem 15.2 $C^*(\pi(A), V(g))$ possesses the property (*). Now, by Corollary 12.9, the mapping

$$b = \sum a_g T_g \to \sum \pi(a_g)V(g) = \tilde{b} \quad ,$$

in view of (15.5), establishes an isomorphism between $C^*(A, T_g)$ and $C^*(\pi(A), V(g))$. □

Corollary 15.5 (The isomorphism theorem for a finite group) *Let* G *be a finite group acting topologically freely on* A. *Then the mapping* $b \to \tilde{b}$ *is an isomorphism of* $C^*(A, T_g)$ *onto* $C^*(\pi(A), V(g))$.

Proof Use the previous corollary and Theorem 12.14. □

Corollary 15.6 *Let* A *be a commutative* C^**–algebra,* G *be a finite group and* $C^*(A, T_g)$ *possesses the property* (*). *Then* $b \in C^*(A, T_g)$ *is invertible if and only if*

$$\det[\tilde{b}(x)] \neq 0 \qquad x \in \text{Prim } A$$

where $\det[\tilde{b}(x)]$ *is the determinant of the matrix* $\tilde{b}(x)$ *defined by* (15.2).

So in the case of a commutative C^*–algebra A this corollary gives us total information about the conditions for invertibility of elements in $C^*(A, T_g)$.

Corollary 15.7 *If* $\tilde{b}$ *defined by* (15.2) *is invertible then* b *defined by* (15.1) *is invertible.*

Proof Since by Theorem 15.2 $C^*(\pi(A), V(g)) \cong A \underset{\hat{T}}{\times} G$ it follows that the mapping

$$\tilde{b} = \sum_{g \in G} \pi(a_g)V(g) \to \sum_{g \in G} a_g T_g = b$$

is a representation for $C^*(\pi(A), V(g))$. So if $\tilde{b}$ is invertible then b is also invertible. □

16. The structure of $C^*(A, T_g)$ in the case of an arbitrary group action

If G acts topologically freely on A the isomorphism theorem (Corollary 12.17) gives us an excellent tool for the description of the structure of $C^*(A, T_g)$. But what shall we do if G does not act topologically freely? We consider this case here. We want to emphasize that up to Theorem 16.14 we do not suppose G to be amenable.

16.1 For any $g \in G$ let us define the set X_g as

$$X_g = \{x \in \text{Prim } A: t_g(x) = x\}$$

where t_g is a homeomorphism of Prim A induced by $\hat{T}_g$.

Let $\mathring{X}_g$ be the interior of X_g and

$$X_\infty = \text{Prim } A \setminus \left[\bigcup_{g \neq e} \mathring{X}_g\right]$$

$$I_g = \{a \in A: \breve{a}(x) = 0 , \quad x \in \overline{\mathring{X}_g}\}$$

$$I^\infty = \{a \in A: \breve{a}(x) = 0 , \quad x \in X_\infty\} .$$

Remarks

(1) If $t_g(x) = x$ then

$$t_{g_1}(x) = \left[t_{g_1} \cdot t_g \cdot t_{g_1^{-1}}\right] (t_{g_1}(x)) = t_{g_1 g g_1^{-1}} (t_{g_1}(x)) .$$

Thus

$$x \in \mathring{X}_g \Rightarrow \forall_{g_1}\ t_{g_1}(x) \in \mathring{X}_{g_1 g g_1^{-1}}$$

and it follows that $\left[\bigcup_{g \neq e} \mathring{X}_g\right]$ is invariant under all t_g, $g \in G$. Consequently X_∞ is also invariant under t_g, that is, the ideal I^∞ is invariant under $\hat{T}_g$.

(2) If G is a commutative group then $t_{g_1 g g_1^{-1}} = t_g$ and consequently (see

the above remark)

$$x \in \overset{\circ}{X}_g \Rightarrow \forall_{g_1}\ t_{g_1}(x) \in \overset{\circ}{X}_g \quad .$$

Thus the ideals I_g are invariant under the action of G.

If $\overset{\circ}{X}_g = \emptyset$ *then* $I_g = \{0\}$ *and we shall exclude such ideals in all the subsequent statements.*

Let us denote by $J^\infty(J_g)$ the ideal in $C^*(A, T_g)$ generated by $I^\infty(I_g)$.

Remarks

(1) $$J^\infty = \overline{\left\{\sum j_g T_g\right\}} \qquad j_g \in I^\infty \tag{16.1}$$

$$J_{g_0} = \overline{\left\{\sum j_g T_g\right\}} \qquad j_g \in S_g \tag{16.2}$$

where S_{g_0} is the ideal in A generated by all the ideals $I_{\tau g_0 \tau^{-1}}$, $\tau \in G$ (see the previous remark). Clearly if G is commutative then $S_{g_0} = I_{g_0}$.

(2) Since $I_{\tau g \tau^{-1}} = T_\tau I_g T_{\tau^{-1}}$ we have that if $\pi(I_g) \neq 0$ then $\pi(I_{\tau g \tau^{-1}}) \neq 0$ for any $\tau \in G$ and $\pi \in \widehat{C^*(A, T_g)}$ where $\widehat{C^*(A, T_g)}$ is the spectrum of $C^*(A, T_g)$.

Consider also the ideals $J^h \subset C^*(A, T_g)$ defined by

$$J^h = \{\cap \text{ Ker } \pi \colon \pi \in \widehat{C^*(A, T_g)},\ \pi(I_h) \neq 0\} \ , \qquad h \in G \ . \tag{16.3}$$

If $\overset{\circ}{X}_g = \emptyset$ *then* $I_g = \{0\}$ *and consequently* $J^g = \emptyset$ *so we exclude such ideals in all the subsequent statements.*

We denote by P_g, P_∞ the canonical homomorphisms

$$C^*(A, T_g) \xrightarrow{P_{g_0}} C^*(A, T_g)/J^{g_0} = C_{g_0} \tag{16.4}$$

$$C^*(A, T_g) \xrightarrow{P_\infty} C^*(A, T_g)/J^\infty = C_\infty \quad . \tag{16.5}$$

16.2 A set $\{\tau_i\}_{i\in I}$ of representations (morphisms) of an algebra B is said to be *sufficient* if an element $b \in B$ is invertible if and only if all $\tau_i(b)$, $i \in I$, are invertible.

Theorem 16.3 $\{P_\infty, P_g\}_{\substack{g\in G\\ g\neq e}}$ *is a sufficient set of morphisms for* $C^*(A, T_g)$.

Proof If there exists a $\tau \in \{\infty\} \cup [G\backslash\{e\}]$ such that $J^\tau = \{0\}$ then $P_\tau(C^*(A, T_g)) = C^*(A, T_g)$ and we are done.

So let us suppose that $J^\tau \neq \{0\}$ for all $\tau \in \{\infty\} \cup G$.

Consider an arbitrary $\pi \in \widehat{C^*(A, T_g)}$. We shall show that there exists a $\tau_0 \in \{\infty\} \cup [G\backslash\{e\}]$ such that $\pi\left(J^{\tau_0}\right) = \{0\}$.

Indeed, suppose on the contrary that

$$\pi(J^\tau\} \neq \{0\} \qquad \text{for all} \quad \tau \in \{\infty\} \cup [G\backslash\{e\}] \quad . \tag{16.6}$$

As $J^\infty = \overline{\left\{\sum a_g T_g\right\}}_{a_g \in I^\infty}$ (see (16.1)) (16.6) implies

$$\pi(I^\infty) \neq \{0\} \quad . \tag{16.7}$$

If $\pi(J^\tau) \neq \{0\}$, $\tau \in [G\backslash\{e\}]$, then in view of (16.3) we conclude

$$\pi(I_\tau) = \{0\} \qquad \tau \in G\backslash\{e\} \quad . \tag{16.8}$$

Consider the ideal $I_\infty = \left\{\bigcup_{g\neq e} I_g\right\}$ generated by all I_g, $g \neq e$. We have

$$I_\infty = \left\{a \in A: \breve{a}(x) = 0 \ , \quad x \in \overline{\left[\bigcup_{g\neq e} \mathring{X}_g\right]}\right\}$$
$$= \{a \in A: \breve{a}(x) = 0 \ , \quad x \in X_\infty\} = I^\infty \ .$$

Consequently in view of (16.8) we conclude

$$\pi(I^\infty) = \pi(I_\infty) = \left\{\bigcup_{g\neq e} \pi(I_g)\right\} = \{0\}$$

and we arrive at a contradiction with (16.7).

Thus there exists a $\tau_0 \in \{\infty\} \cup [G\backslash\{e\}]$ such that $\pi\left(J^{\tau_0}\right) = \{0\}$, that is,

$$J^{\tau_0} \subset \operatorname{Ker} \pi \ . \tag{16.9}$$

In order to show that $\{P_\infty, P_g\}_{\substack{g\in G\\ g\neq e}}$ is a sufficient set of homeomorphisms we have to show that if $\{P_\tau(b)\}$, $\tau \in \{\infty\} \cup G\backslash\{e\}$, are all invertible then $\pi(b)$ is invertible.

In view of (16.9) the canonical mapping

$$C_{\tau_0} = {C^*(A, T_g)}/{J^{\tau_0}} \xrightarrow{\pi_{\tau_0}} \pi(C^*(A, T_g))$$

$(\pi_{\tau_0}[b + J^{\tau_0}] = \pi(b))$ is well defined. Thus as $P_{\tau_0}(b) = [b + J^{\tau_0}]$ is invertible in C_{τ_0}, $\pi(b) = \pi_{\tau_0}(P_{\tau_0}(b))$ is also invertible. The proof is complete. □

16.4 We can also reformulate this result in a different way. Consider the Banach space J_{g_0} (see (16.2)) and the representation $\nu_{g_0}: C^*(A, T_g) \to L(J_{g_0})$ defined by the formula

$$\nu_{g_0}(b)j = bj \ , \qquad b \in C^*(A, T_g) \ , \quad j \in J_{g_0} \ . \tag{16.10}$$

Theorem 16.5 $C_{g_0} \cong \nu_{g_0}(C^*(A, T_g))$ *and the isomorphism is given by*

$$\nu_{g_0}(b) \longleftrightarrow P_{g_0}(b) \qquad b \in C^*(A, T_g)$$

(C_{g_0} *is defined by* (16.4)).

Proof We recall that we are considering only those g_0 for which $\mathring{X}_g \neq \emptyset$ and consequently $J_{g_0} \neq \{0\}$. The proof will be complete if we show that

$$\text{Ker } \nu_{g_0} = J^{g_0} .$$

By $\widehat{C^*(A, T_g)}^{J_{g_0}}$ we denote the set of all $\pi \in \widehat{C^*(A, T_g)}$ such that $\pi(J_{g_0}) \neq \{0\}$. In view of (16.2) and the remark which follows it we have

$$\pi(J_{g_0}) \neq \{0\} \Leftrightarrow \pi(I_{g_0}) \neq \{0\} .$$

Now (16.3) implies

$$J^{g_0} = \left\{ \cap \text{ Ker } \pi\text{: } \pi \in \widehat{C^*(A, T_g)}^{J_{g_0}} \right\} . \tag{16.11}$$

By Theorem 2.83 the mapping

$$\pi \to \pi|_{J_{g_0}} \tag{16.12}$$

is a bijection

$$\widehat{C^*(A, T_g)}^{J_{g_0}} \longleftrightarrow \hat{J}_{g_0} . \tag{16.13}$$

In view of **2.73** (16.12) and (16.13) imply that any non–zero vector $\xi \in H_\pi$ is cyclic for J_{g_0}, that is,

$$\overline{\{\pi(J_{g_0})\xi\}} = H_\pi \tag{16.14}$$

where H_π is the representation space for π. By (16.10)

$$b \in \operatorname{Ker} \nu_{g_0} \Leftrightarrow bj = 0 \qquad \forall j \in J_{g_0} . \tag{16.15}$$

Let us show first that

$$\operatorname{Ker} \nu_{g_0} \subset J^{g_0} . \tag{16.16}$$

In order to prove this it suffices to show (in view of (16.11)) that if $b \in \operatorname{Ker} \nu_{g_0}$ then $\pi(b) = 0$ for any $\pi \in \widehat{C^*(A, T_g)}^{J_{g_0}}$.

Let $b \in \operatorname{Ker} \nu_{g_0}$ and $\pi \in \widehat{C^*(A, T_g)}^{J_{g_0}}$. Since $bj = 0$, $j \in J_{g_0}$ (by (16.15)) we have that

$$0 = \pi(bj)\xi = \pi(b)[\pi(j)\xi] \qquad \forall j \in J_{g_0} , \quad \forall \xi \in H_\pi \tag{16.17}$$

and in view of (16.14), (16.17) implies

$$\pi(b) = 0$$

so (16.16) is proved.

On the other hand let $b \in J^{g_0}$ and take any $j \in J_{g_0}$. In order to prove that

$$J^{g_0} \subset \operatorname{Ker} \nu_{g_0}$$

it suffices (by (16.15)) to prove that $\pi(bj) = 0$ for any $\pi \in \widehat{C^*(A, T_g)}$.

If $\pi \in \widehat{C^*(A, T_g)}^{J_{g_0}}$ then (16.11) implies $\pi(b) = 0$ and so $\pi(bj) = 0$.

If $\pi \bar{\in} \widehat{C^*(A, T_g)}^{J_{g_0}}$ then $\pi(j) = 0$ and so again $\pi(bj) = 0$. The proof is complete. □

Corollary 16.6 $\{P_\infty, \nu_g\}_{\substack{g \in G \\ g \neq e}}$ *is a sufficient set of representations for* $C^*(A, T_g)$.

Proof Use Theorems 16.3 and 16.5. □

Remark 16.7 Since $J_{g_0} = \overline{\left\{\sum j_g T_g\right\}}$, $j_g \in S_{g_0}$ (see (16.2)), we have

$$bj = 0 \qquad \forall j \in J_{g_0} \Leftrightarrow bj = 0 \qquad \forall j \in S_{g_0}$$

and by the definition of S_{g_0} we conclude that

$$bj = 0 \quad \forall j \in J_{g_0} \Leftrightarrow bj = 0 \quad \forall j \in I_{\tau g_0 \tau^{-1}}, \quad \tau \in G . \tag{16.18}$$

16.8 If A is a C^*–subalgebra of $L(H)$ we can describe the representation ν_{g_0} (see (16.10)) in a slightly different way. Let K_{g_0} be the essential subspace for J_{g_0} that is,

$$K_{g_0} = \overline{\{j\xi\}} \qquad j \in J_{g_0}, \quad \xi \in H . \tag{16.19}$$

Proposition 16.9 $\nu_{g_0}(C^*(A, T_g)) \cong C^*(A, T_g)|_{K_{g_0}}$.

Proof K_{g_0} is an invariant subspace for $C^*(A, T_g)$ and in view of (16.19)

$$b \cdot j = 0 \qquad \forall j \in J_{g_0} \Leftrightarrow b|_{K_{g_0}} = 0 \quad .$$

□

Remark 16.10 One can easily see that K_{g_0} is equal to the closed subspace spanned by the vectors

$$\{j\xi\} \quad , \qquad j \in S_{g_0} \quad , \qquad \xi \in H \quad .$$

16.11 Note that

$$A \cap J^{\infty} = I^{\infty} \quad . \tag{16.20}$$

Indeed, for any fixed $a \in A$ and any $\varepsilon > 0$ let $\pi \in \hat{A}$ be an irreducible representation such that Ker $\pi \in X_{\infty}$ and

$$\|\pi(a)\xi\| \geq [\sup_{X_{\infty}} \breve{a}(x)] - \varepsilon$$

where ξ, $\|\xi\| = 1$, is some vector in H_{π}. Let $\nu \in \widehat{C^*(A, T_g)}$ be some extension of π up to an irreducible representation of $C^*(A, T_g)$ (see Theorem 2.81) and let

$$\text{supp } \nu = \{x \in \text{Prim } A \colon \breve{a}(x) = 0 \quad \text{for} \quad a \in \text{Ker } \nu\} \quad .$$

In view of the choice of π and since X_{∞} is invariant under all t_g, $g \in G$, and closed we deduce that supp $\nu \subset X_{\infty}$ (in fact, supp $\nu = \overline{\{t_g(\text{Ker } \pi)\}}_{g \in G}$) and thus by the definition of I^{∞} we have

$$\nu(I^{\infty}) = 0$$

which implies

$$\nu\left(\sum j_g T_g\right) = 0$$

for any finite sum with $j_g \in I^\infty$. Therefore

$$\left\|\nu\left(a + \sum j_g T_g\right)\xi\right\| = \|\nu(a)\xi\| = \|\pi(a)\xi\| \geq \sup_{X_\infty} \breve{a}(x) - \varepsilon$$

which, by the arbitrariness of ε, means that $a \in J^\infty$ only when

$$\breve{a}(x) = 0 \qquad \text{for all} \quad x \in X_\infty$$

and consequently (16.20) is true.

In turn (16.20) implies

$$P_\infty(A) = {}^{A + J^\infty}/_{J^\infty} = {}^{A}/_{A \cap J^\infty} = {}^{A}/_{I^\infty}$$

and it follows (by 2.85) that

$$\text{Prim}[P_\infty(A)] = \text{Prim}\ {}^{A}/_{I^\infty} = X_\infty \quad .$$

We shall say that $\{\hat{T}_g\}_{g\in G}$ acts *topologically freely on* X_∞ if for any $x \in X_\infty$ and for any finite set $g,\dots,g_n$ $(g_i \neq e)$ the set $\left[\bigcup_{i=1}^{n} X_{g_i}\right]$ does not contain any neighbourhood of x

In the case when $\{\hat{T}_g\}_{g\in G}$ acts topologically freely on X_∞ the same argument as that used in the proof of Theorem 12.14 shows that

$$\left\|\sum_F a_g T_g + \sum_M j_h T_h\right\| \geq \sup_{X_\infty} \breve{a}_e(x)$$

where F and M are any finite sets in G and $j_h \in I^\infty$. But this means that

$$\left\|P_\infty\left(\sum_F a_g T_g\right)\right\| = \left\|\sum_F P_\infty(a_g)\, P_\infty(T_g)\right\| \geq \|P_\infty(a_e)\|$$

and thus $P_\infty[C^*(A, T_g)] = C^*(P_\infty(A), P_\infty(T_g))$ possesses the property (*).

So we conclude (in view of the isomorphism theorem) that *if G is amenable and $\{\hat{T}_g\}_{g\in G}$ acts topologically freely on X_∞ then*

$$P_\infty[C^*(A, T_g)] \cong \left({}^{A}/_{I^\infty}\right) \underset{\hat{T}}{\times} G \tag{16.21}$$

and in this situation we can use all the result of Sections 12–15 to work with $P_\infty[C^*(A, T_g)]$.

It should be stressed that if $\{\hat{T}_g\}_{g\in G}$ does not act topologically freely on X_∞ then in general (16.21) is not true.

Example Let us consider the space $\overset{5}{\underset{i=1}{\oplus}} L^2(R_i)$ where $R_i = \mathbb{R}$, $i = \overline{1,5}$. Set A to be the algebra of operators of multiplication by functions

$$a = (a_1(t),\dots,a_5(t))$$

where a_i, $i = \overline{1,5}$, are continuous and have limits at ∞, where all the limits at ∞ are equal:

$$\lim_{t\to\infty} a_i(t) = a(\infty) \qquad i = \overline{1,5} \quad .$$

Clearly

$$\text{Prim}\ A = (\infty) \cup \left(\bigsqcup_{i=1}^{5} R_i\right)$$

with the evident topology.

Let us define the mapping α: $\bigsqcup_{i=1}^{5} R_i \to \bigsqcup_{i=1}^{5} R_i$ by the formulae

$$\left.\begin{aligned}\alpha&: R_1 \to R_2\\ \alpha&: R_2 \to R_1\end{aligned}\right\} \qquad \alpha(t) = t$$

$$\left.\begin{array}{l}\alpha: R_3 \to R_4 \\ \alpha: R_4 \to R_5 \\ \alpha: R_5 \to R_3\end{array}\right\} \qquad \alpha(t) = t$$

and set

$$(Tf)(t) = f(\alpha(t)) \quad , \qquad f \in \bigoplus_{i=1}^{5} L^2(R_i) \quad .$$

We have $T^6 = 1$ and consequently T generates the representation of $\mathbb{Z}_6$. Obviously

$$\overset{\circ}{X}_2 = \overset{\circ}{X}_4 = R_1 \cup R_2$$
$$\overset{\circ}{X}_3 = R_3 \cup R_4 \cup R_5$$
$$X_\infty = (\infty) .$$

Since we have $X_2 \cup X_3 = \text{Prim } A$ we conclude that $\hat{T}$ does not act topologically freely on X_∞.

Observe that the spectrum of T is

$$\sigma(T) = \left\{1, -1, e^{i\frac{2\pi}{3}}, e^{i\frac{4\pi}{3}}\right\} \tag{16.22}$$

and evidently $A/_{I^\infty} \cong \mathbb{C}$ and $\hat{T}$ acts trivially on $A/_{I^\infty}$. But if

$$P_\infty(C^*(A, T)) \cong \left(A/_{I^\infty}\right) \underset{\hat{T}}{\times} \mathbb{Z}_6 \tag{16.23}$$

then

$$\sigma(P_\infty(T)) = \left\{e^{i\frac{2\pi}{6}}\right\}_{i=1}^{6}$$

which contradicts (16.22). Thus (16.23) is not true.

16.12 Observe that if $G = \mathbb{Z}$ then $\{\hat{T}_z\}_{z\in\mathbb{Z}}$ automatically acts topologically freely on X_∞. Indeed if $\bigcup_{i=1}^{n} X_{z_i}$ is a neighbourhood of x then since

$X_{z_1 \cdot \ldots \cdot z_n} \supset \bigcup_{i=1}^{n} X_{z_i}$ we have $x \in \overset{\circ}{X}_{z_1 \cdot \ldots \cdot z_n}$.

The foregoing considerations and the next results help us to understand the structure of $C^*(A, T_g)$ in the case when $\{\hat{T}_g\}_{g \in G}$ does not act topologically freely.

Theorem 16.13 *Let* $C^*(A, T_g)$ *and* $C^*(A_1, V_g)$ *be two* C^**–algebras and let* $\varphi\colon A \to A_1$ *be an isomorphism such that*

$$\varphi(T_g\, aT_{g^{-1}}) = V_g\, \varphi(a)\, V_{g^{-1}} \quad , \qquad g \in G \quad . \tag{16.24}$$

If

$$P_\infty(C^*(A, T_g)) \cong P_\infty(C^*(A_1, V_g)) \tag{16.25}$$

then the mappings $\varphi\colon A \to A_1$, $T_g \to V_g$ *give rise to the isomorphism* $C^*(A, T_g) \cong C^*(A_1, V_g)$ *iff for any* $g_0 \neq e$ *and any finite set* $F \subset G$

$$\left\|\sum_F \varphi(a_g)V_g\right\| = \left\|\sum_F a_gT_g\right\| \quad , \qquad a_g \in S_{g_0} \quad . \tag{16.26}$$

Proof Looking through the proof of Theorem 16.3 and in view of Theorem 16.5 one can easily see that

$$C^*(A, T_g) \cong P_\infty(C^*(A, T_g)) \oplus \bigoplus_{G\setminus\{e\}} \nu_h(C^*(A, T_g)) \quad . \tag{16.27}$$

If (16.25) holds then in order to prove the theorem it suffices to observe that (16.26) implies

$$\nu_{g_0}(C^*(A, T_g)) \cong \nu_{g_0}(C^*(A_1, V_g)) \quad . \tag{16.28}$$

But (16.2) and (16.26) imply

$$J_{g_0} \overset{\gamma}{\cong} C^*(\varphi(S_{g_0}), V_g)$$

and by (16.10) and (16.24)

$$v_{g_0}(a)j = [\gamma^{-1}\varphi(a)\gamma]j$$
$$v_{g_0}(T_g)j = \left[\gamma^{-1}V_g\gamma\right]j \qquad j \in J_{g_0} .$$

Thus (16.28) is true and the proof is complete. □

The following statement is in fact an extension of the isomorphism theorem (see Corollary 12.17) to a rather large class of groups which do not act topologically freely on the whole of Prim A but only on X_∞.

Theorem 16.14 *Let* $C^*(A, T_g)$ *and* $C^*(A_1, V_g)$ *be two* C^**–algebras and* $\varphi: A \to A_1$ *be an isomorphism such that*

$$\varphi(T_g a T_{g^{-1}}) = V_g \varphi(a) V_{g^{-1}} , \qquad g \in G$$

(G *is an amenable group*).

If $\{\hat{T}_g\}_{g\in G}$ *acts topologically freely on* X_∞ *then the mappings* $\varphi: A \to A_1$, $T_g \to V_g$ *give rise to the isomorphism* $C^*(A,T_g) \cong C^*(A_1,V_g)$ *iff for any* $g_0 \neq e$ *and any finite set* $F \subset G$

$$\left\|\sum_F \varphi(a)V_g\right\| = \left\|\sum_F a_g T_g\right\| , \qquad a_g \in S_{g_0} . \tag{16.29}$$

Proof This is evident in view of the previous theorem and (16.21). □

Remarks 16.15

(1) Recall again that if G acts topologically freely (on Prim A) then $S_{g_0} = \{0\}$ for all $g_0 \neq e$ and in this case this statement coincides with Corollary 12.17.

(2) Note that this is the first statement in this section where we used the amenability of G.

(3) Observe that if $G = \mathbb{Z}$ then by 16.12 $\hat{T}$ acts topologically freely on X_∞ and thus this theorem describes the structure of $C^*(A, T)$ in the case of an arbitrary action of $\hat{T}$.

It is clear that the verification of the equality (16.29) is not trivial in general and in fact the level of complexity of this verification corresponds to the level of complexity of $\nu_{g_0}(C^*(A, T_g))$. Nevertheless it is quite reachable in natural situations (see Chapter 6 (in the second part of the book)).

16.16 Now we shall describe one special but still rather common situation when the structure of $C^*(A, T_g)$ can again be described totally only in terms of the action of the homeomorphisms $\{t_g\}_{g\in G}$.

First of all we shall find the condition when $I_g = S_g$ for any g.

For every $x \in \operatorname{Prim} A$ we denote by G_x the subgroup of G consisting of all the elements for which $x \in \overset{\circ}{X}_g$:

$$G_x = \left\{ g \in G\colon x \in \overset{\circ}{X}_g \right\} . \tag{16.30}$$

Evidently if $x \in X_\infty$ then $G_x = \{e\}$.

This implies in particular that the condition for G to act topologically freely is equivalent to the condition $G_x = \{e\}$ for all $x \in \operatorname{Prim} A$.

We recall that a subgroup U of a group V is called *normal* if

$$v^{-1}Uv = U \tag{16.31}$$

for any $v \in V$. Normal subgroups are those which naturally arise when dealing with factor groups.

Now we shall show that $I_g = S_g$ for any g iff G_x is normal for every $x \in \operatorname{Prim} A$.

Lemma 16.17 $\mathrm{G_x}$ *is normal iff for any* $\mathrm{h} \in \mathrm{G}$ *and* $\mathrm{g_1} \in \mathrm{G_x}$ *there exists a neighbourhood* $\mathrm{W(x)} \subset \overset{\circ}{\mathrm{X}}_{\mathrm{g_1}}$ *such that* $\mathrm{t_h(W(x))} \subset \overset{\circ}{\mathrm{X}}_{\mathrm{g_1}}$.

Proof

Necessity. Let G_x be a normal subgroup of G. Then for any $h \in G$ and $g_1 \in G_x$ we have

$$h^{-1}g_1h = g_2 \in G_x \quad ,$$

that is

$$x \in \overset{\circ}{X}_{h^{-1}g_1h} \quad .$$

Thus there exists a neighbourhood $V(x)$ such that $t_{h^{-1}g_1h}$ acts trivially on $V(x)$. Consider some neighbourhood $U(x) \subset \overset{\circ}{X}_{g_1}$ and let

$$W(x) = U(x) \cap V(x) \quad .$$

Then $t_{h^{-1}g_1h}$ acts trivially on $W(x)$ and consequently t_{g_1} acts trivially on $t_h(W(x))$ which means that $t_h(W(x)) \subset \overset{\circ}{X}_{g_1}$.

Sufficiency. Let h, g_1, $W(x)$ be as described in the statement of the lemma. We have to show that $h^{-1}g_1h \in G_x$, that is, $x \in \overset{\circ}{X}_{h^{-1}g_1h}$. But since $t_h(W(x)) \subset \overset{\circ}{X}_{g_1}$ the homeomorphism t_{g_1} acts trivially on $t_h(W(x))$, that is, $t_g[t_h(y)] = t_h(y)$ for every $y \in W(x)$ and it follows that

$$t_{h^{-1}g_1h}(y) = y \qquad y \in W(x)$$

which means that $x \in \overset{\circ}{X}_{h^{-1}g_1h}$ and the proof is complete. □

Corollary 16.18 *The following statements are equivalent*

(i) $I_g = S_g$ *for any* $g \in G$;

(ii) $t_h\left(\overset{\circ}{X}_g\right) = \overset{\circ}{X}_g$ *for any* $h, g \in G$;

(iii) G_x *is normal for every* $x \in \text{Prim } A$.

Proof Clearly (i)⇔(ii). So we have to show that (ii)⇔(iii).

If (ii) is true then $t_h\left(\overset{\circ}{X}_{g_1}\right) \subset \overset{\circ}{X}_{g_1}$ which, by Lemma 16.17, means that G_x is normal.

On the other hand if (iii) is true then (since for any $x \in \overset{\circ}{X}_g$ we have $g \in G_x$) by Lemma 16.17 we have that for any $h \in G$ there exists a $W(x) \subset \overset{\circ}{X}_g$ such that $t_h(W(x)) \subset \overset{\circ}{X}_g$ which implies

$$t_h\left(\overset{\circ}{X}_g\right) \subset \overset{\circ}{X}_g \quad . \tag{16.32}$$

Thus

$$\overset{\circ}{X}_g \subset t_{h^{-1}}\left(\overset{\circ}{X}_g\right) \tag{16.33}$$

and, since h is arbitrary, (16.32) and (16.33) imply $t_h\left(\overset{\circ}{X}_g\right) = \overset{\circ}{X}_g$. The corollary is proved. □

16.19 Now we shall concentrate on the case when all G_x are normal and, under a certain condition on the action of T_g on I_g, construct a special faithful representation of $C^*(A, T_g)$ which will depend only on A and $\{t_g\}_{g \in G}$.

Along with the group G_x we shall consider the group

$$G^x = {}^{G}/_{G_x} \quad .$$

We shall say that G^x acts topologically freely at x if for any finite set $g_1, \ldots, g_n$ such that $[g_i] \neq [e]$ (that is, $g_i \bar{\in} G_x$, $i = \overline{1,n}$) the set $\left[\bigcup_{i=1}^{n} X_{g_i}\right]$ does not contain any neighbourhood of x.

Note that in fact the definition depends not on g_i but only on $[g_i]$. Indeed if, say, $[g_1] = [g_2]$ then there exists on $h \in G_x$ such that $g_1 = g_2 \cdot h$. Since $h \in G_x$ we have that $t_h(y) = y$ for some neighbourhood $O(x)$ and it follows that for every $y \in O(x)$

$$t_{g_1}(y) = t_{g_2 \cdot h}(y) = t_{g_2} \cdot t_h(y) = t_{g_2}(y)$$

and this means that considering sufficiently small neighbourhoods of x we can use $[g_i]$ instead of g_i in the foregoing definition.

16.20 For any $x \in \hat{A}$ consider the space $l^2(G^x, H_x)$ where $G^x = G^{\mathrm{Ker}x}$ and H_x is the representation space for x. Let us define the operators $\pi_x(a)$, $\pi_x(T_{g_0})$ in $l^2(G^x, H_x)$ by the equations

$$\pi_x(a)([g] \otimes \xi) = [g] \otimes x(\hat{T}_g(a))\xi \tag{16.34}$$

$$\pi_x(T_{g_0}) = \pi_x([g_0]) \tag{16.35}$$

and

$$\pi_x([g_0])([g] \otimes \xi) = \left[g \cdot g_0^{-1}\right] \otimes \xi \tag{16.36}$$

where

$$\xi \in H_x\,, \quad [g] \in G^x\,, \quad \text{and} \quad a \in A\,.$$

Theorem 16.21 *Let* A *be a separable* C**–algebra,* G *be an amenable countable group and the following conditions be satisfied:*

(1) G_y *is normal for every* $y \in \mathrm{Prim}\, A$;

(2) G^y *acts topologically freely at every* $y \in \mathrm{Prim}\, A$;

(3) $T_g j = j$ *for any* $j \in I_g$, $g \neq e$.

Then

$$C^*(A, T_g) \cong \bigoplus_{x \in \hat{A}} \pi_x(A) \underset{[\hat{T}_x]}{\times} G^x \cong \bigoplus_{x \in \hat{A}} \pi_x(C^*(A, T_g)) \tag{16.37}$$

where $[\hat{T}_x]$ *is the group of automorphisms of* $\pi_x(A)$ *corresponding to the representation of* G^x *by* $\pi_x([g])$ (see (16.35), (16.36)).

Proof If we fix any $x \in \hat{A}$ then for any finite set $F \subset G$ the mapping

$$\pi_x\left(\sum_F a_g T_g\right) = \sum_F \pi_x(a_g)\, \pi_x(T_g)$$

is clearly algebraic. So in order to prove (16.37) it is enough to prove that:

(i) $$\left\|\sum_F a_g T_g\right\| \geq \left\|\pi_x\left(\sum_F a_g T_g\right)\right\| \qquad (16.38)$$

which means that π_x is a representation for $C^*(A, T_g)$;

and

(ii) for any irreducible representation ν of $C^*(A, T_g)$ there exists a point x such that ν is a representation for $\pi_x(C^*(A, T_g))$;

which along with (i) gives us

$$C^*(A, T_g) \cong \bigoplus_{x \in \hat{A}} \pi_x(C^*(A, T_g)) \quad ;$$

and

(iii) $$\pi_x(A) \underset{[\hat{T}_x]}{\times} G^x \cong \pi_x(C^*(A, T_g)) \quad . \qquad (16.39)$$

We begin with the last statement.

In fact (iii) is evident by Theorem 11.12 since $\pi_x(C^*(A, T_g))$ is simply the right regular representation of $\pi_x(A) \underset{[\hat{T}_x]}{\times} G^x$ (see Remark 15.3).

Now let us prove (ii). For any irreducible representation $\nu \in \widehat{C^*(A, T_g)}$ consider the representation $\nu(A)$ of A and define

$$Y = \text{supp Ker}[\nu(A)] = \{x \in \text{Prim } A\colon x \supset \text{Ker } \nu(A)\} \quad . \tag{16.40}$$

Obviously Y is a closed subset of Prim A and is invariant under all automorphisms $\hat{T}_g$. We can identify Y with Prim $\nu(A)$ (see **2.85**).

The key point of the proof of (ii) is the existence of a point $y_0 \in Y$ with dense orbit, that is, such that $\overline{\{t_g(y_0)\}_{g\in G}} = Y$.

Since A is separable $\nu(A)$ is also separable and the topology on Y is second countable (Dixmier [1], 3.3.4). Let U_k, $k = \overline{1,\infty}$, be a base for the topology on Y. A point $x \in Y$ has a non–dense orbit iff there exists an open set U_j from the base of the topology such that

$$x \in \left[Y \setminus \bigcup_{g\in G} t_g(U_j)\right] \quad .$$

In view of the irreducibility of ν the set

$$V_j = \bigcup_{g\in G} t_g(U_j) \tag{16.41}$$

is dense in Y. As Prim A is a Baire space (Dixmier [1], 3.4.13) the intersection $V = \bigcap_{j=1}^{\infty} V_j$ of all the sets V_j defined by (16.41) is non–empty. Choose any point $y_0 \in V$; its orbit is dense in Y.

Now we have to consider the following cases.

(a) $y_0 \in X_\infty$, that is, $G_{y_0} = \{e\}$.

Since y_0 has dense orbit in Y for any $x_0 \in \hat{A}$ with Ker $x_0 = y_0$

$$\pi_{x_0}(A) \cong \nu(A) \quad .$$

Thus in view of (iii)

$$\pi_{x_0}(C^*(A, T_g)) \cong \nu(A) \underset{\hat{T}}{\times} G$$

and consequently $\nu(C^*(A, T_g))$ is a representation of $\pi_{x_0}(C^*(A, T_g))$ which means that (ii) in this case is true.

(b) $y_0 \bar{\in} X_\infty$ thus $G_{y_0} \neq \{e\}$.

Just as in the previous case we have

$$\pi_{x_0}(C^*(A, T_g)) \cong \nu(A) \underset{[\hat{T}_{x_0}]}{\times} G^{x_0} .$$

(ii) will be proved by the same argument as that used in (a) if we show that

$$\nu(T_h) = 1 \qquad \text{for all} \qquad h \in G_{y_0} . \tag{16.42}$$

By the definition of G_{y_0} (16.30) $y_0 \in \mathring{X}_h$ for all $h \in G_{y_0}$. Note that for every $h \in G_{y_0}$, $\nu(I_h)$ is non–degenerate (i.e. the essential subspace for $\nu(I_g)$ is the whole of H_ν). Indeed, as $y_0 \in \mathring{X}_h$, $\nu(I_h) \neq 0$ and the essential subspace is non–zero and (since $\nu(I_h)$ is the ideal in $\nu(A)$ invariant under $\nu(T_g)$, $g \in G$, in view of the invariance of $\mathring{X}_h$ under t_g) invariant for $\nu(C^*(A, T_g))$. Thus we conclude that it coincides with H_ν.

As $\nu(I_h)$, $h \in G_{y_0}$, is non–degenerate $\nu(I_h)$ is strongly dense in $\nu(A)$ (Dixmier [1], 2.10.4) and consequently the identity $1 \in L(H_\nu)$ lies in the strong closure of $\nu(I_h)$ for all $h \in G_{y_0}$.

Now fix any $h \in G_{y_0}$ and let

$$\nu(i_\lambda) \xrightarrow{\text{strongly}} 1 \qquad i_\lambda \in I_h .$$

Then in view of the assumption of the theorem we have

$$\nu(T_h)\ \nu(i_\lambda) = \nu(i_\lambda) \xrightarrow{\text{strongly}} 1$$

which proves (16.42) and thus the proof of (ii) is complete.

To finish the proof of the theorem we have to prove (16.38).

If $\operatorname{Ker} x \in X_\infty$ then (16.38) is obviously true (see **16.11**). So let us consider the case $\operatorname{Ker} x \bar{\in} X_\infty$.

We have already shown that

$$\pi_x(C^*(A, T_g)) \cong \pi_x(A) \underset{[\hat{T}_x]}{\times} G^x .$$

In view of Theorem 12.8 (16.38) will be proved if we show that for any finite set $F \subset G$

$$\left\| \sum_F a_g T_g \right\| \geq \left\| \sum_{[g]=[e]} \pi_x(a_g) \right\| \tag{16.43}$$

($[g] = [e]$ means $g \in G_y$, $y = \operatorname{Ker} x$).

Evidently

$$\left\| \sum_{[g]=[e]} \pi_x(a_g) \right\| = \sup_{z \in \{G(y)\}} \left[\widecheck{\sum_{[g]=[e]} a_g}(z) \right] \tag{16.44}$$

where $\{G(y)\}$ denotes the orbit of y, that is, the set $\{t_g(y)\}_{g \in G}$.

So let

$$M = \{g \in F\colon [g] = [e]\}$$

$$J = \bigcap_{h \in M} I_h$$

$$W = \bigcap_{h \in M} \overset{\circ}{X}_h , \tag{16.45}$$

$$\beta = \sup_{z \in \{G(y)\}} \left[\widecheck{\sum_{g \in M} a_g}(z) \right] .$$

Fix any $\varepsilon > 0$. By (16.45) there exists a $z_0 \in \{G(y)\}$ such that

$$\left[\widecheck{\sum_{g \in M} a_g(z_0)}\right] > \beta - \varepsilon \quad . \tag{16.46}$$

Since, by the assumption of the theorem, G^y acts topologically freely at y (and thus in z_0) it follows that for any neighbourhood $U(z_0)$ there exists $y' \in U(z_0) \cap W$ such that

$$t_g(y') \neq y' \quad , \qquad g \in F \setminus M \quad . \tag{16.47}$$

In view of the arbitrariness of $U(z_0)$ and the lower semicontinuity of $\vee$ we can assume that

$$\left[\widecheck{\sum_{g \in M} a_g(y')}\right] > \beta - \varepsilon \quad . \tag{16.48}$$

Consider any irreducible representation $\pi \in \hat{A}$ with $\operatorname{Ker} \pi = y'$. By (16.48) there exists $\xi \in H_\pi$, $\|\xi\| = 1$ such that

$$\left\| \pi\left(\sum_{g \in M} a_g\right)\xi \right\| > \beta - \varepsilon \quad . \tag{16.49}$$

Let ν be any extension of π up to an irreducible representation of $C^*(A, T_g)$ (Theorem 2.81). Since $y' \in W$ we conclude that $\nu(J) \neq 0$ (W and J are taken from (16.45)). So (as in the proof of (ii)) the essential subspace for $\nu(J)$ coincide with H_ν, and $\nu(J)$ is strongly dense in $\nu(A)$ which by Kaplansky's density theorem (Theorem 2.97) means that the unit ball of $\nu(J)$ is strongly dense in the unit ball of $\nu(A)$. In particular there exists a sequence $\{j_\lambda\} \in J$, $\|j_\lambda\| \leq 1$, such that

$$\nu(j_\lambda) \xrightarrow{\text{strongly}} 1 \quad . \tag{16.50}$$

Just as was done when proving (16.42) one can show that

$$\nu(T_h) = 1 \quad , \qquad h \in M \quad . \tag{16.51}$$

Now for any j_λ mentioned in (16.50) we have, using (16.51) and (16.49), and the argument used in the proof of Theorem 12.14 (see in fact Lemma 12.15) we have

$$\begin{aligned}
\left\|\nu\left(\sum_F a_g T_g\right)\xi\right\|^2 &\geq \left\|\nu(j_\lambda)\ \nu\left(\sum a_g T_g\right)\xi\right\|^2 \\
&= \left\|\sum_F \nu(j_\lambda a_g T_g)\xi\right\|^2 \\
&= \left\|\sum_M \nu(j_\lambda a_g T_g)\xi\right\|^2 + \left\|\sum_{F\setminus M} \nu(j_\lambda a_g T_g)\xi\right\|^2 \\
&\geq \left\|\sum_M \nu(j_\lambda a_g T_g)\xi\right\|^2 \\
&= \left\|\nu(j_\lambda) \sum_M \nu(a_g)\xi\right\|^2 \\
&= \left\|\nu(j_\lambda) \sum_M \pi(a_g)\xi\right\|^2 \quad .
\end{aligned} \tag{16.52}$$

But in view of (16.50) and (16.49) for some λ_0 we have

$$\left\|\nu(j_{\lambda_0}) \sum_M \pi(a_g)\xi\right\| > \beta - \varepsilon$$

which along with (16.52) means that

$$\left\|\sum a_g T_g\right\| > \beta - \varepsilon \quad .$$

Since ε is arbitrary this proves (16.43) and thus finishes the proof of the theorem. □

Remark 16.22 The example considered in **16.11** shows that if we don't assume G^y to act topologically freely at y then (16.37) is not true.

As an immediate corollary of the foregoing theorem we present here a result which is a generalization of Theorem 14.1 in the case of a separable C^*–algebra A.

Theorem 16.23 *Let* $C^*(A, T_g)$ *be the algebra described in Theorem* 16.21. *If* $J \neq \{0\}$ *is a closed* **–ideal in* $C^*(A, T_g)$ *then* $J \cap A \neq \{0\}$.

Proof Let $A \cap J = \{0\}$. Consider the canonical mapping $\pi\colon C^*(A, T_g) \to C^*(A, Tg)/_J$. We have

$$\pi(A) = A/_J \cong A$$

$$\pi(T_g a T_g^*) = \pi(T_g)\ \pi(a)\ \pi(T_g)^* , \qquad a \in A\ , \quad g \in G$$

and for every $[j] \in \pi(I_g)$ (i.e. $[j] = \pi(j)$ for some $j \in I_g$)

$$\pi(T_g)[j] = \pi(T_g)\ \pi(j) = \pi(T_g j) = \pi(j) = [j]$$

which means that the algebra $\pi(C^*(A, T_g)) = C^*(\pi(A), \pi(T_g))$ satisfies again all the conditions of Theorem 16.21 and thus

$$C^*(A, T_g) \cong C^*(\pi(A), \pi(T_g)) = C^*(A, T_g)/_J .$$

So $J = \{0\}$. □

Remark 16.24 The next example shows that the condition for G^y to act topologically freely at y is essential for the validity of the theorem.

Example Consider the space $H = \left[\bigoplus_{i=1}^{5} L^2(R_i)\right] \oplus \mathbb{C}$ where $R_i = \mathbb{R}$, $i = \overline{1,5}$. Set A to be the algebra of operators of multiplication by functions

$$a = (a_1(t),\dots,a_5(t), \quad a(\infty))$$

where a_i, $i = \overline{1,5}$ are continuous and have limits at ∞ and

$$\lim_{t\to\infty} a_i(t) = a(\infty) \qquad i = \overline{1,5} \quad .$$

Evidently

$$\text{Prim } A = \left(\bigsqcup_{i=1}^{5} R_i\right) \cup (\infty)$$

with the obvious topology.

Define the mapping $\alpha\colon \bigsqcup_{i=1}^{5} R_i \to \bigsqcup_{i=1}^{5} R_i$ by the formulae

$$\left.\begin{array}{l}\alpha\colon R_1 \to R_2\\ \alpha\colon R_2 \to R_1\end{array}\right\} \qquad \alpha(t) = t$$

$$\left.\begin{array}{l}\alpha\colon R_3 \to R_4\\ \alpha\colon R_4 \to R_5\\ \alpha\colon R_5 \to R_3\end{array}\right\} \qquad \alpha(t) = t$$

and consider the operator $T\colon H \to H$ defined by

$$(Tf)(t) = f(\alpha(t)) \quad , \qquad t \in \bigsqcup_{i=1}^{5} R_i$$

$$(Tf)(\infty) = e^{\frac{i\pi}{3}} f(\infty) \quad .$$

We have $T^6 = 1$ and consequently T generates a representation of $\mathbb{Z}_6$.

Obviously $G_\infty = \{e\}$, thus $G^\infty = \mathbb{Z}^6$ and since Prim $A = X_2 \cup X_3$ G^∞ does not act topologically freely at ∞.

By the structure of the topology Prim A does not contain isolated points so A does not contain compact operators besides zero (by Theorem 14.19) but $C^*(A, T)$ does contain non–zero compact operators! Indeed let us consider the operator

$$b = (T^2 - 1)(T^3 - 1) \quad .$$

In view of the definition of T we have

$$T^2\Big|_{\bigoplus_{i=1}^{2} L^2(R_i)} = 1$$

$$T^3\Big|_{\bigoplus_{i=3}^{5} L^2(R_i)} = 1$$

and consequently

$$b\Big|_{\bigoplus_{i=1}^{5} L^2(R_i)} = 0$$

while

$$b\Big|_{\mathbb{C}} = \left(e^{i\frac{2\pi}{3}} - 1\right)(e^{i\pi} - 1) \neq 0 \quad .$$

Thus b is a one–dimensional operator.

Notes and remarks

Section 11 C^*–algebras generated by dynamical systems and concrete classes of such algebras have been considered in connection with problems of general operator algebra theory and applications first of all to representation theory and quantum mechanics. The investigation of such algebras was started in the prominent work by Murray and Von Neumann [1]. By now it is absolutely impossible to cover all the sources connected with this theme;

the bibliography is vast. The most detailed exposition and bibliography up to the end of the 1970s is presented by Pedersen [1]. Here we sketched only a few (necessary for our purposes) facts concerning amenability. An up–to–date description of the amenable world is given by Peterson [1].

Section 12 Results interrelated with the isomorphism theorem (Corollary 12.17) have been obtained by a number of authors in connection with the treatment of different problems in analysis.

The mapping $N\colon C^*(A, T_g) \to A$ (see **12.2**) was introduced by Von Neumann [1]. It was investigated by many authors and, in the theory of operator algebras, plays a role analogous to that of conditional mathematical expectation of a random variable in probability theory.

It may be worth remarking that properties similar to the property (*) (see **12.2**) almost automatically arise when we consider objects which may be expanded into some sort of series. For example if we consider a function

$$f(z) = \sum_{n=0}^{\infty} c_n z^n$$

which is holomorphic in the unit circle $|z| \le 1$ then its Taylor coefficients c_n satisfy the Cauchy inequalities

$$|c_n| \le \max_{|z|=1} |f(z)|$$

(see for example Shabat [1], Chapter II, Section 6) which simply express the property (*) in this case if one considers the algebra $C^*(\mathbb{C}, T)$ with $T = z$.

One particular isomorphism theorem was proved by Arveson [2] and Arveson, Josephson [1] while considering the problem of the classification of measure–preserving automorphisms. O'Donovan [1] treated the subject while investigating the structure of ideals in cross–products. Olesen and Pedersen [1] brought to light the correlation between the Connes spectrum, the Borches and the Arveson spectrum, and the topologically free action (see **12.13**) of the group of automorphisms (in the case of $G = \mathbb{Z}$). Their results are also interrelated with the description of ideals in cross–products and finding the criteria for automorphisms to be outer. Perhaps the first article where it was observed that the algebras generated by unitary groups possessing the property

(*) (see **12.2**) possess also some isomorphism properties and the criteria for groups to be such (the non–degenerate action) was brought to light in the article by Arveson [1].

It is worth mentioning that as a rule other sorts of isomorphism theorems have been obtained while investigating the problem of the classification of C^*–algebras generated by dynamical systems, see, for example, De Brabanter [1, 2], De Brabanter, Zettl [1], Brebken [1], Pimsner, Voiculescu [1], Riedel [1, 2], Rieffel [1], Putnam, Schmidt, Skau [1] and others.

In connection with Theorem 12.8 and Corollary 12.9 it is worth mentioning the paper by Landstad [1] where the criteria for W^* and C^*–algebras to be isomorphic to a certain cross–product (with an arbitrary locally compact group) were found. These criteria are stated in terms of a duality theory. Some of the results of this article (concerning discrete groups) may be deduced from Theorem 12.8.

The isomorphism theorem for $C^*(A, T)$ in the case when A is commutative was obtained by Lebedev [4, 7]. The proof was different but the vital importance of the topologically free action and the property (*) has been established. This work initiated the systematic investigation of the subject in the work by Antonevich [11, 14], Antonevich, Brenner [2], Antonevich, Lebedev [1], Brenner [2, 4], Karlovich [1], Lebedev [11, 13] and Batkhin [2, 4].

We want to distinguish in this connection the work by Brenner [2, 4] where the isomorphism theorem for subexponential groups of automorphisms was established in an original way using the spectral radius formulae (see Section 22). The idea of this proof was generalized further by Antonevich [11, 14] up to admissible groups. Karlovich [1] obtained the isomorphism theorem for some local–type algebras (see Notes and Remarks to Sections 34 and 35). The isomorphism theorem in its present form was proved by Lebedev [13].

In the case of Von Neumann algebras the topologically free action of a group can be formulated in a different way (which is reasonable to call a metrically free action). Namely, let A be a Von Neumann subalgebra of $L(H)$ and α: $G \to \mathrm{Aut}(A)$ be a representation of a discrete group G by automorphisms of A. We say that G acts *metrically freely* on A if for any finite set $\{g_1,\dots,g_n\} \subset G$ and any non–zero projection $P \in A \cap A'$ there exists a non–zero subprojection $P_1 \in A \cap A'$ (that is $P_1P = PP_1 = P_1$)

such that $\alpha(g_i)(P_1)\cdot\alpha(g_j)(P_1) = 0$, $i,j \in \overline{1,h}$, $i \neq j$.

Under this condition the isomorphism theorem for Von Neumann algebras can be proved (see for example Golodets [1] for some particular results).

Section 13 As we have already mentioned (see Notes and Remarks concerning Section 8) the spectral symmetry of weighted shifts was observed by many authors. The results of this section are straightforward corollaries of the isomorphism theorem and they were obtained in various forms in the work of Lebedev [2, 13], Brenner [2], and Antonevich [11, 14].

The problem of the coincidence of spectra defined by the same expression in different functional spaces was settled apparently by Burnat [1] (the history of this is presented by Kozlov, Shubin [1]). The examples in **13.11**, **13.12** belong to Antonevich [14]. In connection with this problem, it is worth mentioning the papers by Biktashev [1], Biktashev, Mischenko [1], Mischenko, Sharipov [1], and Sharipov [1] where, for example, Biklashev [1] and Mischenko, Sharipov [1] used the isomorphism theorem from Antonevich, Brenner [2].

Section 14 In the Notes and Remarks concerning Section 8 we noted that the coincidence of the conditions for invertibility and Fredholmity of particular operators generated by weighted composition operators was observed by many authors.

Propositions 14.4 and 14.5 describe the general C^* picture of such situations. Together with the results of **14.6–14.27** they give explicit criteria which bring to light the internal structure and give the solution to the problem.

A different approach is presented by Krupnik, Feldman [1].

In connection with Theorem 14.19, we should also mention the paper by Kamowitz [1] treating the same problem for commutative semisimple Banach algebras.

Propositions 14.4, 14.5 and 14.6 in the case of a commutative algebra A and $G = \mathbb{Z}$ where proved by Lebedev [1, 2, 7]. Further developments were obtained by Brenner [2], Antonevich [11, 14], and Lebedev [12, 13, 16].

When considering the structure of ideals in cross–products and in connection with Theorem 14.1 we should mention the work by O'Donovan [1], Olesen and Pedersen [1] (see Notes and Remarks for Section 12), and Kawamura, Tomiyama [1].

In this section we follow the article by Lebedev [16].

Section 15 Algebras generated by weighted composition operators in the case when composition operators form a finite group where treated by a number of authors. In this connection we have to distinguish the work of Litvinchuk [1, 2], Antonevich [2, 3]. Przeworska–Rolewicz [1], and Antonevich, Brenner [1].

The results of this section follow the structure developed by Antonevich [2, 3]. The idea of this structure arose earlier in the work of a number of authors and in particular in the work by Babbage [1].

Section 16 Here we follow the work by Lebedev [16].

Chapter 3: Invertibility

'Доказательство – набор слов, убеждающих нас в том, что очевидное утверждение верно' О. Слин
'Зто правильно, но неверно' А. Платонов 'Деревянное растение'*

The conditions for invertibility of elements in $C^*(A, T_g)$ in the case of a finite group G were considered in Section 15. If G is infinite the situation is much more complicated.

The main purpose of this chapter is to obtain some general results about the conditions for invertibility of elements in $C^*(A, T_g)$. We begin with a description of conditions for the invertibility of operators of the form $a_0 + a_1 T$ in the case of a commutative algebra A (Sections 17 and 18).

Passing from the commutative case to the general case we find that the conditions for the invertibility of two–term operators are naturally linked with the hyperbolicity conditions for the corresponding linear extension of the initial dynamical system induced by T (Sections 19 and 20).

Then we find a number of formulae and estimates for the spectral radius of elements in $C^*(A, T_g)$ and provide criteria for invertibility of certain classes of group actions Sections 22–24.

It happens that the structure of the spectrum of a weighted shift operator aT has deep interrelations with the Lyapunov exponents of a C^*–dynamical system (Section 25).

The chapter also contains a description of some full subalgebras in $C^*(A, T_g)$ and the corresponding Wiener–type theorems (Section 26), and we present two more conceptually distinct methods for investigating the conditions for invertibility of elements in $C^*(A, T_g)$: one based on the trajectorial decomposition of $\hat{A}$ (Section 21); and the other on an approximation of the

---

*'A proof is a set of words convincing us that an obvious statement is true'

O. Slin

'This is true but not correct' A. Platonov *Wooden plant*

group of automorphisms $\{\hat{T}_g\}$ by some other group $\{\hat{U}_g\}$ which makes it possible in a number of situations to simplify the group action.

17. Operators of the form $a_0 + a_1T$. The commutative algebra case.

Throughout this section A is a commutative C^*–algebra, with identity, of operators acting on a Hilbert space H, and T is a unitary operator such that $TAT^* = A$ and $\{\hat{T}^z\}$ acts topologically freely on A.

We recall (see Proposition 13.8) that in this situation the operator $a_0 + a_1T$ is invertible if and only if $|a_0| + |a_1|T$ is invertible.

We shall denote by M the maximal ideal space of A and by $M(\hat{T})$ the maximal ideal space of the C^*–subalgebra of all the $\hat{T}$–invariant elements in A. Let Λ be the set of all ergodic measures with respect to t, where t is the homeomorphism of M induced by $\hat{T}$.

Theorem 17.1 *Let* $\mathrm{M}(\hat{\mathrm{T}})$ *be a connected space. If* a_0 *and* a_1 *are not invertible then* $\mathrm{b} = \mathrm{a}_0 + \mathrm{a}_1\mathrm{T}$ *is not invertible.*

Proof By the foregoing remark without loss of generality it can be assumed that $a_i(x) \geq 0$, $x \in M$. Since a_i are not invertible there exist points x_0, x_1 such that $a_i(x_i) = 0$.

We consider the following possible cases.

I. There exists a measure μ ergodic with respect to the homeomorphism t such that $x_i \in \operatorname{supp} \mu$, $i = 0,1$. In this case we shall use the following lemma.

Lemma 17.2 *Let* $\mathrm{x}_0 \in \operatorname{supp} \mu$, $\mathrm{a}(\mathrm{x}_0) = 0$ *and* $\mathrm{a}(\mathrm{x}) \geq 0$ *for* $\mathrm{x} \in \mathrm{M}$. *Then for any* $\varepsilon > 0$ *there exists a positive number* $\Delta(\varepsilon)$ *such that for any* δ *satisfying* $0 < \delta < \Delta(\varepsilon)$ *there exists an invertible operator* $\mathrm{a}_{\varepsilon,\delta} \in \mathrm{A}$ *such that*

$$\|\mathrm{a}_{\varepsilon,\delta} - \mathrm{a}\| < \varepsilon \quad , \qquad \mathrm{S}_\mu(\mathrm{a}_{\varepsilon,\delta}) = \delta$$

(where S_μ *is defined in* 4.3).

Proof Let $a_\varepsilon = a + \varepsilon I$ and φ be a continuous function satisfying the following conditions:

(i) $\frac{1}{2} \le \varphi(x) \le 1$, $x \in M$;

(ii) $$\varphi(x) = \begin{cases} \frac{1}{2} , & a(x) \le \frac{\varepsilon}{2} \\ 1 , & a(x) \ge \varepsilon \end{cases} .$$

Such a function exists by the Urysohn Lemma. Obviously for any non–negative c

$$\left\| \varphi^c a_\varepsilon - a \right\| < \varepsilon \tag{17.1}$$

and $\varphi^c a_\varepsilon$ is an invertible operator.

We also have that

$$S_\mu(\varphi^c a_\varepsilon) = [S_\mu(\varphi)]^c \cdot S_\mu(a_\varepsilon) \le S_\mu(a_\varepsilon) \tag{17.2}$$

and clearly

$$S_\mu(\varphi^0 a_\varepsilon) = S_\mu(a_\varepsilon) \tag{17.3}$$

and the function $S_\mu(\varphi^c a_\varepsilon)$ depends continuously on c in $[0, +\infty]$ and tends to zero as c tends to $+\infty$.

Now choose $\Delta(\varepsilon) = S_\mu(a_\varepsilon)$ and a suitable operator of the form $\varphi^c a_\varepsilon$ to be $a_{\varepsilon,\delta}$. By (17.1), (17.2) and (17.3) the lemma is proved. □

Let us continue the proof of the theorem. Since $x_0, x_1 \in \operatorname{supp} \mu$ by Lemma 17.2 for any $\varepsilon > 0$ there exist a number $\delta > 0$ and invertible operators $a^0_{\varepsilon,\delta}$ and $a^1_{\varepsilon,\delta}$ such that

$$\left\| a^i_{\varepsilon,\delta} - a_i \right\| < \varepsilon \quad \text{and} \quad S_\mu(a^i_{\varepsilon,\delta}) = \delta \qquad i = 0,1 . \tag{17.4}$$

Note that the operator $a^0_{\varepsilon,\delta} + a^1_{\varepsilon,\delta}T$ is not invertible.

Indeed, the operator $a^0_{\varepsilon,\delta} + a^1_{\varepsilon,\delta}T$ is invertible if and only if $I + \left(a^0_{\varepsilon,\delta}\right)^{-1} a^1_{\varepsilon,\delta}T$ is invertible. By Theorem 13.10 the spectrum $\sigma\left(\left(a^0_{\varepsilon,\delta}\right)^{-1} a^1_{\varepsilon,\delta}T\right)$ is the annulus

$$\max_{\nu\in\Lambda} S_\nu\left[\left(a^0_{\varepsilon,\delta}\right)^{-1} a^1_{\varepsilon,\delta}\right] \geq |\lambda| \geq \min_{\nu\in\Lambda} S_\nu\left[\left(a^0_{\varepsilon,\delta}\right)^{-1} a^1_{\varepsilon,\delta}\right]$$

but in view of (17.4)

$$S_\mu\left[\left(a^0_{\varepsilon,\delta}\right)^{-1} a^1_{\varepsilon,\delta}\right] = \frac{S_\mu\left(a^1_{\varepsilon,\delta}\right)}{S_\mu\left(a^0_{\varepsilon,\delta}\right)} = 1$$

and so $I + \left(a^0_{\varepsilon,\delta}\right)^{-1} a^1_{\varepsilon,\delta}T$ is not invertible. Consequently $a^0_{\varepsilon,\delta} + a^1_{\varepsilon,\delta}T$ is also not invertible.

By the construction of the operators $a^i_{\varepsilon,\delta}$, $i = 0,1$, we have that

$$\left\|(a^0_{\varepsilon,\delta} + a^1_{\varepsilon,\delta}T) - (a_0 + a_1T)\right\| \leq \left\|a^0_{\varepsilon,\delta} - a_0\right\| + \left\|a^1_{\varepsilon,\delta} - a_1\right\| < 2\varepsilon \ .$$

Thus we have shown that in any neighbourhood of the operator $a_0 + a_1T$ there are non–invertible operators. Hence, $a_0 + a_1T$ is not invertible and in case I the theorem is proved.

II(a) Suppose that some ergodic measure μ with respect to t satisfies the following conditions:

(i) $x_0 \in \text{supp}\ \mu$; and

(ii) the set of zeros of $a_1(x)$ is separated from supp μ.

By Lemma 17.2 for any $\varepsilon > 0$ one can choose an invertible operator $a^0_{\varepsilon,\delta}$ such that

$$\left\|a^0_{\varepsilon,\delta} - a_0\right\| < \varepsilon \qquad \text{and} \qquad S_\mu\left(a^0_{\varepsilon,\delta}\right) = \delta \leq S_\mu(a_1) \ .$$

Note that the operator $a^0_{\varepsilon,\delta} + a_1T$ is not invertible.

Indeed, the operator $a^0_{\varepsilon,\delta} + a_1T$ is invertible if and only if the operator $I + \left(a^0_{\varepsilon,\delta}\right)^{-1} a_1T$ is invertible.

By Theorem 13.10 (since $\left(a^0_{\varepsilon,\delta}\right)^{-1} a_1$ is not invertible) the spectrum $\sigma\left(\left(a^0_{\varepsilon,\delta}\right)^{-1} a_1T\right)$ the disk

$$|\lambda| \leq \max_{\nu\in\Lambda} S_\nu\left[\left(a^0_{\varepsilon,\delta}\right)^{-1} a_1\right] \quad .$$

But $S_\mu\left[\left(a^0_{\varepsilon,\delta}\right)^{-1} a_1\right] \geq 1$ hence $I + \left(a^0_{\varepsilon,\delta}\right)^{-1} a_1T$ and consequently $a^0_{\varepsilon,\delta} + a_1T$ are not invertible. Just as before this implies that $a_0 + a_1T$ is not invertible.

II(b) Suppose that some ergodic measure μ satisfies:

(i) $x_1 \in \operatorname{supp} \mu$; and

(ii) the set of zeros of $a_0(m)$ is separated from supp μ.

Then the proof is carried out in a similar way to II(a) by passing to the operator

$$a_0T^{-1} + a_1 = (a_0 + a_1T)T^{-1} \quad .$$

III Finally, suppose that the set of zeros of $a_0(x)$ and $a_1(x)$ is separated from the supports of all the measures ergodic with respect to t.

For each measure μ we have either

$$S_\mu(a_0) \leq S_\mu(a_1) \tag{17.5}$$

or

$$S_\mu(a_0) \geq S_\mu(a_1) \quad . \tag{17.6}$$

Assume that (17.5) holds. Since the zero set of $a_0(x)$ is separated from supp μ, there exists an $\varepsilon > 0$ such that the set $V_\varepsilon = \{x \in M: a_0(x) \le \varepsilon\}$ is separated from supp μ. Hence $S_\mu(a_0) = S_\mu\left(a_\varepsilon^0\right)$ where

$$a_\varepsilon^0(x) = \begin{cases} a_0(x) , & a_0(x) \ge \varepsilon \\ \varepsilon , & a_0(x) \le \varepsilon \end{cases} .$$

Obviously, $\left\|a_\varepsilon^0 - a_0\right\| \le \varepsilon$. Moreover, as $S_\mu\left[\left(a_\varepsilon^0\right)^{-1} a_1\right] \ge 1$ and the operator $\left(a_\varepsilon^0\right)^{-1} a_1$ is not invertible, neither is $a_\varepsilon^0 + a_1 T$. Since this holds for all arbitrarily small ε the operator $a_0 + a_1 T$ is not invertible.

If (17.6) holds, then the non–invertibility of $a_0 + a_1 T$ can be proved similarly by passing to the operator $a_0 T^{-1} + a_1$. The proof is complete. □

Theorem 17.3 *Let* $M(\hat{T})$ *be a connected space. The operator* $b = a_0 + a_1 T$ *is invertible iff one of the following conditions holds:*

(1) a_0 *is an invertible operator, and*

$$S_\nu(a_0) > S_\nu(a_1)$$

for any measure $\nu \in \Lambda$;

(2) a_1 *is an invertible operator, and*

$$S_\nu(a_1) > S_\nu(a_0)$$

for any measure $\nu \in \Lambda$.

Proof According to Theorem 17.1 if b is invertible then at least one of the operators a_i, $i = 0,1$, is invertible.

If a_0 is invertible but a_1 is not then the operator $b = a_0 + a_1T$ is invertible if and only if $b' = I + (a_0^{-1}a_1)T$ is invertible, that is (see Theorem 13.10), if and only if condition (1) holds.

The case when a_1 is invertible but a_0 is not is handled similarly and leads to condition (2).

If both a_0 and a_1 are invertible then case (1) of Theorem 13.10 must be applied to the operator b'. This proves the theorem. □

Remark The results of this section can also be obtained by means of essentially different ideas (see Section 19, and in particular, Corollary 19.31).

18. Examples.

Now we shall apply the results of Section 17 to the computation of invertibility conditions for concrete operators of the form $b = a_0 + a_1T$.

18.1 Consider the Hilbert space $l^2(\mathbb{Z})$ and the algebra A of operators of multiplication by functions $a \in l^\infty(\mathbb{Z})$. Let T be the bilateral shift operator

$$(Tf)(z) = f(z-1) \qquad f \in l^2(\mathbb{Z}) \quad .$$

The subalgebra of all the elements in A which are invariant under $\hat{T}$ is the subalgebra of operators of multiplication by constant functions so $M(\hat{T})$ consists of a single point. By Theorem 17.1 if

$$\inf_{\mathbb{Z}} |a_0(z)| = \inf_{\mathbb{Z}} |a_1(z)| = 0$$

then the operator $b = a_0 + a_1T$ is not invertible. Unfortunately if a_0 and a_1 are arbitrary elements of A then in view of the fact that the maximal ideal space of $l^\infty(\mathbb{Z})$ is 'very large' we are not able to use Theorem 17.3 effectively. But if a_0 and a_1 have 'reasonable' behaviour at infinity we can consider an algebra which is narrower than $l^\infty(\mathbb{Z})$ and obtain effective results. For example, let there exist the limits

$$\lim_{z\to+\infty} a_i(z) = a_i(+\infty)$$

$$i = 0,1 \quad .$$

$$\lim_{z\to-\infty} a_i(z) = a_i(-\infty)$$

Then we can consider the C^*–algebra $A_1 \subset l^\infty(\mathbb{Z})$ of the functions having limits at infinity. The maximal ideal space of A_1 is $\{-\infty\} \cup \mathbb{Z} \cup \{+\infty\}$ and the only ergodic measures under the homeomorphism induced by $\hat{T}$ are $\delta(-\infty)$ and $\delta(+\infty)$. Hence by Theorem 17.3 the operator $b = a_0 + a_1 T$ is invertible if and only if one of the following conditions holds:

(1) $a_0(z) \neq 0$ and $|a_0(\pm\infty)| > |a_1(\pm\infty)|$;

(2) $a_1(z) \neq 0$ and $|a_1(\pm\infty)| > |a_0(\pm\infty)|$.

We recall here that the conditions for Fredholmity and invertibility may be different for such operators (see **8.9**).

18.2 In the same manner we can consider the space $L^2(\mathbb{R})$ and an operator of the form $b = a_0 + a_1 T_h$ where a_i, $i = \overline{0,1}$, are the operators of multiplication by continuous bounded functions and T_h is the shift operator

$$(Tf)(x) = f(x + h)$$

where h is a given number (cf. **8.4**).

So, for example, if a_i have limits at infinity

$$\lim_{x\to\pm\infty} a_i(x) = a_i(\pm\infty)$$

we conclude that the operator $b = a_0 + a_1 T$ is invertible if and only if

(1) $a_0(x) \neq 0$ and $|a_0(\pm\infty)| > |a_1(\pm\infty)|$,

(2) $a_1(x) \neq 0$ and $|a_1(\pm\infty)| > |a_0(\pm\infty)|$.

By Theorem 14.26 and Remark 14.18 the operator b is invertible iff it is Fredholm.

18.3 Let X be a compact space and $\alpha\colon X \to X$ be a homeomorphism acting topologically freely on X such that the set $\Omega(\alpha)$ of all the non–wandering points consists only of periodic points. Let μ be some Borel measure on X such that supp $\mu = X$ and α preserves the equivalence class of μ. In the space $L^2_\mu(X)$ consider the operator

$$b = a_0 + a_1 T$$

where T and A are defined in 3.3.

Using the arguments of Example 5.11(a) and Theorem 17.3 we conclude that if X is a connected space then the operator b is invertible if and only if one of the following conditions holds:

(1) $a_0(x) \neq 0$ for $x \in X$ and

$$\prod_{k=0}^{p(x)-1} \left|a_0(\alpha^k(x))\right| > \prod_{k=0}^{p(x)-1} \left|a_1(\alpha^k(x))\right| \qquad \text{for all} \quad x \in \Omega(\alpha) \ ;$$

(2) $a_1(x) \neq 0$ for $x \in X$ and

$$\prod_{k=0}^{p(x)-1} \left|a_1(\alpha^k(x))\right| > \prod_{k=0}^{p(x)-1} \left|a_0(\alpha^k(x))\right| \qquad \text{for all} \quad x \in \Omega(\alpha)$$

where $p(x)$ is the period of $x \in \Omega(\alpha)$.

By Theorem 14.26 and Remark 14.18 if X does not contain an isolated point then b is invertible iff it is Fredholm.

18.4 Consider the operator b acting on the space $L^2(S^1)$ by the formula

$$[bf](x) = a_0(x)f(x) + a_1(x)|\alpha'(x)|^{\frac{1}{2}} f(\alpha(x))$$

where $a_i \in C(S^1)$, $i = 0,1$, and α is a homeomorphism of S^1 acting topologically freely and preserving the equivalence class of the Lebesgue probability measure dx, and $\alpha'(x)$ is the Radon–Nikodym derivative of the measure $\alpha(\mathrm{d}x)$ with respect to the measure dx.

We have to consider the following cases (cf. Example 5.11).

(a) The rotation number $\tau(\alpha)$ is rational: $\tau(\alpha) = {}^{M}/_{N}$ an irreducible fraction. Here we find ourselves under the condition of the previous example 18.3 and consequently b is invertible if and only if one of the following conditions holds:

(1) $a_0(x) \neq 0$ for $x \in S^1$ and

$$\prod_{k=0}^{N-1} \left|a_0(\alpha^k(x))\right| > \prod_{k=0}^{N-1} \left|a_1(\alpha^k(x))\right| \qquad \text{for all } x \in \Omega(\alpha)\ ;$$

or

(2) $a_1(x) \neq 0$ for $x \in S^1$ and

$$\prod_{k=0}^{N-1} \left|a_1(\alpha^k(x))\right| > \prod_{k=0}^{N-1} \left|a_0(\alpha^k(x))\right| \qquad \text{for all } x \in \Omega(\alpha)\ .$$

(b) α is a transitive homeomorphism. Using the notation introduced in Example 5.11(b) along with Theorem 17.3 we deduce that b is invertible if and only if one of the following conditions holds:

(1) $a_0(x) \neq 0$ for $x \in S^1$ and

$$\int_{S^1} \ln\left|a_0(h^{-1}(x))\right| \mathrm{d}x > \int_{S^1} \ln\left|a_1(h^{-1}(x))\right| \mathrm{d}x\ ;$$

(2) $a_1(x) \neq 0$ for $x \in S^1$ and

$$\int_{S^1} \ln\left|a_1(h^{-1}(x))\right| \mathrm{d}x > \int_{S^1} \ln\left|a_0(h^{-1}(x))\right| \mathrm{d}x\ .$$

(c) α is an intransitive homeomorphism. Using the notation introduced in Example 5.11(c) along with Theorem 17.3 we conclude that b is invertible if and only if one of the following conditions holds:

(1) $a_0(x) \neq 0$ for $x \in S^1$ and

$$\int_{S^1} \ln|\tilde{a}_0(x)|dx > \int_{S^1} \ln|\tilde{a}_1(x)|dx \ ;$$

(2) $a_1(x) \neq 0$ for $x \in X$ and

$$\int_{S^1} \ln|\tilde{a}_1(x)|dx > \int_{S^1} \ln|\tilde{a}_0(x)|dx \ .$$

(d) The homeomorphism α changes orientation. Now we are again under the condition of **18.3** and using the notation of Example 5.11(d) and Theorem 17.3 we have that b is invertible if and only if one of the following conditions holds:

(1) $a_0(x) \neq 0$ for $x \in S^1$ and

$$|a_0(x_k)| > |a_1(x_k)| \qquad k = 1,2 \ (x_1, x_2 \text{ are two fixed points})$$
$$|a_0(x)a_0(\alpha(x))| > |a_1(x)\cdot a_1(\alpha(x))| \qquad \text{for} \quad x \in \Omega'(\alpha) \ ;$$

(2) $a_1(x) \neq 0$ for $x \in S^1$ and

$$|a_1(x_k)| > |a_0(x_k)| \qquad k = 1,2$$
$$|a_1(x)a_1(\alpha(x))| > |a_0(x)\cdot a_0(\alpha(x))| \qquad \text{for} \quad x \in \Omega'(\alpha) \ .$$

By Theorem 14.26 and Remark 14.18 in all the cases considered b is invertible iff it is Fredholm.

18.5 Let G be a compact abelian group with a countable base of topology and let μ be the Haar measure on G. Consider an operator b on $L^2_\mu(G)$

defined by

$$[bf](g) = a_0(g)f(g) + a_1(g)f(g_0 + g)$$

where $g_0 \in G$ and $a_i \in C(G)$, $i = 0,1$.

I. Suppose that g_0 is an element of finite order n, that is, $ng = 0$ and $kg \neq 0$, $k = \overline{1,n-1}$.

Then by Corollary 15.6 b is invertible iff

$$\det[b_{k,j}(g)] \neq 0 \qquad \text{for} \quad g \in G \quad . \tag{18.1}$$

where $k,j \in \mathbb{Z}(\mathrm{mod}\ n)$ and

$$b_{k,j}(g) = a_{j-k}(g + kg_0)$$

(here $a_s \equiv 0$ when $s \,\bar{\in}\, \{0, 1\}$. Hence b is invertible iff

$$\prod_{k=0}^{n-1} a_0(g + kg_0) \neq \prod_{k=0}^{n-1} a_1(g + kg_0) \qquad \text{for} \quad g \in G \quad .$$

Remark In fact, by Corollary 15.6 the condition (18.1) is a necessary and sufficient condition for the invertibility of the operator b defined by

$$(bf)(g) = \sum_{k=0}^{n-1} a_k(g)f(kg_0 + g)$$

and in this case we do not need G to have a countable base of the topology.

II. Suppose now that g_0 is an element of infinite order and ${}^{G}/_{G_0}$ is connected (where G_0 is the closed subgroup generated by g_0). In view of Theorem 17.3 and Theorem 8.7 we conclude that b is invertible if and only if one of the following conditions holds:

(1) $a_0(g) \neq 0$ for $g \in G$ and

$$\int_{G_0} \ln|a_0(g + \gamma)|d\gamma > \int_{G_0} \ln|a_1(g + \gamma)|d\gamma \ ;$$

for all $g \in G$;

(2) $a_1(g) \neq 0$ for $g \in G$ and

$$\int_{G_0} \ln|a_1(g + \gamma)|d\gamma > \int_{G_0} \ln|a_0(g + \gamma)|d\gamma \ ;$$

for all $g \in G$.

Here $d\gamma$ is the normalized Haar measure on G_0.

In both cases I and II if G does not contain an isolated point then by Theorem 14.26 b is invertible iff it is Fredholm.

18.6 For any continuous mapping α: $X \to X$ of a topological space X, X is said to be α*–irreducible* if it cannot be represented as a union of two non–void disjoint closed subsets invariant under α.

18.7 Let X be a compact space with a countable base of the topology and μ be a Borel measure on X such that supp $\mu = X$. Let α be a homeomorphism acting topologically freely on X, preserving the equivalence class of μ, and such that the action of α on $\Omega(\alpha)$ is included in an action of a compact group G (see 5.21). (This is actually a generalization of all the previously considered examples.)

Consider the operator b acting on $L^2_\mu(X)$ by the formula

$$[bf](x) = a_0(x)f(x) + a_1(x)|\alpha'(x)|^{\frac{1}{2}}f(\alpha(x))$$

where $a_i \in C(X)$, $i = 0{,}1$, and α' is the Radon–Nikodym derivative of the measure $\alpha(\mu)$ with respect to the measure μ.

In view of Theorem 17.3 and Theorem 5.22 we conclude that if X is α–irreducible then b is invertible if and only if one of the following conditions holds:

(1) $a_0(x) \neq 0$ for $x \in X$ and

$$\int_{G_0} \ln|a_0(g(x))|\,dg > \int_{G_0} \ln|a_1(g(x))|\,dg \qquad \text{for all } x \in \Omega(\alpha) ;$$

(2) $a_1(x) \neq 0$ for $x \in X$ and

$$\int_{G_0} \ln|a_1(g(x))|\,dg > \int_{G_0} \ln|a_0(g(x))|\,dg \qquad \text{for all } x \in \Omega(\alpha) ;$$

where G_0 is the closed subgroup of G generated by α and dg is the normalized Haar measure on G_0.

By Theorem 14.26 if X does not contain an isolated point then b is invertible iff it is Fredholm.

19. Operators of the form $a_0 + a_1T$. The general case. Hyperbolicity.

In this section we consider operators of the form $a_0 + a_1T$ in the case when A is an arbitrary C^*–algebra with an identity and T is a unitary element such that $TAT^* = A$. We recall once more that when $\{\hat{T}^z\}$ acts topologically freely on A the algebra $C^*(A, T)$ possesses the property (*) (Theorem 12.14).

When A is not commutative the conditions for invertibility of operators of the form $a_0 + a_1T$ cannot be obtained in such an effective manner as in Section 17 (and the reason for this will be examined below). The investigation of operators of the form $a_0 + a_1T$ leads to a correlation between the invertibility of the such operators and the hyperbolicity of the corresponding linear extensions of the dynamical system induced by $\hat{T}$.

We want to stress that a number of ideas and a lot of terminology in this and the subsequent sections: linear extensions, hyperbolicity, stable and unstable bundles, dynamical spectra, exponential dichotomy, Lyapunov exponents, and so on, go back to the theory of differential equations and dynamical systems.

Our first main result here is the following lemma which is also interesting for its own sake.

Lemma 19.1 *Let* $C^*(A, T)$ *possesses the property* (*). *Suppose that* 1 *does not belong to the spectrum of the operator* aT. *Then the whole circle* $|\lambda| = 1$ *does not meet the spectrum and the spectral projection*

$$P = \frac{1}{2\pi i} \int_{|\lambda|=1} (\lambda - aT)^{-1} d\lambda$$

belongs to A.

Proof As the spectrum is invariant under rotations (Corollary 13.6) and 1 does not belong to the spectrum it follows that no point of the circle $|\lambda| = 1$ belongs to the spectrum.

Since $C^*(A, T)$ possesses the property (*) we can identify $C^*(A, T)$ with the regular representation $\tilde{\pi}(A \underset{\hat{T}}{\times} \mathbb{Z})$ (see 11.11, Theorems 11.12 and 12.8) in $L(l^2(\mathbb{Z}, H))$. So hereafter we think of the elements of $C^*(A, T)$ as elements of $\tilde{\pi}(A \underset{\hat{T}}{\times} \mathbb{Z})$.

Consider the projection

$$P = \frac{1}{2\pi i} \int_{|\lambda|=1} (\lambda - aT)^{-1} d\lambda$$

and its coefficients $a_n = N(PT^{-n})$ (see 12.2). Let P_k, $k \in \mathbb{Z}$, be the projection in $l^2(\mathbb{Z}, H)$ defined by

$$P_k\xi = \{0, \xi_k, 0, \dots\} \qquad \xi \in l^2(\mathbb{Z}, H)$$

where

$$\xi = \{\xi_n\}_{n\in\mathbb{Z}}, \quad \xi_n \in H.$$

Clearly

$$P_k \in A'$$
$$P_k = T^k P_0 T^{-k}$$

and the subspaces Im P_k are pairwise orthogonal. $P \in A$ iff $a_n = 0$, $n \neq 0$, and this equality is valid if

$$P_0 a_n = 0 \qquad n \neq 0$$

so we shall prove the last one.

Just as in step I of Theorem 7.2 one can verify that $H_1 = \text{Im } P$ and $H_2 = \text{Ker } P$ are invariant under P_0. Thus

$$P_0 P = P P_0$$

and in terms of the coefficients a_n of P this means that

$$(P_0 - P_n) a_n = 0 \qquad n \in \mathbb{Z} \ .$$

Since Im P_0 and Im P_n are orthogonal when $n \neq 0$ we conclude

$$P_0 a_n = 0 \qquad n \neq 0$$

which finishes the proof. □

As an immediate corollary of the lemma we have

Corollary 19.2 *If an algebra* A *contains no non–zero projections besides* I *then the spectrum* σ(aT) *is connected.*

Now we consider the main objects of this section.

19.3 Denote by $H(\hat{A})$ the *universal space* $\oplus_{\pi \in \hat{A}} H_\pi$. A *linear extension* of a mapping $\alpha: \hat{A} \to \hat{A}$ is a mapping $\beta: H(\hat{A}) \to H(\hat{A})$ such that the fibre H_π is mapped linearly into the fibre $H_{\alpha(\pi)}$.

19.4 A linear extension β is said to be *hyperbolic* (with respect to the algebra A) if there exists a projection $P \in A$ such that the subspaces $H^s = \bigoplus_{\pi\in\hat{A}} \operatorname{Im} \pi(P)$ and $H^u = \bigoplus_{\pi\in\hat{A}} \operatorname{Ker} \pi(P)$ are invariant under β, β maps H^u onto H^u and there exist constants $C > 0$, $0 < \gamma < 1$ such that

$$\|\beta^n(y)\| \le C\gamma^n\|y\| \quad , \qquad y \in H^s \quad , \qquad n = 1,2,\ldots \tag{19.1}$$

$$\|\beta^n(y)\| \ge C\gamma^{-n}\|y\| \quad , \qquad y \in H^u \quad , \qquad n = 1,2,\ldots \quad . \tag{19.2}$$

The subspace H^s is called stable and H^u is called unstable.

Remarks

(1) It follows from the definition that $H(\hat{A}) = H^s \oplus H^u$.

(2) Estimate (19.2) and the condition that β is onto on H^u imply that β is bijective on H^u.

19.5 Since the mapping t of $\hat{A}$ (induced by $\hat{T}$) is defined by the equation

$$t(\pi)[a] = \pi[\hat{T}^{-1}(a)] \qquad \pi \in \hat{A} \quad , \qquad a \in A$$

the spaces $H_{t(\pi)}$ and H_π are isomorphic. Let us denote by $\theta_\pi\colon H_\pi \to H_{t(\pi)}$ the operator satisfying the equation

$$\theta_\pi \pi[\hat{T}^{-1}(a)](\theta_\pi^{-1} h) = t(\pi)[a](h) \quad , \qquad h \in H_{t(\pi)} \quad . \tag{19.3}$$

19.6 A linear bounded operator is said to be *hyperbolic* if its spectrum does not meet the unit circle $|\lambda| = 1$.

Lemma 19.7 *An operator* b *is hyperbolic iff there exists a decomposition* $\mathrm{H} = \mathrm{E}^s \oplus \mathrm{E}^u$ *such that both* E^s *and* E^u *are invariant under* b, b *is invertible on* H^u *and there exist constants* $\mathrm{C} > 0$, $0 < \gamma < 1$, *such that*

$$\|\mathrm{b}^n(\mathrm{h})\| \le \mathrm{C}\gamma^n\|\mathrm{h}\| \quad , \qquad \mathrm{h} \in \mathrm{E}^s \quad , \qquad \mathrm{n} = 1,2,\ldots \tag{19.4}$$

$$\|\mathrm{b}^n(\mathrm{h})\| \ge \mathrm{C}\gamma^{-n}\|\mathrm{h}\| \quad , \qquad \mathrm{h} \in \mathrm{E}^u \quad , \qquad \mathrm{n} = 1,2,\ldots \quad . \tag{19.5}$$

Proof This follows easily from the spectral decomposition $H = P(H) \oplus (I - P)H = E^s \oplus E^u$ where

$$P = \frac{1}{2\pi i} \int_{|\lambda|=1} (\lambda - b)^{-1} d\lambda$$

and the fact that both the spectral radii of $b|_{P(H)}$ and $b^{-1}|_{(I-P)H}$ are less than $\gamma < 1$. □

19.8 It may be useful to present a geometric description of E^s and E^u. To do this let us introduce, for any vector $h \neq 0$, the numbers

$$\bar{\gamma}^{+}(h) = \overline{\lim_{n\to\infty}} \|b^n h\|^{\frac{1}{n}}$$

$$\underline{\gamma}^{+}(h) = \underline{\lim}_{n\to\infty} \|b^n h\|^{\frac{1}{n}}$$

and the numbers

$$\bar{\gamma}^{-}(h) = \overline{\lim_{n\to\infty}} \|h_n\|^{-\frac{1}{n}}$$

$$\underline{\gamma}^{-}(h) = \underline{\lim}_{n\to\infty} \|h_n\|^{-\frac{1}{n}}$$

where $\{h_n\}_{n=1}^{\infty}$ is a sequence such that $h_0 = h$ and $bh_n = h_{n-1}$ (we set $\underline{\gamma}^{-} = \bar{\gamma}^{-} = 0$ if there is no such sequence). Clearly $\bar{\gamma}^{-}$ and $\underline{\gamma}^{-}$ in general depend on the sequence $\{h_n\}_{n=1}^{\infty}$ with the described property.

Lemma 19.9 *Let* b *be a hyperbolic operator. Then*:

(a) h $\in$ E^s *iff* $\bar{\gamma}^{+}$ *and* $\underline{\gamma}^{+}$ *are less than* 1;

(b) *if* h $\in$ E^s *then* $\bar{\gamma}^{-}$(h) < 1 *for any sequence* $\{h_n\}$ *mentioned in* **19.8**;

(c) $h \in E^u$ *iff there exists a sequence* $\{h_n\}$ *mentioned in* **19.8** *such that* $\gamma^-(h) > 1$;

(d) *if* $h \in E^u$ *then* $\gamma^+ > 1$.

Moreover $\overline{\gamma}^+(h)$ *and* $\gamma^+(h)$ *both lie in the same connected component of* $|\sigma(b)|$ *and also* $\gamma^-(h)$ *and* $\overline{\gamma}^-(h)$ *for any particular* $\{h_n\}$ *defined in* **19.8** *both lie in the same connected component of* $|\sigma(b)|$.

Proof (a) Follows easily from Lemma 19.7 and the argument used in the proof of Theorem 7.2 (see Step I).

Let us prove (b). If $h \in E^s$ and there is no sequence $\{h_n\}$ with the property mentioned in **19.8** then $\overline{\gamma}^-(h) = 0 < 1$.

Now let $\{h_n\}$ be any such sequence. There exists a unique decomposition

$$h_n = h_n^s + h_n^u \ , \qquad h_n^s \in E^s \ , \qquad h_n^u \in E^u \ .$$

By the invariance of E^s and E^u under b one has

$$b^n h_n^s \in E^s$$

$$b^n h_n^u \in E^u$$

which means that

$$b^n h_n^s = h \qquad \text{and} \qquad h_n^u = 0$$

and consequently (19.4) implies (b).

(d) Follows immediately from (19.5).

So let us verify (c). To see this consider an arbitrary vector

$$h = h_0 = h_0^s + h_0^u$$

and let $bh_n = h_{n-1}$. Thus we have $b^n h_n = h_0$ and if

$$h_n = h_n^s + h_n^u$$

we obtain

$$b^n h_n^s = h_0^s$$

$$b^n h_n^u = h_0^u$$

which implies, by Lemma 19.7, that

$$\left\|h_n^s\right\| \geq C_1 \left[\gamma_1^{-1}\right]^n \left\|h_0^s\right\|$$

$$\left\|h_n^u\right\| \leq C_2 [\gamma_1]^n \left\|h_0^u\right\|$$

where $0 < \gamma_1 < 1$ and C_1, C_2 are non–zero constant. So we conclude that

$$\|h_n\| \geq \left\|h_n^s\right\| - \left\|h_n^u\right\| \geq C_1 \left[\gamma_1^{-1}\right]^n \left\|h_0^s\right\| - C_2 [\gamma_1]^n \left\|h_0^u\right\| .$$

Now if $h_0^s \neq 0$ the last inequality shows that

$$\|h_n\|^{\frac{1}{n}} \geq \left[\gamma_1^{-1}\right] \cdot \sqrt[n]{\omega}$$

where ω is some non–zero constant. In turn this implies

$$\|h_n\|^{-\frac{1}{n}} \leq \gamma_1 \cdot \sqrt[n]{\omega}$$

and thus

$$\bar{\gamma}^{-}(h) \le \gamma_1 < 1 \quad .$$

So we obtain the next statement: *if* $h_0^s \neq 0$ *then* $\bar{\gamma}^{-}(h) < 1$ *for any sequence* $\{h_n\}$ *mentioned in* **19.8.**

On the other hand if $h_0^s = 0$ then there exists a sequence $\{h_n\} \in E^u$ such that

$$b^n h_n = h$$

and for this sequence (19.5) implies

$$\gamma^{-}(h) > 1 \quad .$$

So (c) is proved.

To finish the proof let us first of all check that

$$\bar{\gamma}^{+}(h) \in |\sigma(b)| \quad .$$

Indeed, if $\bar{\gamma}^{+}(h) = 0$ then b is not invertible (or it will be $\bar{\gamma}^{+}(h) \ge \frac{1}{\|b^{-1}\|}$) and so $\bar{\gamma}^{+}(h) = 0 \in \sigma(b)$.

So let us consider the situation when $\bar{\gamma}^{+}(h) > 0$. Note first that clearly

$$\bar{\gamma}^{+}(h) \le r(b) \quad .$$

Now suppose that $\bar{\gamma}^{+}(h) = \lambda \,\bar{\in}\, |\sigma(b)|$.

Without loss of generality we can assume that $\lambda < 1$ and

$$[\lambda - \varepsilon, 1] \cap |\sigma(b)| = \emptyset$$

for some $\varepsilon > 0$.

The same argument as the one used in (a) shows that in this case

$h \in E^s$ and since

$$\sigma\left(b\big|_{E^s}\right) = \{t\colon |t| < 1\} \cap \sigma(b)$$

we have

$$r\left(b\big|_{E^s}\right) < \lambda - \varepsilon < \bar{\gamma}^{+}$$

so

$$\bar{\gamma}^{+}(h) \le r\left(b\big|_{E^s}\right) < \bar{\gamma}^{+}(h)$$

which is a contradiction. Thus $\bar{\gamma}^{+}(h) \in |\sigma(b)|$.

To see that $\bar{\gamma}^{+}(h)$ and $\underline{\gamma}^{+}(h)$ lie in the same connected component of $|\sigma(b)|$ let us suppose on the contrary that there exists a $\omega \notin |\sigma(b)|$ such that

$$\underline{\gamma}^{+}(h) < \omega < \bar{\gamma}^{+}(h) \quad .$$

As in the foregoing argument we can set $\omega = 1$. Then

$$\underline{\gamma}^{+}(h) < 1 \Rightarrow h \in E^s \Rightarrow \bar{\gamma}^{+}(h) < 1$$

and the desired statement is proved.

Finally let us verify that $\underline{\gamma}^{-}(h)$ and $\bar{\gamma}^{-}(h)$ lie in the same connected component of $|\sigma(b)|$.

Consider any point $\omega \notin |\sigma(b)|$. Again we can assume that $\omega = 1$. For any h one can consider the following situations.

(1) $h_0^s \ne 0$. In this case, as we know, $\bar{\gamma}^{-}(h) < 1$.

(2) $h_0^s = 0$. Here we can come to the next two cases.

(i) There exists no n_0 such that $h_{n_0}^s \ne 0$. In this case (see the argument in the proof of (c)) we again have

$$\bar{\gamma}(h) < 1 \quad .$$

(ii) $h_n^s = 0$, $n = 1,2,\ldots$. Here (see again the proof of (c)) we have

$$\gamma^-(h) > 1.$$

Thus we conclude that in any case $\bar{\gamma}(h)$ and $\gamma^-(h)$ belong to $|\sigma(b)|$. (It is worth mentioning that

$$\bar{\gamma}(h) \le r(b)$$

which follows from the inequalities

$$\|h\| = \left\|b^n h_n\right\| \le C[r(b)]^n \|h_n\|$$

for some constant C.)

To see that both $\gamma^-(h)$ and $\bar{\gamma}(h)$ lie in the same connected component let us assume that for some sequence $\{h_n\}$ mentioned in **19.8** we have

$$\gamma^-(h) < 1$$

where $1 \bar{\in} |\sigma(b)|$.

As we have already checked, this may occur only when $h_{n_0}^s \ne 0$ for some n_0. But in this case

$$\bar{\gamma}(h) < 1$$

so $\bar{\gamma}(h)$ and $\gamma^-(h)$ should lie in the same connected component of $|\sigma(b)|$. The proof is complete. □

19.10 For any two positive numbers $t_1 \neq t_2$ not belonging to the spectrum $\sigma(b)$ let Δ_{t_1,t_2} be the part of the spectrum lying in the ring $\{\lambda : t_1 < |\lambda| < t_2\}$ and P_{t_1,t_2} be the corresponding Riesz projection, $E_{t_1,t_2} = \operatorname{Im} P_{t_1,t_2}$.

For any positive t not lying in $\sigma(b)$ by Δ_t we denote the part of $\sigma(b)$ lying in the circle $\{\lambda : |\lambda| < t\}$, by P_t the corresponding Riesz projection, and $E_t = \operatorname{Im} P_t$.

To any connected component $\Delta \subset |\sigma(b)|$ there corresponds a subspace E_Δ defined by

$$E_\Delta = \begin{cases} \displaystyle\bigcap_{\substack{t_1,t_2 \notin |\sigma(b)| \\ (t_1,t_2) \supset \Delta}} E_{t_1,t_2} , & \text{if } \Delta \not\ni 0 , \\[2ex] \displaystyle\bigcap_{\substack{t \notin |\sigma(b)| \\ [0,t) \supset \Delta}} E_t , & \text{if } \Delta \ni 0 . \end{cases}$$

The previous lemma gives us a tool for the geometric description of E_t, E_{t_1,t_2}, and E_Δ.

Lemma 19.11

(1) E_{t_1,t_2} *coincides with the closed linear span of vectors* $\mathrm{h} \neq 0$ *such that* $\gamma^+, \bar{\gamma}^+ \in (t_1,t_2)$, $\bar{\gamma}^-(\mathrm{h}) < t_2$ *for every* $\{\mathrm{h}_n\}$ *defined in* 19.8 *and* $\gamma^-(\mathrm{h}) > t_1$ *for some* $\{\mathrm{h}_n\}$.

(2) E_t *coincides with the closed linear span of vectors* $\mathrm{h} \neq 0$ *such that* $\gamma^\pm$, $\bar{\gamma}^\pm < t$.

(3) E_Δ *coincides with the closed linear span of vectors* $\mathrm{h} \neq 0$ *such that*:

(i) $\gamma^\pm, \bar{\gamma}^\pm \in \Delta$ *if* $\Delta \ni 0$;

(ii) $\overset{+}{\gamma}^+, \overset{-}{\bar{\gamma}} \in \Delta$, $\bar{\gamma}(h) \le t_2$ *for every* $\{h_n\}$ *defined in* **19.8** *and* $\gamma^-(h) \ge t_1$ *for some* $\{h_n\}$ *when* $\Delta = [t_1,t_2]$, $t_1 > 0$.

Proof Obvious by the previous lemma and the definitions of E_t, E_{t_1,t_2}, and E_Δ. □

19.12 (Recall 19.5) With the operator aT we shall associate a linear extension β of t (t is the mapping of $\hat{A}$ induced by $\hat{T}$) by the formula

$$\beta(h) = t(\pi)[a](\theta_\pi h) \quad , \qquad h \in H_\pi \quad . \tag{19.6}$$

We can easily link the hyperbolicity of β with the hyperbolicity of aT.

Theorem 19.13 *Let* $\{\hat{T}^z\}$ *act topologically freely on* A. *The operator* aT *is hyperbolic iff the corresponding linear extension is hyperbolic.*

Proof Consider the enveloping algebra $\tilde{A}$ of operators on $H(\hat{A})$ defined by the formula

$$\tilde{a}(h_\pi) = \pi(a)h_\pi \quad , \qquad h_\pi \in H_\pi \quad , \qquad a \in A$$

and the operator θ: $H(\hat{A}) \to H(\hat{A})$:

$$\theta(h_\pi) = \theta_\pi(h_\pi) \quad , \qquad h_\pi \in H_\pi \quad .$$

We have that

$$A \cong \tilde{A}$$

and the formulae (19.3) imply

$$(\theta\tilde{a}\theta^*)h_\pi = \pi(\hat{T}(a))h_\pi \quad .$$

Thus

$$\theta\tilde{a}\theta^* = (\widetilde{TaT^*})$$

and in view of Corollary 12.17 we conclude that

$$C^*(A, T) \cong C^*(\tilde{A}, \theta) \quad . \tag{19.7}$$

But the formula (19.6) for the linear extension β is simply the formula defining the operator $\tilde{a}\theta$ and the condition for β to be hyperbolic is (in view of Lemmas 19.1 and 19.7) just the condition for $\tilde{a}\theta$ to be hyperbolic. Finally by (19.7) $\tilde{a}\theta$ is hyperbolic iff aT is hyperbolic. □

Corollary 19.14 *If* a_0 *is an invertible element in* A *then the operator* $\mathrm{b} = \mathrm{a}_0 + \mathrm{a}_1\mathrm{T}$ *is invertible iff the linear extension* β *of* t *defined by the formula*

$$\beta(\mathrm{h}) = \mathrm{t}(\pi)\left[\mathrm{a}_0^{-1}\mathrm{a}_1\right](\theta_\pi \mathrm{h}) \qquad \mathrm{h} \in \mathrm{H}_\pi$$

is hyperbolic ($\{\hat{\mathrm{T}}^{\mathrm{z}}\}$ *acts topologically freely*).

Proof $b = a_0 + a_1T = a_0\left[I + a_0^{-1}a_1T\right]$ is invertible iff $a_0^{-1}a_1T$ is hyperbolic. □

Remarks

(1) Since aT is hyperbolic iff $(aT)^*$ is hyperbolic we can also obtain the conditions for the hyperbolicity of aT in terms of the corresponding linear extension of t defined similarly to (19.6) by $(aT)^* = \hat{T}^{-1}(a^*)T^*$.

(2) If a_1 is invertible one can pass from b to $bT^* = a_0T^* + a_1$ and find a condition for the invertibility of b analogous to that formulated in Corollary 19.14.

19.15 Now we consider the general case when both a_0 and a_1 may not be invertible. First of all we have to obtain a technical result (Lemma 19.16) which is a generalization of the Riesz spectral decomposition (Theorem 2.14).

Let A and B be linear bounded operators acting from a Banach space H into a Banach space E. Consider a linear operator bundle $F(\lambda) = \lambda A + B$. The *spectrum* of $F(\lambda)$ is the set $\sigma(F)$ of all numbers in the extended complex plane $\overline{\mathbb{C}} = \mathbb{C} \cup \infty$ for which $F(\lambda)$ is not invertible (we mean that $F(\infty)$ is not invertible iff A is not invertible).

Evidently $\sigma(F)$ is a closed non–empty set.

Lemma 19.16 *Let σ_1 and σ_2 be two isolated and mutually supplementary parts of* $\sigma(F)$ *and* $\sigma_1 \not\ni \infty$. *Then there exist the decompositions* $H = H_1 \oplus H_2$ *and* $E = E_1 \oplus E_2$ *and the decomposition* $F(\lambda) = F_1(\lambda) \oplus F_2(\lambda)$ *(where* $F_i(\lambda) = \lambda A_i + B_i$ *and* $A_i\colon H_i \to E_i$, $B_i\colon H_i \to E_i$, $i = 1,2$) *such that* $\sigma(F_i) = \sigma_i$, $i = 1,2$. *The corresponding projections are given by the formulae*

$$P_{H_1} = CA, \qquad P_{E_1} = AC, \qquad P_{H_2} = I - P_{H_1}, \qquad P_{E_2} = I - P_{E_1}$$

where

$$C = \frac{1}{2\pi i} \int_{\partial D} (\lambda A + B)^{-1} d\lambda \quad .$$

Here ∂D *is the oriented boundary (consisting of a finite number of closed rectifiable curves) of an open set* D *of the complex plane for which*

$$D \cap \sigma(F) = \sigma_1 \ ; \qquad [\overline{\mathbb{C}} \setminus \overline{D}] \cap \sigma(F) = \sigma_2 \quad .$$

Moreover we have that A_1 *is invertible and if* 0 *does not belong to* σ_2 *then* B_2 *is invertible and if* 0 *does not belong to* σ_1 *then* B_1 *is invertible.*

Proof First of all we shall prove that

$$CAC = C \quad . \tag{19.8}$$

For any $\lambda \not\in \sigma(F)$ we define C_λ by the formula

$$C_\lambda = (\lambda A + B)^{-1} \quad .$$

Note that for any $\lambda,\gamma \,\bar{\in}\, \sigma(F)$ we have

$$C_\lambda A C_\gamma = \frac{C_\lambda - C_\gamma}{\gamma - \lambda} \quad . \tag{19.9}$$

Indeed

$$\begin{aligned} C_\lambda &= C_\lambda(\gamma A + B)C_\gamma \\ &= C_\lambda((\gamma - \lambda)A + \lambda A - B)C_\gamma \\ &= (\gamma - \lambda)C_\lambda A C_\gamma + C_\gamma \end{aligned}$$

which implies (19.9).

As the distance from σ_1 to ∂D is positive there exists another open set D' such that $\sigma_1 \subset D' \subset D$ and $\partial D'$ is in D. Since C_λ is an analytic function in $\mathbb{C}\backslash\sigma(F)$ we have that

$$\frac{1}{2\pi i}\int_{\partial D'} C_\gamma d\gamma = \frac{1}{2\pi i}\int_{\partial D} C_\lambda d\lambda = C \quad .$$

Now

$$CAC = \frac{1}{(2\pi i)^2}\int_{\partial D}\int_{\partial D'} C_\lambda A C_\gamma d\lambda d\gamma = \frac{1}{(2\pi i)^2}\int_{\partial D}\int_{\partial D'} \frac{C_\lambda - C_\gamma}{\gamma - \lambda}\, d\lambda d\gamma$$

$$= \frac{1}{(2\pi i)^2}\left[\int_{\partial D} C_\lambda d\lambda \int_{\partial D'} \frac{1}{\gamma - \lambda}\, d\gamma - \int_{\partial D'} C_\gamma d\gamma \int_{\partial D} \frac{1}{\gamma - \lambda}\, d\lambda\right]$$

$$= C$$

where we used (19.9) and the fact that

$$\int_{\partial D'} \frac{1}{\gamma - \lambda}\, d\gamma = 0 \qquad \text{and} \qquad \int_{\partial D} \frac{1}{\gamma - \lambda}\, d\lambda = -2\pi i$$

since λ is not in $\overline{D'}$ and γ is in D.

So (19.8) is proved. It implies that

$$CACA = CA \qquad \text{and} \qquad ACAC = AC \quad .$$

Thus $P = CA$ and $Q = AC$ are the projections. Now let us denote

$$H_1 = P(H) , \qquad H_2 = (I-P)H , \qquad E_1 = Q(E) , \qquad E_2 = (I-Q)E .$$

The definition of P and Q gives us

$$AP = QA \qquad \text{and} \qquad CQ = PC \quad .$$

Consequently

$$H_1 \xrightarrow{A} E_1 \xrightarrow{C} H_1 \xrightarrow{A} E_1$$

and as AC and CA are bijections it follows that A and C are both bijections here. This means that $A_1 = AP$ is invertible (as an operator from H_1 to E_1) and CQ is invertible (as an operator from E_1 to H_1).

Since

$$A(I - P) = (I - Q)A \qquad \text{and} \qquad C(I - Q) = (I - P)C$$

we conclude also that $A_2 = A(I - P)$ acts from H_2 into E_2 and $C(I - Q)$ acts from E_2 into H_2.

Note now that for an arbitrary ξ and for any $\eta \,\bar{\in}\, \sigma(F)$ we have

$$(\xi A + B)C_\eta = [(\xi - \eta)A + \eta A + B]C_\eta = (\xi - \eta)AC_\eta + I \quad (19.10)$$

and in an analogous way

$$C_\eta(\xi A + B) = (\xi - \eta)C_\eta A + I \quad . \quad (19.11)$$

For any $\xi \,\bar{\in}\, \partial D$ let us define the operator Φ_ξ by the formula

$$\Phi_\xi = \frac{1}{2\pi i} \int_{\partial D} \frac{C_\lambda}{\lambda - \xi} \, d\lambda \quad . \quad (19.12)$$

We have

$$(\xi A + B)\Phi_\xi = \frac{1}{2\pi i}\int_{\partial D} \frac{(\xi A + B)C_\lambda}{\lambda - \xi}\, d\lambda$$

$$= \frac{1}{2\pi i}\int_{\partial D} \frac{(\xi - \lambda)AC_\lambda + I}{\lambda - \xi}\, d\lambda = -AC + \frac{1}{2\pi i}\int_{\partial D} \frac{d\lambda}{\lambda - \xi} \quad (19.13)$$

where we used (19.10).

Thus (19.13) implies

$$(\xi A + B)\Phi_\xi = \begin{cases} -Q, & \text{when } \xi \text{ is not in } \overline{D} \\ I - Q, & \text{when } \xi \text{ is in } D. \end{cases} \quad (19.14)$$

Similarly

$$\Phi_\xi(\xi A + B) = \begin{cases} -P, & \text{when } \xi \text{ is not in } \overline{D} \\ I - P, & \text{when } \xi \text{ is in } D. \end{cases} \quad (19.15)$$

Note that as $[\mathbb{C}\backslash\sigma(F)]$ is an open set we can always assume that $0 \,\bar{\in}\, \partial D$, and it follows from (19.14) and (19.15) that

$$B\Phi_0 = \begin{cases} -Q, & \text{when } 0 \text{ is not in } \overline{D} \\ I - Q, & \text{when } 0 \text{ is in } D. \end{cases} \quad (19.16)$$

$$\Phi_0 B = \begin{cases} -P, & \text{when } 0 \text{ is not in } \overline{D} \\ I - P, & \text{when } 0 \text{ is in } D. \end{cases} \quad (19.17)$$

From (19.16) and (19.17) we conclude that

$$BP = QB, \qquad B(I - P) = (I - Q)B$$

and

$$\Phi_0 Q = P\Phi_0, \qquad \Phi_0(I - Q) = (I - P)\Phi_0.$$

This means that B_i acts from H_i into E_i, $i = 1,2$ (where $B_1 = BP$, $B_2 = B(I - P)$), and Φ_{0i} acts from E_i into H_i, $i = 1,2$ (where $\Phi_{01} = \Phi_0 Q$, $Q_{02} = \Phi_0(I - Q)$).

All these arguments and the formulae (19.14) and (19.15) imply that

$$\sigma(F_i) = \sigma_i \qquad i = 1,2$$

where $F_i(\lambda) = \lambda A_i + B_i$.

Moreover, if 0 does not belong to σ_2 we can always choose D in such a way that $0 \in D$ and in view of (19.16), (19.17) B_2 is invertible in this case. If 0 does not belong to σ_1 then we can always choose D such that $0 \bar{\in} \bar{D}$ and again by (19.16), (19.17) we have that in this case B_1 is invertible. The proof is complete. □

19.17 A pair (A, B) of operators (acting from H to E) is said to be *hyperbolic* if the spectrum $\sigma(F)$ $(F(\lambda) = \lambda A + B)$ does not meet the unit circle $|\lambda| = 1$.

Remarks

(1) Clearly (A, B) is hyperbolic iff (B, A) is hyperbolic.

(2) An operator b is hyperbolic (see **19.6**) iff the pair (I, b) is hyperbolic.

The following lemma is a generalization of Lemma 19.1 and in fact is the main technical result of this section.

Lemma 19.18 *Let* $C^*(A, T)$ *possesses the property* (*). *Suppose that the operator* $b = a_0 + a_1T$ *is invertible. Then the pair* (a_0, a_1T) *is hyperbolic and the projections*

$$P = Ca_0 \qquad \textit{and} \qquad Q = a_0C$$

where

$$C = \frac{1}{2\pi i} \int_{|\lambda|=1} (\lambda a_0 + a_1T)^{-1} d\lambda$$

belong to A.

Proof Since b is invertible and $C^*(A, T)$ possesses the property (*) it follows (in view of Corollary 13.6) that the pair (a_0, a_1T) is hyperbolic.

By Lemma 19.16 P and Q are projections and $a_{01}(a_{0i}: H_i \to E_i$ where $H_1 = PH$, $H_2 = (I - P)H$, $E_1 = QH$, $E_2 = (I - Q)H$, $a_{01} = a_0 P$, $a_{02} = a_0(I - P))$ is invertible and $a_{12}T_2(a_{1i}T_i: H_i \to E_i$, where $a_{11}T_1 = (a_1 T)P$ and $(a_{12}T_2) = (a_1 T)(I - P))$ is invertible.

Since the spectrum $\sigma(F_1)$ $(F_1(\lambda) = \lambda a_{01} + a_{11}T)$ lies in the unit circle and a_{01} is invertible one can describe the subspace H_1 in an invariant geometric way. Here is the description.

Let $\xi_0 \in H_1$. As a_{01} is invertible there exists a unique vector $\xi_1 \in H_1$ such that

$$a_0 \xi_1 = a_1 T \xi_0 \quad .$$

Furthermore there exists a unique vector $\xi_2 \in H_1$ such that

$$a_0 \xi_2 = a_1 T \xi_1$$

and so on: for any $n = 0,1,\ldots$ there exists a unique vector $\xi_{n+1} \in H_1$ such that

$$a_0 \xi_{n+1} = a_1 T \xi_n \quad .$$

Clearly

$$\xi_n = \left(a_{01}^{-1}\, a_{11} T_1\right)^n \xi_0 \quad . \tag{19.18}$$

Since the spectrum $\sigma(F_1)$ lies in the unit circle we have that the spectrum $\sigma(a_{01}^{-1}\, a_{11} T_1)$ also lies in the unit circle. Thus

$$\|\xi_n\| \to 0 \quad . \tag{19.19}$$

Now let ξ_0 be an arbitrary vector in H. Consider the equation

$$a_0\xi_1 = a_1T\xi_0 \quad . \tag{19.20}$$

We have the following possibilities:

(1) equation (19.20) has no solution;
(2) the equation has more than one solution;
(3) there exists a unique solution ξ_1 of equation (19.20).

In both cases (1) and (2) (in view of the previous argument) $\xi_0 \bar{\in} H_1$.

Consider case (3) and the next equation:

$$a_0\xi_2 = a_1T\xi_1 \quad . \tag{19.21}$$

Just as in equation (19.20) we may obtain the same three possibilities. Again in cases (1) and (2) we conclude

$$\xi_0 \bar{\in} H_1 \quad .$$

So, continuing this procedure we have to consider the following case: for any $n = 0,1,...$ there exists a unique vector ξ_{n+1} such that

$$a_0\xi_{n+1} = a_1T\xi_n \quad .$$

Let $\xi_n = \xi_n^1 + \xi_n^2$ where $\xi_n^i \in H_i$ (this decomposition is unique). Since a_{01} and $a_{12}T_2$ are invertible we have that

$$\xi_n^1 = \left(a_{01}^{-1}\, a_{11}T_1\right)^n \xi_0^1 \tag{19.22}$$

and

$$\left[(a_{12}T_2)^{-1}a_{02}\right]^n \xi_n^2 = \xi_0^2 \quad . \tag{19.23}$$

As in the foregoing argument (see (19.18) and (19.19)) using the fact that $\sigma(F_1)$ lies in the unit circle and $\sigma(F_2)$ lies outside of the unit circle we conclude

$$\left\|\xi_n^1\right\| \to 0 \tag{19.24}$$

and (if $\xi_0^2 \neq 0$)

$$\left\|\xi_n^2\right\| \to \infty \quad . \tag{19.25}$$

Thus if $\xi_0^2 \neq 0$ we have

$$\|\xi_n\| \geq \left\|\xi_n^2\right\| - \left\|\xi_n^1\right\| \to \infty \quad .$$

Finally we have the following description of H_1: $\xi_0 \in H_1$ iff the equation

$$a_0\xi_{n+1} = a_1 T\xi_n \tag{19.26}$$

has a unique solution for any $n = 0,1,\ldots$ and

$$\|\xi_n\| \to 0 \quad .$$

Since $C^*(A, T)$ possesses the property (*) we may assume (as we did in the proof of Lemma 19.1) that $C^*(A, T) = \tilde{\pi}(A \underset{\hat{T}}{\times} \mathbb{Z}) \subset L(l^2(\mathbb{Z}, H))$.

Let P_k, $k \in \mathbb{Z}$, be the projections introduced in the proof of Lemma 19.1. In order to prove that $P = Ca_0$ belongs to A it suffices (as in the proof of Lemma 19.1) to show that

$$P_0 P = PP_0 \quad . \tag{19.27}$$

Let $\overline{H}_n = \operatorname{Im} P_n$, $n = 0,\pm 1,\ldots$. Then we have that

$$a_0\colon \ \overline{H}_n \to \overline{H}_n \tag{19.28}$$

$$a_1T\colon \overline{H}_n \to \overline{H}_{n+1} \quad . \tag{19.29}$$

Let $\xi_n = \{\xi_{nk}\}$, $n = 0,1,\ldots$, $k = 0,\pm 1,\ldots$ ($\xi_{nk} \in \overline{H}_k$ are the coordinates of ξ_n). In view of (19.28) and (19.29), equation (19.26) means that

$$a_0\xi_{(n+1)(k+1)} = a_1T\ \xi_{nk} \quad .$$

Thus (by the properties of P_k) the equation

$$a_0y = a_1T\ P_k\xi_n$$

has a unique solution for any n and k (namely $y = P_{k+1}\xi_{k+1} = \xi_{(n+1)(k+1)}$) and since $\|\xi_n\| \to 0$ implies that

$$\|\xi_{nk}\| \to 0 \qquad n \to \infty$$

we conclude that if $\eta_0 \in H_1$ then $P_k\eta_0 \in H_1$ for any k, that is

$$P_kP = PP_kP \quad . \tag{19.30}$$

In an analogous way one can verify that

$$P_k(I - P) = (I - P)P_k(I - P) \quad . \tag{19.31}$$

(In order to make the argument verifying (19.31) completely analogous to that in the proof of (19.30) one can pass first to the operator $bT^* = a_0T^* + a_1$ and then after going through the previous argument return back to b.)

Now (19.30) and (19.31) imply that

$$P_kP = PP_k \quad .$$

Thus (19.27) is proved and consequently $P \in A$.

In order to prove that $Q = a_0C$ belongs to A recall that

$$E_1 = QH = a_0H_1 \tag{19.32}$$

(see the proof of Lemma 19.16). In view of (19.32) we have that $\xi \in E_1$ iff $\xi = a_0\eta$ for some $\eta \in H_1$, and so (since $P_k \in A'$)

$$P_k\xi = P_ka_0\eta = a_0P_k\eta \quad .$$

But we have already proved that $P_k\eta \in H_1$, therefore we conclude

$$P_k\xi \in E_1 \qquad \text{if} \quad \xi \in E_1 \quad .$$

This means that

$$P_kQ = QP_kQ \quad . \tag{19.33}$$

In a similar way one can find that

$$P_k(I - Q) = (I - Q)P_k(I - Q) \tag{19.34}$$

and (19.33) and (19.34) imply, as (19.30) and (19.31), that $Q \in A$. We have finished. □

19.19 With the operator $b = a_0 + a_1T$ we shall associate the *relation* ρ on $H(\hat{A})$ (that is, a set of pairs $(z,y) \in H(\hat{A}) \times H(\hat{A})$) in the following way: if $z \in H_{\pi_1}$ and $y \in H_{\pi_2}$ then

$$(z,y) \in \rho \Leftrightarrow \begin{cases} \pi_1 = t^{-1}(\pi_2) \\ \pi_2(a_0)y = \pi_2(a_1)(\theta_{\pi_1}z) \end{cases}$$

where θ_π is defined by (19.3)

Remarks

(1) If a_0 is invertible then this relation is simply the graph of the linear extension β defined in Corollary 19.14.

(2) The relation ρ corresponding to the operator b does not change if one multiplies b by an invertible element from A.

(3) The relation ρ is linear in the following way: the intersection of ρ with every product of fibres $H_\pi \times H_{t(\pi)}$ is a linear subspace in this product.

19.20 The relation ρ is said to be *hyperbolic* (with respect to the algebra A) if there exists a projection $P \in A$ such that for the subspaces $H^s = \bigoplus_{\pi\in\hat{A}} \operatorname{Im} \pi(P)$ and $H^u = \bigoplus_{\pi\in\hat{A}} \operatorname{Ker} \pi(P)$ there exist mappings $\beta_s\colon H^s \to H^s$ and $\beta_u\colon H^u \to H^u$ satisfying the following conditions:

(1) β_s is a linear extension of t and H^s is stable under β_s (see (19.1));

(2) β_u is a linear extension of t^{-1} and H^u is stable under β_u;

(3) the relation ρ is the direct sum of graphs β_s and β_u; that is, a pair (z,y), where $z = (z_s,z_u)$ and $y = (y_s,y_u)$, belongs to ρ iff $y_s = \beta_s z_s$ and $z_u = \beta_u y_u$;

(4) for any $\pi \in \hat{A}$

$E^s_\pi = \pi(a_0)H^s_\pi$ is a closed subspace in H_π

$$\left(H^s_\pi = \pi(P)H_\pi\right)$$

$E^u_\pi = \pi(a_1)\theta_{t^{-1}(\pi)} H^u_{t^{-1}(\pi)}$ is a closed subspace in H_π

$$\left(H^u_\pi = \pi(I - P)H_\pi\right) \quad ;$$

(5) $E^u_\pi + E^s_\pi = H_\pi$ that is, for every $\xi \in H_\pi$ there exist $x_s \in H^s_\pi$ and $x_u \in H^u_{t^{-1}(\pi)}$ such that $\pi(a_0)x_s + \pi(a_1)\,\theta_{t^{-1}(\pi)}x_u = \xi$.

Theorem 19.21 *Let* $\{\hat{T}^z\}$ *act topologically freely on* A. *The operator* $b = a_0 + a_1T$ *is invertible iff the corresponding relation* ρ *is hyperbolic.*

Proof b is invertible iff the operator $\tilde{b} = \tilde{a}_0 + \tilde{a}_1\theta$ (see the proof of Theorem 19.13) is invertible and $\tilde{b}$ is invertible iff the pair $(\tilde{a}_0, \tilde{a}_1\theta)$ is hyperbolic. Now we have to prove that the conditions (1)–(5) are the conditions of the hyperbolicity of $(\tilde{a}_0, \tilde{a}_1\theta)$.

I. Suppose (1)–(5) are satisfied. Note first that

$$E^s_\pi \cap E^u_\pi = \{0\} \quad . \tag{19.35}$$

Indeed if $E^s_\pi \cap E^u_\pi \ni v \neq 0$ then there exist $x_s \in H^s_\pi$ and $x_u \in H^u_{t^{-1}(\pi)}$ such that

$$\pi(a_0)x_s = v = \pi(a_1)\theta_{t^{-1}(\pi)}x_u \quad .$$

As $v \neq 0$ both x_s and x_u are non–zero and by the definition of ρ, $(x_u, x_s) \in \rho$. The decomposition $H = H^s \oplus H^u$ gives us the decomposition $x_s = (x_s, 0)$ and $x_u = (0, x_u)$. Now condition (3) implies $x_s = \beta_s 0 = 0$ and $x_u = \beta_u 0 = 0$ and consequently $v = 0$. Thus we have arrived at a contradiction.

From (19.35) and condition (5) we conclude

$$E^s_\pi \oplus E^u_\pi = H_\pi \quad . \tag{19.36}$$

Now let us show that $\pi(a_0)$: $H^s_\pi \to E^s_\pi$ is an invertible operator. Indeed let $x_s \in H^s_\pi$ and $\pi(a_0)x_s = 0$. Then $(x_s, 0) \in \rho$ and it follows from (3) that $x_s = 0$. Thus $\pi(a_0)$: $H^s_\pi \to E^s_\pi$ is injective and in view of the definition of E^s_π, $\pi(a_0)$ is in fact bijective. Similarly one can show that $\pi(a_1)\theta_{t^{-1}(\pi)}: H^u_{t^{-1}(\pi)} \to E^u_\pi$ is invertible.

Denote $E^s = \underset{\pi}{\oplus} E^s_\pi$, $E^u = \underset{\pi}{\oplus} E^u_\pi$. The foregoing argument proves that

$$H = E^s \oplus E^u$$

and $\tilde{a}_0$: $H^s \to E^s$ and $\tilde{a}_1\theta$: $H^u \to E^u$ are invertible.

Let us verify that $\tilde{a}_0$ maps H^u into H^u and $\tilde{a}_1\theta$ maps H^s into H^s.

Indeed take any $y_u \in H^u_\pi$ and let

$$\pi(a_0)y_u = x_s + x_u \qquad x_s \in E^s_\pi \ , \quad x_n \in E^n_\pi \quad .$$

Then there exists a $y_s \in H^s_\pi$ such $\pi(a_0)y_s = -x_s$ and consequently

$$\pi(a_0)(y_u + y_s) = x_u = \pi(a_1)\theta_{t^{-1}(\pi)}z_u$$

where z_u is some element in $H^u_{t^{-1}(\pi)}$.

Thus $(z_u, y_s + y_u) \in \rho$ and in view of condition (3) we conclude (as $z_s = 0$) that

$$y_s = \beta_s z_s = 0$$

and so $x_s = -\pi(a_0)y_s = 0$ which means that

$$\pi(a_0): H^u_\pi \to E^u_\pi \quad .$$

In an analogous way one can show that $\pi(a_1)\theta_{t^{-1}(\pi)}$: $H^s_{t^{-1}(\pi)} \to E^s_\pi$

which implies

$$\tilde{a}_0: \quad H^u \to E^u$$

$$\tilde{a}_1\theta: H^s \to E^s \quad .$$

So we obtain two decompositions $H = H^s \oplus H^n$ and $H = E^s \oplus E^u$ and the corresponding decomposition of $\tilde{b}(\lambda) = \lambda\tilde{a}_0 + \tilde{a}_1\theta$, $\tilde{b}(\lambda) = \tilde{b}_1(\lambda) \oplus \tilde{b}_2(\lambda)$ where

$$\tilde{b}_1(\lambda) = (\lambda\tilde{a}_{01} + \tilde{a}_{11}\theta_1): H^s \to E^s$$

$$\tilde{b}_2(\lambda) = (\lambda\tilde{a}_{02} + \tilde{a}_{12}\theta_2): H^u \to E^u$$

and $\tilde{a}_{01}$ and $\tilde{a}_{12}\theta_2$ are invertible.

Note that

$$(z,y) \in \rho \Leftrightarrow \tilde{a}_0 y = \tilde{a}_1\theta z$$

and if $z = (z_s, z_u)$, $y = (y_s, y_u)$ then in view of the foregoing argument we have

$$y_s = \tilde{a}_{01}^{-1}\, \tilde{a}_{11}\theta_1 z_s \tag{19.37}$$

$$z_u = (\tilde{a}_{12}\theta_2)^{-1} a_{02} y_u \quad . \tag{19.38}$$

In fact one can choose z_s and y_u to be arbitrary elements in H^s and H^u and then y_s and z_u should be defined by (19.37) and (19.38). Using condition (3) we conclude

$$\beta_s = \tilde{a}_{01}^{-1}\, \tilde{a}_{11}\theta_1$$

$$\beta_u = (\tilde{a}_{12}\theta_2)^{-1} a_{02}$$

and conditions (1) and (2) imply that both the spectral radii of $\tilde{a}_{01}^{-1}\tilde{a}_{11}\theta_1$ and $(\tilde{a}_{12}\theta_2)^{-1}a_{02}$ are less than 1, which means that the spectrum of $\tilde{b}_1(\lambda)$ lies in the unit circle while the spectrum of $\tilde{b}_2(\lambda)$ lies outside of the unit circle. Thus $(\tilde{a}_0, \tilde{a}_1\theta)$ is a hyperbolic pair.

II. Now let the pair $(\tilde{a}_0, \tilde{a}_1\theta)$ be hyperbolic. Consider the spectral projections $\tilde{P} = \tilde{C}\tilde{a}_0$ and $\tilde{Q} = \tilde{a}_0\tilde{C}$ defined by Lemma 19.18 (both of them are in $\tilde{A}$). Denote

$$H^s = \tilde{P}H\ , \qquad H^u = (I-\tilde{P})H\ , \qquad E^s = \tilde{Q}H\ , \qquad E^u = (I-\tilde{Q})H$$

$$\beta^s = \tilde{a}_{01}^{-1}\tilde{a}_{11}\theta_1\ , \qquad \beta^u = (\tilde{a}_{12}\theta_2)^{-1}a_{02}\ .$$

Routine computation now shows that all the conditions (1)–(5) are satisfied. □

Remark Along with the operator $b = a_0 + a_1T$ one can also consider the operator $bT^* = a_0T^* + a_1$ and the operator $b^* = a_0^* + \hat{T}^{-1}(a_1^*)T^*$ and since all of these operators are invertible or not invertible simultaneously we can define the conditions for invertibility of b in terms of the relation ρ_1 corresponding to bT^* or ρ_2 corresponding to b^*.

The idea of hyperbolicity is actually very useful when investigating the spectrum of $b(\lambda) = \lambda a_0 + a_1T$ and in the rest of the paragraph with the help of this idea we shall describe some spectral properties of $b(\lambda)$.

Theorem 19.22 *Suppose that* $C^*(A, T)$ *possesses the property* (*). *Let* $\sigma(b)$ *be the spectrum of* $b(\lambda)$. *To any connected component* Δ *of* $\sigma(b)$ *there correspond the subspaces* H_Δ *and* E_Δ *such that*

$$b(\lambda):\ H_\Delta \to E_\Delta$$

and

(1) *if* $H_\Delta \neq \{0\}$ *then* $\sigma(b)|_{H_\Delta} \subset \Delta$;

(2) *if* Δ *is an open and closed set (in the relative topology on* $\sigma(b)$*) then both the projections*

$$P_\Delta: H \to H_\Delta \qquad \textit{and} \qquad Q_\Delta: H \to E_\Delta$$

belong to A *and* $\sigma(b)|_{H_\Delta} = \Delta$;

(3) $H = \underset{\Delta\in\Lambda}{\oplus} H_\Delta = \underset{\Delta\in\Lambda}{\oplus} E_\Delta$ *where* Λ *is the set of all connected components of* $\sigma(b)$.

Proof Let Δ be some connected component of $\sigma(b)$. Then (as $\sigma(b)$ is closed and invariant under rotations) Δ has the form

$$\Delta = \{\lambda \in \mathbb{C}: \alpha \leq |\lambda| \leq \beta\}$$

where $\alpha,\beta \in [0, +\infty]$.

I. Suppose first that $\alpha \neq 0$, $\beta \neq +\infty$. One can obviously, choose closed intervals $[\alpha_i, \beta_i]$, $i = 1,2,\ldots$, such that the points α_i, β_i do not belong to $\sigma(b)$ and

$$\Delta = \bigcap_{i\geq 1} \Delta_i \tag{19.39}$$

where $\Delta_i = \{\lambda: \alpha_i \leq |\lambda| \leq \beta_i\}$.

For any $\gamma \in (0, +\infty)$, $\gamma \,\overline{\in}\, \sigma(b)$, we denote by C_γ the operator

$$C_\gamma = \frac{1}{2\pi i} \int_{|\lambda|=\gamma} b(\lambda)^{-1} d\lambda \quad .$$

In view of Lemmas 19.16 and 19.18 the projections

$$P_\gamma = C_\gamma a_0 \qquad \text{and} \qquad Q_\gamma = a_0 C_\gamma \tag{19.40}$$

both belong to A and for the subspaces

$$H_{1\gamma} = P_\gamma H\ , \qquad H_{2\gamma} = (I-P_\gamma)H\ , \qquad E_{1\gamma} = Q_\gamma H\ , \qquad E_{2\gamma} = (I-Q_\gamma)H \tag{19.41}$$

we have

$$b(\lambda):\ H_{k\gamma} \to E_{k\gamma} \qquad k = 1,2 \tag{19.42}$$

and

$$\sigma(b)|_{H_{k\gamma}} = \sigma(b) \cap D_\gamma^k \qquad k = 1,2 \tag{19.43}$$

where $D_\gamma^1 = \{\lambda\colon |\lambda| < \gamma\}$, $D_\gamma^2 = \{\lambda\colon |\lambda| > \gamma\}$.

Let $H_{\Delta_i} = (P_{\beta_i} - P_{\alpha_i})H$ and $E_{\Delta_i} = (Q_{\beta_i} - Q_{\alpha_i})H$. If Δ is an open and closed set then the distance from Δ to $\{\sigma(b)\backslash\Delta\}$ is positive and one can choose Δ_1 in such a way that $\Delta_1 \cap \sigma(b) = \Delta$.

Put $H_\Delta = H_{\Delta_1}$, $E_\Delta = E_{\Delta_1}$ and the foregoing argument shows that (2) is true.

In general put

$$H_\Delta = \bigcap_{i\ge1} H_{\Delta_i}\ , \qquad E_\Delta = \bigcap_{i\ge1} E_{\Delta_i}$$

(clearly H_Δ and E_Δ do not depend on the particular choice of Δ_i in (19.39)).

Suppose that $H_\Delta \neq \{0\}$. Via the construction we have

$$b(\lambda):\ H_\Delta \to E_\Delta\ .$$

By Lemma 19.18 we have that both

$$a_0:\ H_{\Delta_i} \to E_{\Delta_i}$$

$$a_1T\colon H_{\Delta_i} \to E_{\Delta_i}$$

are bijections and

$$\sigma(b)|_{H_{\Delta_i}} = \sigma(b) \cap \Delta_i \quad . \tag{19.44}$$

Thus both

$$a_0\colon \ H_\Delta \to E_\Delta$$

$$a_1T\colon H_\Delta \to E_\Delta$$

are bijections.

Consider any $\lambda_0 \bar{\in} \Delta$ and let $|\lambda_0| > \beta$. Then there exists a β_{i_0} such that $\beta < \beta_{i_0} < |\lambda_0|$. Without loss of generality we can assume that $\beta_{i_0} = 1$. In view of (19.44) the operator $\left(a_0^{-1}a_1T\right)\colon H_{\Delta_{i_0}} \to H_{\Delta_{i_0}}$ is hyperbolic and in fact (by Lemma 19.7)

$$\left\|\left(a_0^{-1}a_1T\right)^n h\right\| \le C\ \gamma^n\|h\| \qquad h \in H_{\Delta_{i_0}}, \quad n = 1,2,\ldots$$

where $\gamma < 1$. So we conclude that (since $H_\Delta \subset H_{\Delta_{i_0}}$) the spectral radius of $\left(a_0^{-1}a_1T\right)\Big|_{H_\Delta}$ is less than 1 and consequently $\lambda_0 \bar{\in} \sigma(b)|_{H_\Delta}$. In an analogous way (using the invertibility of $a_1T\colon H_\Delta \to E_\Delta$) it can be shown that for any $|\lambda_0| < \alpha$, $\lambda_0 \bar{\in} \sigma(b)|_{H_\Delta}$. Thus (1) is true.

II. Now let $\alpha = 0$, $\beta < +\infty$. In this case we can choose points β_i, $i = 1,2,\ldots$, such that $\beta_i \bar{\in} \sigma(b)$ and

$$\Delta = \bigcap_{i\geq 1} \Delta_i$$

where $\Delta_i = \{\lambda : |\lambda| \leq \beta_i\}$.

Let

$$H_{\Delta_i} = P_{\beta_i}H , \qquad E_{\Delta_i} = Q_{\beta_i}H \qquad \text{and} \qquad H_\Delta = \bigcap_{i\geq 1} H_{\Delta_i} .$$

This case can now be considered in an analogous way to the previous one.

III. $\alpha > 0$, $\beta = +\infty$ is similar to the foregoing.

IV. $\alpha = 0$, $\beta = +\infty$ is trivial with $H_\Delta = H$.

(3) is evident via the construction of H_Δ and E_Δ.

The proof of the theorem is finished. □

Remark If Δ is not an open and closed set in $\sigma(b)$ then it may happen that

(1) $H_\Delta = \{0\}$

(2) $\sigma(b)|_{H_\Delta} \neq \Delta$.

Examples

(1) Consider the space $H = l^2(\mathbb{Z}, l^2)$. Let A be the algebra of operator–valued functions $a = \{a(z)\}_{z\in\mathbb{Z}}$, $a_z \in L(l^2)$, having (the uniform) operator limits at $\pm\infty$. Define the operator T by

$$(Tf)(z) = f(z - 1) \quad , \qquad f \in H \ , \quad z \in \mathbb{Z} \ .$$

Thus

$$(TaT^*)(z) = a(z - 1) \quad , \qquad a \in A$$

and clearly $C^*(A, T)$ possesses the property (*).

Consider the operator $b = a_0 + a_1T$ where $a_0 = I$ and $a_1 = a_1(z)$ is defined in the following way: all $a_1(z)$, $z \in \mathbb{Z}$, are operators on l^2 of multiplication by the functions $a_{1z}(n)$, $n \in \mathbb{N}$, where

(1) $\forall z, \forall n \ a_{1z}(n) \neq 0$,

(2) $a_{10}(n) \xrightarrow[n\to\infty]{} 0$,

(3) $a_{1z}(n) = \alpha_n$, $z \neq 0$,

where α_n is a decreasing sequence of positive numbers and $\alpha_n \searrow \alpha$, $\alpha > 0$.

Consider now the space $l^2(\mathbb{N} \times \mathbb{Z}) \cong l^2(\mathbb{Z}, l^2)$ and the algebra $\mathcal{A}$ of operators of multiplication by functions $a(n, z)$ having limits

$$\lim_{n\to\infty} a(n, z) = a(\infty, z)$$

$$\lim_{z\to\pm\infty} a(n, z) = a(n, \pm\infty)$$

$$\lim_{\substack{n\to\infty \\ z\to\pm\infty}} a(n, z) = a(\infty, \pm\infty) \quad .$$

Clearly $a_1 \in \mathcal{A}$ and $T\mathcal{A}T^* = \mathcal{A}$. But $\mathcal{A}$ is a commutative C^*–algebra and one can easily compute the spectrum $\sigma(b) = \sigma(a_1T)$ with the help of the results of Sections 4, 7 and 17.

Actually

$$\sigma(b) = \bigcup_{k\in N\cup\infty} \Delta_k \tag{19.45}$$

where

$$\Delta_k = \{\lambda: |\lambda| = \alpha_k\} \qquad k \in \mathbb{N}$$

$$\Delta_\infty = \{\lambda: |\lambda| \leq \alpha\} \quad .$$

Evidently $H_{\Delta_k} = \{f \in l^2(\mathbb{N} \times \mathbb{Z}): f(n, z) = 0 \ n \neq k\}$. Consequently $\bigoplus_{k\in\mathbb{N}} H_{\Delta_k} = H$ and thus $H_{\Delta_\infty} = \{0\}$.

(2) Consider $H' = H \oplus l^2(\mathbb{Z})$ where H is the space from the previous example. The action T on $l^2(\mathbb{Z})$ will be defined as

$$(Tf)(z) = f(z - 1)$$

and let $a_1 = 0$ on $l^2(\mathbb{Z})$. Now let $b = b_1 \oplus b_2$ where b_1 is the operator on H described in the previous example and $b_2 = I + a_1T = I + 0$ on $l^2(\mathbb{Z})$. Obviously $\sigma(b)$ is described by (19.45) and $H_{\Delta_\infty} = l^2(\mathbb{Z})$ while $\sigma(b)|_{\Delta_\infty} = 0 \neq \Delta_\infty$. □

19.23 In the case when all $\pi(a_0)$, $\pi(a_1)$, $\pi \in \hat{A}$, are compact operators we can effectively describe the dimensions of the spectral subspaces (arising from Theorem 19.22). If Δ is an open and closed set in $\sigma(b)$ then by $H_{\Delta\pi}$, $E_{\Delta\pi}$ we shall denote the subspaces of H_π defined by

$$H_{\Delta\pi} = \pi(P_\Delta)H_\pi \quad , \qquad E_{\Delta\pi} = \pi(Q_\Delta)H_\pi \quad .$$

Theorem 19.24 *Suppose that* $C^*(A, T)$ *possesses the property* (*). *If all* $\pi(a_0)$, $\pi(a_1)$, $\pi \in \hat{A}$, *are compact operators and* $\sigma(b) \neq \mathbb{C}$ *then* $\dim H_\pi < \infty$ *for every* $\pi \in \hat{A}$ *and* $\dim H_{\Delta_\pi}$ *and* $\dim E_{\Delta_\pi}$ *are* t*–invariant functions on* $\hat{A}$.

Proof Take any connected component in $\mathbb{C}\backslash\sigma(b)$ and any circle $\{|\lambda| = \alpha\}$ lying in it. Let Δ_1 be the part of $\sigma(b)$ lying in this circle and Δ_2 be its complement (Δ_1 or Δ_2 may be empty).

Let $\pi \in \hat{A}$ be any irreducible representation of A with a representation space H_π and $\nu \in \widehat{C^*(A, T)}$ be its extension (Theorem 2.81) with $H_\nu \supset H_\pi$.

By Lemmas 19.16 and 19.18 we have that

$$\nu(a_0P_{\Delta_1}): \mathrm{Im}\ \nu(P_{\Delta_1}) \to \mathrm{Im}\ \nu(Q_{\Delta_1})$$

is a bijection.

Since

$$\nu(A)|_{H_\pi} = \pi(A)|_{H_\pi}$$

and a_0, P_{Δ_1}, $Q_{\Delta_1} \in A$ it follows that $\pi(a_0)$ is a bijection of $H_{\Delta_1\pi}$ onto $E_{\Delta_1\pi}$ (or $H_{\Delta_1\pi} = E_{\Delta_1\pi} = 0$).

Moreover

$$\nu(T^* aT)|_{H_\pi} = \pi(T^* aT)|_{H_\pi}$$

which means (see 19.5) that

$$\nu(T^* aT)|_{H_\pi} = t(\pi)[a] \quad .$$

Consider the projection $\overline{P}$: $H_\nu \to H_\pi$. As H_π is an invariant space for $\nu(A)$, we have that $\overline{P} \in \nu(A)'$. Evidently the projection $\overline{\overline{P}} = \nu(T)\overline{P}\nu(T^*)$ also belongs to $\nu(A)'$ and the equality

$$\overline{\overline{P}}\nu(T) = \nu(T)\overline{P}$$

along with the foregoing argument shows that we can identify Im $\overline{\overline{P}}$ with $H_{t(\pi)}$ and under this identification $\nu(T)$ establishes the isomorphism

$$\nu(T)\colon H_\pi \to H_{t(\pi)} \quad .$$

By Lemmas 19.16 and 19.18 we have that

$$\nu(a_1 T(I - P_{\Delta_1}))\colon \text{Im } \nu(I - P_{\Delta_1}) \to \text{Im } \nu(I - Q_{\Delta_1})$$

is a bijection which (having in mind the foregoing consideration) means now that

$$t(\pi)[a_1]\nu(T)\colon H_{\Delta_2\pi} \to E_{\Delta_2 t(\pi)}$$

is a bijection (or $H_{\Delta_2\pi} = E_{\Delta_2 t(\pi)} = 0$).

Since both $\pi(a_0)$ and $t(\pi)[a_1]$ are compact and $\nu(T)$ is unitary it follows that $\dim H_{\Delta_1\pi}$ and $\dim H_{\Delta_2\pi}$ are finite and consequently $\dim H_\pi = \dim H_{\Delta_1\pi} + \dim H_{\Delta_2\pi} < \infty$. As $\nu(T)\colon H_\pi \to H_{t(\pi)}$ is a bijection $\dim H_\pi$ is t–invariant.

Now let Δ be any open and closed set in $\sigma(b)$ and $\Delta_1' = \Delta \cap \Delta_1$, $\Delta_2' = \Delta \cap \Delta_2$. Let $\dim H_\pi = K$ and $\dim H_{\Delta_2'\pi} = m$. By the foregoing argument

$$\dim H_{\Delta_2'\pi} = \dim E_{\Delta_2't(\pi)} = m \quad .$$

As $t(\pi)[a_0]$ is a bijection of $H_{\Delta_1't(\pi)}$ onto $E_{\Delta_1't(\pi)}$ we have

$$\begin{aligned}\dim H_{\Delta_2't(\pi)} &= \dim H_{t(\pi)} - \dim H_{\Delta_1't(\pi)} \\ &= K - \left(K - \dim E_{\Delta_2't(\pi)}\right) = K - (K - m) = m\end{aligned}$$

so $\dim H_{\Delta_2'\pi}$ is t–invariant.

In an analogous manner one can verify that $\dim H_{\Delta_1'\pi}$, $\dim E_{\Delta_1'\pi}$, $\dim E_{\Delta_2'\pi}$ are all t–invariant. □

Remark 19.25 If we consider an operator aT, such that $\pi(a)$ are compact for all $\pi \in \hat{A}$, and the spectral subspace $H_{\Delta\pi}$ corresponding to an open and closed set $\Delta \subset \sigma(aT)$, then if $\Delta \not\ni 0$ one can see that the function $\dim H_{\Delta\pi}$ is finite and t–invariant. This is true since in this case $H_{\Delta\pi} = E_{\Delta\pi}$ and $t(\pi)[a]\nu(T)$ is a bijection of $H_{\Delta\pi}$ onto $E_{\Delta t(\pi)}$.

Corollary 19.26 *If* $\dim \mathrm{H}_\pi < \infty$ *for every* $\pi \in \hat{\mathrm{A}}$ *then* $\dim \mathrm{H}_{\Delta\pi}$ *and* $\mathrm{E}_{\Delta\pi}$ *are* t–*invariant functions of* $\hat{\mathrm{A}}$ *for every open and closed set* $\Delta \subset \sigma(\mathrm{b})$.

Proof If $\sigma(b) = \mathbb{C}$ then $\Delta = \mathbb{C}$ and $H_\pi = H_{\Delta\pi}$. Since $\nu(T)$: $H_\pi \to H_{t(\pi)}$ is a bijection dim $H_{\Delta\pi}$ is t–invariant.

The other cases are covered by Theorem 19.24. □

19.27 Recall (see **18.6**) that a topological space X is called α–irreducible under the mapping α: $X \to X$ if it cannot be decomposed into two non–empty closed sets which are invariant under α.

The foregoing results lead to the following lemma.

Lemma 19.28 *If* dim $\mathrm{H}_\pi < \infty$ *for every* $\pi \in \hat{\mathrm{A}}$ *and* Tr $\pi(\mathrm{a})$ *is a continuous function for any* $\mathrm{a} \in \mathrm{A}$ *then* dim $\mathrm{H}_{\Delta\pi}$ *and* dim $\mathrm{E}_{\Delta\pi}$ *are constants on any* t*–irreducible subset of* $\hat{\mathrm{A}}$.

Proof Since dim $H_\pi < \infty$ for every $\pi \in \hat{A}$ we have that Prim $A = \hat{A}$ (Dixmier [1], 4.3.7). In view of Corollary 19.26 and the continuity of Tr we conclude that dim $H_{\Delta\pi} = \mathrm{Tr}\ \pi(P_\Delta)$ and dim $E_{\Delta\pi} = \mathrm{Tr}\ \pi(Q_\Delta)$ are continuous t–invariant functions on $\hat{A}$ taking values in $\mathbb{N}$. Thus they are constant on any t–irreducible subset of $\hat{A}$. □

Remarks 19.29

(1) If Tr is not a continuous function on $\hat{A}$ then it may happen that dim $H_{\Delta\pi}$ and dim $E_{\Delta\pi}$ are not constant on t–irreducible subsets of $\hat{A}$.

Example Let A be the algebra of functions defined on the interval [0, 2], taking values in the C^*–algebra M_2 of all complex 2×2 matrices and such that:

(i) on the interval [0, 1] $a(x)$ is a continuous matrix function and $a(1)$ is diagonal;

(ii) on the interval (1, 2], $a(x)$ is a continuous matrix function having the limit

$$\lim_{x \to 1+0} a(x) = a(1+0)$$

where $a(1+0)$ is a diagonal matrix and $a_{11}(1+0) = a_{11}(1)$.

Clearly $\text{Prim}\, A = \hat{A} = ([0, 1) \cup (1, 2] \cup \{1\} \cup \{1-0\} \cup \{1+0\})$ where for $\pi \in [0, 1) \cup (1, 2]$ H_π is two–dimensional and $\pi(a) = a(\pi)$, and for all the other points H_π is one–dimensional and $1(a) = a_{11}(1)$, $(1-0)(a) = a_{22}(1)$, $(1+0)(a) = \lim_{x \to 1+0} a_{22}(x)$. The Jacobson topology on $[0, 1) \cup (1, 2]$ is the usual one, whereas basic neighbourhoods for the remaining points are

$(1-\varepsilon, 1+h)$	for	$\{1\}$
$(1-\varepsilon, 1) \cup \{1-0\}$	for	$\{1-0\}$
$(1, 1+h) \cup \{1+0)$	for	$\{1+0\}$.

Thus in fact $\hat{A}$ is connected.

Consider the space $H = L^2([0, 2], \mathbb{C}^2)$ and the operator T defined by

$$(Tf)(x) = |\alpha'(x)|^{\frac{1}{2}} f(\alpha(x)) \qquad f \in H$$

where $\alpha\colon [0, 2] \to [0, 2]$ is a diffeomorphism having 0, 1 and 2 as the only fixed points.

Let

$$(af)(x) = a(x)f(x) \qquad f \in H\,, \quad a \in A \quad .$$

Then obviously $TAT^* = A$, T acts topologically freely (thus $C^*(A, T)$ possesses the property (*)), and $\hat{A}$ is t–irreducible.

Consider the matrix function a which is constant on $[0, 1]$, $a_{11}(1) = 1$, $a_{22}(1) = 2$, and constant on $(1, 2]$ and $a_{22}(1+0) = 3$. Then the operator aT has the spectrum

$$\sigma(aT) = \Delta_1 \cup \Delta_2 \cup \Delta_3$$

where $\Delta_1 = \{\lambda\colon |\lambda| = 1\}$, $\Delta_2 = \{\lambda\colon |\lambda| = 2\}$ and $\Delta_3 = \{\lambda\colon |\lambda| = 3\}$. Considering the function $b(\lambda) = \lambda a_0 + a_1 T = \lambda I + aT$ we have, for example, for Δ_2

$$H_{\Delta_2\pi} = E_{\Delta_2\pi}$$

and dim $H_{\Delta_2\pi} = 1$ on $[0, 1)$ which dim $H_{\Delta_2\pi} = 0$ on $(1, 2]$.

(2) If dim $H_\pi = \infty$ then it may even happen that dim $H_{\Delta\pi}$ is not t–invariant.

Example Consider the space $H = l^2(\mathbb{Z}, H_1)$ where H_1 is a Hilbert space such that $H_1 = H_2 \oplus H_3$ and $H_3 \cong H_1$ (that is, dim $H_1 = \infty$ and card H_3 = card H_1). Let A be the algebra of all the operators having the form

$$(af)(z) = a(z)\, f(z) \qquad f \in H$$

where $a(z)$ are operators on H_1 such that there exist uniform limits $\lim_{z\to\pm\infty} a(z) = a(\pm\infty)$. Define the operator T by

$$(Tf)(z) = f(z - 1) \qquad f \in H \quad .$$

Obviously $\hat{A} \supset \Omega = \{-\infty \cup \mathbb{Z} + \infty\}$ with the corresponding topology and $H_\pi \cong H_1$ for all $\pi \in \Omega$ and $\pi(a) = a(\pi)$. Consider the operator a acting as follows:

$$[ah_2](z) = \begin{cases} \frac{1}{2}h_2(z) , & z < 0 \\ 0 , & z = 0 \end{cases} \qquad \text{where } h_2(z) \in H_2 , \quad z \le 0$$

$$(ah_3)(z) = \begin{cases} 2h_3(z) , & z < 0 \\ 2U[h_3(0)] , & z = 0 \end{cases} \qquad \text{where } h_3(z) \in H_3 , \; z \le 0$$

and U is an isomorphism of H_3 onto H_1

$$(ah)(z) = 2h(z) \qquad \text{where} \quad z > 0, \quad h(z) \in H_1 \ .$$

Denote by

$$H^s = \bigoplus_{-\infty}^{-1} H_2 \qquad \text{and} \qquad H^u = \bigoplus_{-\infty}^{-1} H_3 \bigoplus_{0}^{+\infty} H_1 \ .$$

Evidently we have

$$aT\colon H^s \to H^s$$
$$aT\colon H^u \to H^u$$

and

$$\sigma(aT)|_{H^s} \subset \{\lambda\colon |\lambda| < 1\}$$
$$\sigma(aT)|_{H^u} = \{\lambda\colon |\lambda| = 2\} \ .$$

Consider $H_{\Delta_1\pi}$ where $\Delta_1 = \sigma(aT)|_{H^s}$. By the construction of a and T we have

$$H_{\Delta_1\pi} = \begin{cases} H_2 \ , & \pi < 0 \\ 0 \ , & \pi \geq 0 \ . \end{cases}$$

So dim $H_{\Delta_1\pi}$ is not t–invariant.

The foregoing examples show that if Tr $\pi(a)$ is not a continuous function or dim H_π is infinite the behaviour of $H_{\Delta\pi}$ can be rather pathological. But if Tr $\pi(a)$ is continuous and dim H_π is finite the description of the spectrum of the function $b(\lambda)$ can be obtained in a rather clear form which is shown in the following.

Theorem 19.30 *Suppose that* $C^*(A, T)$ *possesses the property* (*). *Let there exist a natural number* n *such that* dim $H_\pi \leq n$, *for all* $\pi \in \hat{A}$ *and let* Tr $\pi(a)$ *be continuous for any* $a \in A$. *If* Prim A *is* t*–irreducible then the spectrum of the function* $b(\lambda) = \lambda a_0 + a_1T$ *(and in particular the spectrum of* aT*) has no more than* n *connected components and is such that there exist numbers*

$$0 \le \alpha_1 \le \beta_1 < \alpha_2 \le \beta_2 < \ldots < \alpha_p \le \beta_p \le +\infty \qquad \textit{where} \quad p \le n$$

such that

$$\sigma(b) = \bigcup_{i=1}^{p} \Delta_i \quad , \qquad \Delta_i = \{\lambda \in \mathbb{C} : \alpha_i \le |\lambda| \le \beta_i\}$$

and $\alpha_1 = 0$ *iff* a_1 *is not invertible,* $\beta_p = \infty$ *iff* a_0 *is not invertible.*

Proof Evidently $0 \in \sigma(b)$ iff a_1 is not invertible and $\infty \in \sigma(b)$ iff a_0 is not invertible. Since $C^*(A, T)$ possesses the property (*) $\sigma(b)$ is invariant under rotations. Let $0 \le \alpha < \beta \le +\infty$ be any points lying in different connected components of $\mathbb{C}\backslash\sigma(b)$. Consider the open and closed set

$$\Delta = \sigma(b) \cap \{\alpha \le |\lambda| \le \beta\} \quad .$$

In view of Theorem 19.22, Lemma 19.28, and the t–irreducibility of Prim A

$$0 < \dim H_{\Delta\pi} \le n$$

and $\dim H_{\Delta\pi}$ is constant on Prim A.

Since $\dim H_{(\Delta_1 \cup \Delta_2)\pi} = \dim H_{\Delta_1\pi} + \dim H_{\Delta_2\pi}$ for any two non–intersecting non–void open and closed sets Δ_1 and Δ_2 the set $\mathbb{C}\backslash\sigma(b)$ cannot have more than $(n + 1)$ connected components and thus $\sigma(b)$ cannot have more than n connected components. The theorem is proved. □

Remark If Tr $\pi(a)$ is not continuous it may happen that the spectrum of $b(\lambda)$ has more than n connected components while $\dim H_\pi \le n$ for all $\pi \in \hat{A}$. See the example in Remark 19.29 (1).

As an immediate corollary we can again obtain the results of Section 17.

Corollary 19.31 (cf. Theorem 17.1) *Let the conditions of Theorem* 19.30 *be satisfied and* $n = 1$ (*thus* A *is commutative*). *If both* a_0 *and* a_1 *are not* *invertible then* $b = a_0 + a_1T$ *is not invertible.*

Proof Since a_0 and a_1 are not invertible the spectrum of $b(\lambda) = \lambda a_0 + a_1T$ contains 0 and ∞. If b is invertible then the spectrum of $b(\lambda)$ does not meet the unit circle and thus has more than one connected component. We arrive at a contradiction. □

If $t\colon \hat{A} \to \hat{A}$ has periodic points we can deduce some simple necessary conditions for $b = a_0 + a_1T$ to be invertible.

Let π be a periodic point of period n, that is,

$$t^n(\pi) = \pi \qquad \text{and} \qquad t^j(\pi) \neq \pi \qquad 0 < j < n \quad .$$

With the operator b we can associate the matrix

$$F_\pi(\lambda) = \begin{pmatrix} \lambda\pi[a_0] & \pi(a_1)\theta_{t^{-1}(\pi)} & 0 & \dots & 0 \\ 0 & \lambda t^{-1}(\pi)\ [a_0] & t^{-1}(\pi)[a_1]\theta_{t^{-2}(\pi)} & \dots & 0 \\ \dots & \dots & \dots & \dots & \dots \\ t^{-n+1}(\pi)[a_1]\theta_\pi & 0 & 0 & \dots & \lambda t^{-n+1}(\pi)[a_0] \end{pmatrix}$$

defining an operator in $\bigoplus_{i=0}^{-n+1} H_{t^i(\pi)}$ (θ is defined in **19.5**).

Theorem 19.32 *Suppose that* $C^*(A, T)$ *possesses the property* (*), $\pi(a_0)$, $\pi(a_1)$ *are compact for all* $\pi \in \hat{A}$ *and* $b = a_0 + a_1T$ *is invertible. Then*

$$\dim H_\pi < \infty \qquad \text{for all} \quad \pi \in \hat{A} \tag{19.46}$$

and for any periodic point π_0

$$\det[F_{\pi_0}(\lambda)] \neq 0 \quad , \qquad |\lambda| = 1 \quad . \tag{19.47}$$

If, moreover, $\operatorname{Tr} \pi(a)$ *is a continuous function on* $\hat{A}$ *for any* $a \in A$ *then for any periodic points* π_0 *and* π_1 *lying in the same* t*–irreducible component* *of* $\hat{A}$

$$\omega(\pi_0) = \omega(\pi_1) \tag{19.48}$$

where

$$\omega(\pi) = \frac{1}{2\pi n(\pi)} \left[\arg \det F_{\pi}(\lambda)\right]_{|\lambda|=1}$$

and where $n(\pi)$ *is the period of* π.

Proof Since b is invertible the spectrum of $b(\lambda)$ is not the whole of $\mathbb{C}$ and in view of Theorem 19.24 (19.46) is true.

For any periodic $\pi_0 \in \hat{A}$ let us consider the representation $\nu(C^*(A, T))$ on the space $L_{\pi_0} = \bigoplus_{i=0}^{-n+1} H_{t^i(\pi_0)}$ defined by the formulae

$$\nu(a)\Big|_{H_{t^i(\pi_0)}} = t^i(\pi_0)[a] , \qquad \nu(T)\Big|_{H_{t^i(\pi_0)}} = \theta_{t^i(\pi_0)} .$$

ν is really a representation of $C^*(A, T)$ since $C^*(A, T) \cong A \underset{\hat{T}}{\times} \mathbb{Z}$ by Theorem 12.8. Obviously $F_{\pi_0}(\lambda)$ is simply $\nu(b(\lambda))$ written in matrix form. Thus if b is invertible then $b(\lambda)$ and $F_{\pi_0}(\lambda)$ are also invertible and so (19.47) is true.

Let Δ be the part of $\sigma(b)$ lying in the unit circle. Clearly

$$\dim H_{\Delta\pi_0} = \frac{1}{n(\pi_0)} \dim L_{\Delta\pi_0}$$

where $L_{\Delta\pi_0}$ is the subspace of L_{π_0} corresponding to the part of the spectrum of $F_{\pi_0}(\lambda)$ lying in the unit circle. In fact $\dim L_{\Delta\pi_0}$ is the number of zeros of $\det[F_{\pi_0}(\lambda)]$ lying in the unit circle. Thus

$$\dim H_{\Delta\pi_0} = \frac{1}{n(\pi_0)2\pi} \left[\arg \det F_{\pi_0}(\lambda)\right]_{|\lambda|=1} = \omega(\pi_0) \quad .$$

If Tr $\pi(a)$ is continuous then by Lemma 19.28 dim $H_{\Delta\pi}$ is constant on any t–irreducible component of $\hat{A}$ and (19.48) is true. □

19.33 We shall now make a few final remarks concerning the so–called dynamical spectrum of a linear extension or a relation. The *dynamical spectrum* $\sum_D(\beta)$ *of the linear extension* β defined by (19.6) is the set of all $\omega \in \mathbb{R}$ such that the extension β_ω:

$$\beta_\omega(h) = e^{-\omega}t(\pi)[a](\theta_\pi h) \quad , \qquad h \in H_\pi$$

is not hyperbolic, and the *dynamical spectrum* $\sum_D(\rho)$ *of the relation* ρ defined in **19.17** is the set of all $\omega \in \mathbb{R}$ such that the relation ρ_ω:

$$(z,\ y) \in \rho_\omega \Leftrightarrow (e^{-\omega}z,\ y) \in \rho$$

is not hyperbolic.

In an obvious way using the foregoing results one can easily obtain corresponding statements for the description of the dynamical spectrum. For example

Lemma 19.34 *Let* $\hat{\mathrm{T}}$ *act topologically freely on* A. *Then* $\omega \in \sum_D(\beta)$ $\left(\sum_D(\rho)\right)$ *iff* $e^\omega \in \sigma(\mathrm{aT})$ $(\sigma(\mathrm{b}))$.

This lemma combined with Theorem 19.22 and Theorem 19.30 contains in particular the Sacker–Sell spectral theorem (see Sacker, Sell [1, 2] and Chicone, Swanson [1]).

One more statement which is also worth mentioning is the following.

Theorem 19.35 *Let* $C^*(\mathrm{A},\ \mathrm{T})$ *possess the property* (*) *and let the operator* $\mathrm{b} = \mathrm{a}_0 + \mathrm{a}_1\mathrm{T}$ *be invertible. Then* b *is homotopically equivalent (in the set*

of invertible operators) to the direct sum of invertible operators

$$a_0P \oplus a_1T(I - P)$$

where

$$a_0P\colon PH \to QH$$

$$a_1T(I - P)\colon (I - P)H \to (I - Q)H$$

and the projections P, Q *are defined in Lemma* 19.18.

Proof Since $C^*(A, T)$ possesses the property (*) and b is invertible we have by Corollary 13.6 that $b(\lambda) = \lambda a_0 + a_1T$ is invertible when $|\lambda| = 1$.

So Lemmas 19.16 and 19.18 imply the invertibility of the operator–valued functions

$$b'_{1-t} = a_0P + ta_1TP\colon PH \to QH$$

$$b''_{1-t} = ta_0(I - P) + a_1T(I - P)\colon (I - P)H \to (I - Q)H$$

for all $0 \le t \le 1$.

Thus the set $\{b_\tau\}_{0\le\tau\le 1} = \{b'_\tau \oplus b''_\tau\}$ of invertible operators implement the desired homotopy between $b = b_0$ and $a_0P \oplus a_1T(I - P) = b_1$. □

Remark 19.36 In this section we used $H(\hat{A})$ as a universal space characterizing A and $C^*(\tilde{A}, \theta)$ (see the proof of Theorem 19.13) as a faithful representation of $C^*(A, T)$ (under the condition that $\hat{T}$ acts topologically freely). But in fact Lemmas 19.1 and 19.18 show that any other faithful representations of $C^*(A, T)$ will do the job (for some particular examples see Antonevich [10, 13], and Latuskin, Stepin [1, 2]).

20. Operators of the form $\sum a_iT^i$. Examples.

Here we show that the investigation of operators of the form $b = \sum_{i=-k}^{s} a_iT^i$ in the case when $C^*(A, T)$ possesses the property (*) can be reduced to that of a two–term operator $\bar{\bar{b}} = d_0 + d_1S$, where d_0, d_1 belong to a C^*–algebra

$D \subset \mathrm{Mat}_m(\mathbb{C}) \otimes A$ and S is some unitary operator (induced by T) such that $C^*(D, S)$ again possesses the property (*), so one can use the ideas considered in the previous section.

20.1 Let us first fix any integer $m \geq 2$ and let H be the space where $C^*(A, T)$ acts and $\overline{H} = \bigoplus_{i=1}^{m} H_i$, $H_i = H$, $i = \overline{1,m}$. We denote by $S: \overline{H} \to \overline{H}$ a unitary operator

$$Sv = \left(T^m v_0, \dots, T^m v_{m-1}\right) \tag{20.1}$$

where $v = (v_0, \dots, v_{m-1}) \in \overline{H}$.

With any operator $b = \sum a_k T^k \in C^*(A, T)$ we shall associate the element $\overline{b} \in D \times_{\hat{S}} \mathbb{Z}$ of the form

$$\overline{b} = \sum_l d_l \otimes l \quad , \qquad l \in \mathbb{Z} \tag{20.2}$$

where $d_l \in L(\overline{H})$ is defined by the operator matrix

$$[d_l]_{ij} = \hat{T}^i(a_{lm+j-i}) \qquad i,j = 0,\dots,m-1 \quad . \tag{20.3}$$

D is a C^*–subalgebra of $L(\overline{H})$ generated by operators of the form d_l, and $\hat{S}$ is the group of automorphisms of D defined by $\hat{S}^z(d) = S^z d S^{z*}$, $d \in D$, $z \in \mathbb{Z}$.

Theorem 20.2 *Let* $\mathrm{C}^*(\mathrm{A}, \mathrm{T})$ *possesses the property* (*). *Then the mapping* ψ: $\mathrm{b} \to \overline{\mathrm{b}}$ *defined by* (20.2) *establishes an isomorphism between* $\mathrm{C}^*(\mathrm{A}, \mathrm{T})$ *and* $\mathrm{D} \times_{\hat{\mathrm{S}}} \mathbb{Z}$. *Thus* b *is invertible iff* $\overline{\mathrm{b}}$ *is invertible.*

Proof Consider the natural decomposition

$$\mathbb{Z} \cong \mathbb{Z}_m \oplus \mathbb{Z} \tag{20.4}$$

induced by the uniqueness of the expression $z = k + lm$, $0 \le k < m - 1$, $l \in \mathbb{Z}$, for every $z \in \mathbb{Z}$. Since $C^*(A, T)$ possesses the property (*) it is isomorphic to $A \times_{\hat{T}} \mathbb{Z}$ (by Theorem 12.8). Now the result follows immediately from Lemma 11.16 and Theorem 15.2 (here we have $D \cong A \times_{[\hat{T}]_m} \mathbb{Z}_m$ where $[\hat{T}]_m$ is the representation of $\mathbb{Z}_m$ by automorphisms induced by the decomposition (20.4) and clearly S^z generates the automorphism $\hat{S}^z$ of D which corresponds to τ_2 mentioned in Lemma 11.16). □

Corollary 20.3 *If an operator* b *has the form* $b = \sum_0^p a_k T^k$ *then taking* $m = p$ *we have*

$$\bar{b} = d_0 \otimes 0 + d_1 \otimes 1$$

where d_i, $i = 0,1$, *are described by* (20.3). *Thus we need to consider only two–term elements.*

Corollary 20.4 *If* T *acts topologically freely then the mapping* $b \to \bar{\bar{b}} \in L(\bar{H})$ *where*

$$\bar{\bar{b}} = \sum d_l S^l$$

(d_l *are defined by* (20.3), S *by* (20.1)) *establishes an isomorphism between* $C^*(A, T)$ *and* $C^*(D, S)$.

In particular under this isomorphism an operator $b = \sum_0^m a_k T^k$ *is mapped into*

$$\bar{\bar{b}} = d_0 + d_1 S \quad .$$

Proof Since Prim A = Prim[$\mathrm{Mat}_m(\mathbb{C}) \otimes A$] and in view of the definition of S, the homeomorphism of Prim[$\mathrm{Mat}_m(\mathbb{C}) \otimes A$] corresponding to the automorphism (of $\mathrm{Mat}_m(\mathbb{C}) \otimes A \subset L(\overline{H})$)$S(\quad\cdot\quad)\,S^*$ is simply t^m (where t is the homeomorphism of Prim A induced by $\hat{T}$). As T acts topologically freely it follows that S acts topologically freely. Thus

$$C^*(\mathrm{Mat}_m(\mathbb{C}) \otimes A,\ S) \cong [\mathrm{Mat}_m(\mathbb{C}) \otimes A] \underset{\hat{S}}{\times} \mathbb{Z} \tag{20.5}$$

and since $D \subset \mathrm{Mat}_m(\mathbb{C}) \otimes A$ (20.5) implies

$$C^*(D,\ S) \cong D \underset{\hat{S}}{\times} \mathbb{Z}$$

which (see the proof of Theorem 20.2) proves the corollary. □

Since the conditions of invertibility for the operator $b = \sum_{i=-k}^{s} a_i T^i$ are the same as those for $bT^k = \sum_{j=0}^{s+k} a_{j-k} T^j$, Corollaries 20.3 and 20.4 give us the result we wanted to obtain.

These corollaries actually describe a procedure analogous to that of passing from a differential equation of high order to a system of equations of first order. Notice that if the coefficient a_0 of b is invertible then the coefficient d_0 of $\overline{b}(\overline{\overline{\mathrm{b}}})$ is invertible as well, since d_0 is a triangular operator matrix with invertible elements on the diagonal. Similarly if the coefficient a_p of b is invertible then the coefficient d_1 of $\overline{b}(\overline{\overline{\mathrm{b}}})$ is invertible.

If A is a commutative algebra then Section 17 gives us the conditions of invertibility for a two–term operator. If A is not commutative or an operator b has more than two terms then the results of Section 19 and Corollaries 20.3 and 20.4 establish a relation between the conditions of invertibility of b and hyperbolicity of the corresponding linear extension (or relation). The last problem is a classical one for the theory of dynamical systems and, as is known, cannot be solved effectively in general.

Here we present an example where the idea of hyperbolicity works well and gives us a numerical result.

Example 20.5 Consider the space $L^2(\mathbb{R})$ and an operator b of the form

$$b = 1 + a_1 T + a_2 T^2$$

where a_1, a_2 are operators of multiplication by continuous functions $a_i(x)$, $i = 1,2$, having limits at $\pm\infty$:

$$\lim_{x \to \pm\infty} a_i(x) = a_i(\pm\infty)$$

and T is defined by

$$(Tf)(x) = f(x + h) \quad , \qquad f \in L^2(\mathbb{R}) \quad .$$

Let us investigate the conditions for invertibility of b. Without loss of generality we can assume that $h = 1$.

Following the ideas of this section first of all we must consider the operator $\bar{\bar{b}} = d_0 + d_1 S$ where S and d_i, $i = 0,1$, are given by (20.1) and (20.3). We have

$$d_0(x) = \begin{bmatrix} \hat{T}^0(a_0) & \hat{T}^0(a_1) \\ \hat{T}(a_{-1}) & \hat{T}(a_0) \end{bmatrix} = \begin{bmatrix} 1 & a_1(x) \\ 0 & 1 \end{bmatrix}$$

$$d_1(x) = \begin{bmatrix} \hat{T}^0(a_2) & \hat{T}^0(a_3) \\ \hat{T}(a_1) & \hat{T}(a_2) \end{bmatrix} = \begin{bmatrix} a_2(x) & 0 \\ a_1(x + 1) & a_2(x + 1) \end{bmatrix}$$

$$(Sv)(x) = v(x + 2)$$

where $v(x) = [v_0(x), v_1(x)]$, $v_i(x) \in L^2(\mathbb{R})$.

In this example $\hat{A}$ = Prim A = Prim[$\mathrm{Mat}_2(\mathbb{C}) \otimes A$] = $\mathbb{R}$ where $\mathbb{R}$ is the real line compactificated by $+\infty$ and $-\infty$. The universal space (see 19.3) $H(\widehat{\mathrm{Mat}_2(\mathbb{C}) \otimes A}) = \bigoplus_{\overline{\mathbb{R}}} H_x$ where $H_x = \mathbb{C}^2$. Since d_0 is invertible we can use Corollary 19.14 with

$$C(x) = \left[d_0^{-1} \cdot d_1\right](x) = \begin{bmatrix} a_2(x) - a_1(x)a_1(x+1) & -a_1(x)a_1(x+1) \\ a_1(x+1) & a_2(x+1) \end{bmatrix}, \quad x \in \mathbb{R}$$

and $t(x) = x - 2$, $t(\pm\infty) = \pm\infty$ (here t is a homeomorphism of $\overline{\mathbb{R}}$ corresponding to $\hat{S}$), and

$$\theta_x\colon H_x \to H_{t(x)}, \qquad x \in \overline{\mathbb{R}}$$

where

$$\theta_x(\xi) = \xi, \qquad \xi \in \mathbb{C}^2 \cong H_x .$$

By Corollary 19.14 $\bar{\bar{b}}$ is invertible if the linear extension

$$\begin{aligned} \beta(x, \xi) &= (x - 2,\ C(x-2)\xi), \qquad (x, \xi) \in \mathbb{R} \times \mathbb{C}^2 \\ \beta(\pm\infty, \xi) &= (\pm\infty,\ C(\pm\infty)\xi) \end{aligned} \tag{20.6}$$

is hyperbolic (here by (x, ξ) we denote the vector in H_x equal to $\xi \in \mathbb{C}^2$).

Since $+\infty$ and $-\infty$ are the only fixed points for t we can also use Theorem 19.32. Let $\lambda_1(x)$, $\lambda_2(x)$, $x \in \overline{\mathbb{R}}$ be the eigenvalues of $C(x)$. We shall assume that $|\lambda_1(x)| \le |\lambda_2(x)|$.

As $\mathbb{R}$ is t–irreducible in view of Theorem 19.32 we conclude that if $\bar{b}$ is invertible then

$$|\lambda_i(\pm\infty)| \neq 1, \qquad i = 1,2$$

and the number of eigenvalues lying in the unit circle should be the same for $+\infty$ and $-\infty$. Thus $\bar{\bar{b}}$ can be invertible only when one of the following conditions is satisfied:

$$\begin{aligned}
&(1) \quad |\lambda_i(\pm\infty)| < 1 , \quad i = 1,2 ; \\
&(2) \quad |\lambda_i(\pm\infty)| > 1 , \quad i = 1,2 ; \\
&(3) \quad |\lambda_1(-\infty)| < 1 < |\lambda_2(-\infty)| ; \\
&\qquad |\lambda_1(+\infty)| < 1 < |\lambda_2(-\infty)| .
\end{aligned} \tag{20.7}$$

Let us consider each of these situations in turn.

I. In the first case one can easily verify that the whole of the universal space is stable under β. Thus β is hyperbolic and consequently $\bar{\bar{b}}$ (or b) is invertible.

II. In the second case since the unstable fibre over $+\infty$ and $-\infty$ is the whole of $\mathbb{C}^2$ and since the projection mentioned in definition 19.4 belongs to $\mathrm{Mat}_2(\mathbb{C}) \otimes A$ it follows that in this situation the whole of the universal space should coincide with the unstable bundle H^u. Moreover by definition 19.4 $C(x)$ should be invertible for every $x \in \mathbb{R}$ which is clearly equivalent to the condition of the invertibility of $d_1(x)$

$$a_2(x) \neq 0 \qquad x \in \mathbb{R} . \tag{20.8}$$

On the other hand if (20.8) is satisfied then one can observe that condition (2) in (20.7) implies the hyperbolicity of β and thus the invertibility of $\bar{\bar{b}}$ (or b).

So if condition (2) in (20.7) is satisfied then b is invertible iff $a_2(x) \neq 0$, $x \in \mathbb{R}$.

III. Finally let us consider the third case in (20.7). This is the most complicated one. Here we shall investigate it under the additional assumption that a_1 and a_2 are constants in some neighbourhoods $O(+\infty)$ and $O(-\infty)$, that is, there exists $N > 0$ such that

$$\begin{aligned}
a_i(x) &= a_i(+\infty) , \qquad x \geq N \quad i = 1,2 \\
a_i(x) &= a_i(-\infty) , \qquad x \leq -N \quad i = 1,2 .
\end{aligned}$$

Let $l_j(\pm\infty)$, $i = 1,2$, be the eigenvectors corresponding to the eigenvalues $\lambda_j(\pm\infty)$. If β is hyperbolic then the stable fibre H_x^s for every $x \le -N$ consists of vectors of the form $\gamma l_1(-\infty)$, $\gamma \in \mathbb{C}$ (this is in view of the fact that H^s can be described in an invariant geometric way as all the vectors $l(x) \in H_x$ tending to zero under the mapping β). The corresponding argument for $+\infty$ shows that for every $x \ge N$ the unstable fibre H_x^u consists of the vectors of the form $\gamma l_2(+\infty)$.

Consider some even number $N_0 = 2k \ge N$ and the interval $\Delta = [N_0, N_0 + 2]$. For any $x_0 \in \Delta$ the vector $\xi = l_2(+\infty) \in H_x$ belongs to the unstable fibre H_x^u. Since the unstable bundle is invariant under β it follows that $\beta^{2k+1}(\xi) \in H_y^u$, where $y \in -\Delta$. As we have already noticed $H_y^s = \{\gamma l_1(-\infty)\}$; thus a necessary condition for β to be hyperbolic is that for every $x \in \Delta$ the vectors

$$\begin{aligned} &v(x) = C(x - (2k + 1)2) \cdot \ldots \cdot C(x - 4)C(x - 2)l_2(+\infty) \\ \text{and} \quad & \qquad\qquad\qquad\qquad\qquad\qquad\qquad\qquad\qquad\qquad (20.9) \\ &l_1(-\infty) \end{aligned}$$

are linearly independent.

In fact we shall now show that if b satisfies condition (3) in (20.7) and a_1 and a_2 are constants for some neighbourhoods of $+\infty$ and $-\infty$ then condition (20.9) is necessary and sufficient for b to be invertible.

We need to prove only sufficiency. To prove this we simply construct the stable and unstable bundles H^s and H^u. Thus the hyperbolicity of β and consequently the invertibility of b will be established.

We have already observed that for every $x > N$ the unstable fibre consists of the vectors proportional to $l_2(+\infty)$. Consider the bundle over $\mathbb{R} \cup +\infty$, that is, the set of images of H_x^u, $x \ge N$ under all the mappings β^n, $n > 0$, and let us set $H_{-\infty}^u = \{\gamma l_2(-\infty)\}$. In view of (20.9) $H_x^u \ne \{0\}$ for every $x \in \mathbb{R}$. This bundle will satisfy all the conditions for the unstable bundle H^u (from definition 19.4) if we show that the projection on it belongs to $\mathrm{Mat}_2(\mathbb{C}) \otimes A$ or equivalently if we show that it is a continuous vector bundle over $\mathbb{R}$.

Clearly it is a continuous vector–bundle over $\mathbb{R} \cup +\infty$ by construction. So we have to check only the point $-\infty$.

Consider any point $x \leq -N$ and any vector $\xi \in H_x^u$ where H_x^u is the above–defined fibre. In view of (20.9) ξ is linearly independent with $l_1(-\infty)$. Thus $\xi = \mu_1 l_1(-\infty) + \mu_2 l_2(-\infty)$ where $\mu_2 \neq 0$. For any $k > 0$ we have

$$\beta^k(\xi) = [C(-\infty)]^k \xi = \mu_1[\lambda_1(-\infty)]^k l_1(-\infty) + \mu_2[\lambda_2(-\infty)]^k l_2(-\infty) \quad .$$

Dividing this equality by $\mu_2[\lambda_2(-\infty)]^k$ we observe that

$$\frac{\beta^k(\xi)}{\mu_2[\lambda_2(-\infty)]^k} \to l_2(-\infty)$$

which proves the continuity of the constructed bundle at $-\infty$.

Now let us construct the stable bundle. As we know, for every $x \leq -N$, $H_x^s = \{\gamma l_1(-\infty)\}$. For any $x \in (-N, -N + 2]$ let us define H_x^s as

$$H_x^s = \beta^{-1}\left(H_{x-2}^s\right) \quad . \tag{20.10}$$

In view of the definition of β (see (20.6)) $H_x^s = H_{x-2}^s$ here. Now for any $x \in (-N + 2, -N + 4]$ define H_x^s again by (20.10). The condition (20.9) implies $\dim H_x^s = 1$.

Let us check that the above–defined bundle is continuous over $[-\infty, -N + 4]$. It is obviously so for $[-\infty, -N + 2]$. For $x \in (-N + 2, -N + 4]$ we define the base of $H_x = \mathbb{C}^2$ by setting $w_1(x) \in H_x^u$ where H_x^u is the fibre already constructed. Evidently one can choose $w_1(x)$ to form a continuous vector section. For $x \in (-N, -N + 2]$ we take $w_1(x) = \lambda(x + 2)\ \beta(w_1(x + 2))$ where $\lambda(x) \neq 0$ is taken in such a way that $w_1(x)$ forms a continuous vector section over $(-N, -N + 4]$. The second vector

$w_2(x)$ of the base we take in an arbitrary way such that $w_2(x)$ forms a continuous vector section over $(-N, -N + 4]$. If this base is fixed then the matrix of the mapping $\beta_x\colon H_x \to H_{x-2}$ for all $x \in (-N + 2, -N + 4]$ has the form

$$\beta_x = \begin{pmatrix} \lambda^{-1}(x) & g(x) \\ 0 & h(x) \end{pmatrix}$$

where $g(x)$ and $h(x)$ are continuous functions. β_x is invertible iff $h(x) \neq 0$.

If $h(x) = 0$ then by (20.10)

$$H_x^s = \operatorname{Ker} \beta_x = \left\{ \gamma \begin{pmatrix} -\lambda(x) g(x) \\ 1 \end{pmatrix} \right\} . \tag{20.11}$$

If $h(x) \neq 0$ then again by (20.10) a routine computation shows that

$$H_x^s = \left\{ \gamma \begin{pmatrix} -\lambda(x) g(x) + \lambda(x) h(x) f(x) \\ 1 \end{pmatrix} \right\} \tag{20.12}$$

where $f(x)$ is a continuous function.

Comparing (20.11) and (20.12) we conclude that if $h(x) \to 0 = h(x_0)$ then H_x^s tends continuously to $H_{x_0}^s$ which proves the continuity of the bundle H_x^s over $[-\infty, -N + 4]$.

Continuing this procedure (using (20.10)) we construct a continuous vector bundle H^s over $-\infty \cup \mathbb{R}$. Now let us set $H_{+\infty}^s = \{\gamma l_1(+\infty)\}$.

If $\lambda_1(+\infty) \neq 0$ then the continuity of the constructed vector bundle can be verified by the same argument as the one used for H^u at $-\infty$. If $\lambda_1(+\infty) = 0$ then β_x is not invertible in a neighbourhood of $+\infty$ and in this neighbourhood we have $H_x^s = \operatorname{Ker} C(+\infty) = \{\gamma l_1(+\infty)\}$. Thus the constructed bundle H^s is continuous over the whole of $\overline{\mathbb{R}}$.

Thus we have constructed two bundles H^u and H^s over $\overline{\mathbb{R}}$ which are invariant under β, and β is invertible on H^u. If we choose any base $w_1(x)$, $w_2(x)$ of H_x in such a way that $w_1(x) \in H_x^u$, $w_2(x) \in H_x^s$ and both w_1 and w_2 are continuous vector sections then the mapping β in this base has the form (cf. (20.6))

$$\beta(x, \xi) = \left(x - 2\ ,\ \begin{pmatrix} C_1(x - 2) & 0 \\ 0 & C_2(x - 2) \end{pmatrix}\xi\right)$$

where C_1, C_2 are continuous functions over $\mathbb{R}$, $C_2(x) \neq 0$ and

$$\lim_{x\to\pm\infty} C_i(x) = \lambda_i(\pm\infty) \quad , \qquad i = 1,2 \quad .$$

Obviously H^u and H^s satisfy all the conditions of definition 19.4 and thus β is hyperbolic. The proof is complete. □

21. The trajectorial approach.

Though hyperbolic ideas are very fruitful when one considers operators of the form $\sum a_i T^i$, they have natural boundaries and can hardly be used in the general case of the investigation of elements in $C^*(A, T_g)$ with arbitrary amenable G. In this section we present a method of finding the conditions of invertibility of any $b \in C^*(A, T_g)$ which is based on the construction of a suitable family of representations of $C^*(A, T_g)$ on the discrete Hilbert spaces associated with the trajectories of points in the spectrum $\hat{A}$ of the algebra A.

21.1 Throughout this section we shall assume that A is a separable C^*–algebra and $C^*(A, T_g)$ possesses the property (*) (recall that this is always so if $\{\hat{T}_g\}$ acts topologically freely).

For each point $x \in \hat{A}$ we define a representation π_x of $C^*(A, T_g)$ in the space $l^2(G, H_x)$ by the equations:

$$\pi_x(a)(g \otimes h) = g \otimes x(\hat{T}_g(a))h \quad , \qquad a \in A \qquad (21.1)$$

$$\pi_x(T_{g_0})(g \otimes h) = (g \cdot g_0^{-1}) \otimes h \quad , \qquad g_0 \in G \qquad (21.2)$$

where $g \in G$, $h \in H_x$.

Since $C^*(A, T_g)$ possesses the property (*) in view of Theorems 12.8 and

11.12 and Remark 15.3 one can see that the representation $\pi = \oplus_{x \in \hat{A}} \pi_x$ of $C^*(A, T_g)$ on $\oplus_{x \in \hat{A}} H_x$ is faithful and thus $b \in C^*(A, T_g)$ is invertible iff $\pi(b)$ is invertible. Consequently, if b is invertible then all $\pi_x(b)$, $x \in \hat{A}$, are invertible.

The question arises: is the converse true? It may be supposed that it can happen, say, that b is not invertible while all $\pi_x(b)$, $x \in \hat{A}$, are invertible, but the function $\left\|[\pi_x(b)]^{-1}\right\|$ is not bounded on $\hat{A}$ (and in this case $\pi(b)$ is not invertible).

The next theorem shows that this situation is impossible.

Theorem 21.2 *An element* $b \in C^*(A, T_g)$ *is invertible if and only if all the elements* $\pi_x(b)$, $x \in \hat{A}$, *are invertible.*

Proof We have already observed that if b is invertible then all $\pi_x(b)$, $x \in \hat{A}$, are invertible.

Now suppose that b is not invertible. We shall prove that in this case there exists an $x \in \hat{A}$ such that $\pi_x(b)$ is not invertible.

As b is not invertible there exists a $\tau \in \widehat{C^*(A, T_g)}$ such that $\tau(b)$ is not invertible (Naimark [1], §25.3). Consider the representation $\tau(A)$ of the algebra A and define

$$Y = \text{supp Ker}[\tau(A)] = \{p \in \text{Prim } A\colon p \supset \text{Ker } \pi(A)\} \quad .$$

Y is a closed subset of Prim A and is invariant under all automorphisms $\hat{T}_g$. We can identify Y with Prim $\tau(A)$ (see 2.85).

By the same argument as the one used in proving (ii) in the proof of Theorem 16.21 we can prove the existence of a point $y_0 \in Y$ with dense orbit.

Now let x_0 be an irreducible representation of A with Ker $x_0 = y_0$. Observe that $\pi_{x_0}(b)$ is not invertible.

Indeed, by construction $\pi_{x_0}[C^*(A, T_g)]$ possesses the property (*) (π_{x_0} is simply the right regular representation for $\pi_{x_0}(A) \underset{\hat{T}_{x_0}}{\times} G$ where $\hat{T}_{x_0}$ is the group of automorphisms of $\pi_{x_0}(A)$ induced by $\pi_{x_0}(T_g)$, $g \in G$, so we can apply Theorems 11.12 and 12.8) and by the choice of x_0 $\pi_{x_0}(A) \cong \tau(A)$. So by Theorem 12.8

$$\pi_{x_0}(C^*(A, T_g)) \cong \tau(A) \underset{\hat{T}}{\times} G$$

and consequently $\tau(C^*(A, T_g))$ is a representation of $\pi_{x_0}(C^*(A, T_g))$. Thus $\pi_{x_0}(b)$ is not invertible (as $\tau(b)$ is not invertible) and the proof is complete. □

Theorem 16.21 deals with the situation when $C^*(A, T_g)$ does not possess the property (*) but the corresponding trajectorial representations defined by (16.34)–(16.36) completely describe the conditions for the invertibility of elements in $C^*(A, T_g)$ which is stated in the next theorem.

Theorem 21.3 *If* $\mathrm{C}^*(\mathrm{A}, \mathrm{T}_g)$ *is the* C^**–algebra described in Theorem* 16.21 *then an element* $\mathrm{b} \in \mathrm{C}^*(\mathrm{A}, \mathrm{T}_g)$ *is invertible if and only if all the elements* $\pi_{\mathrm{x}}(\mathrm{b})$, $\mathrm{x} \in \hat{\mathrm{A}}$, *are invertible.*

Proof The proof of (i) and (ii) in the proof of Theorem 16.21 in fact show that the statement is true. □

Remark 21.4 If Ker $x_0 = y_0$ and Ker $x_1 = y_1$ where y_0 and y_1 lie in the same orbit (that is, there exists a $g_0 \in G$ such that $t_{g_0}(y_0) = y_1$) then clearly $\pi_{x_0}(b)$ is invertible iff $\pi_{x_1}(b)$ is invertible. So it suffices to consider in Theorems 21.2 and 21.3 only one (any) point in each orbit.

21.5 Though the objects $\pi_x(b)$ are simpler than b they are still too complicated in the case when A and G are arbitrary. But if, for example, $\dim x(A) < \infty$ and $G = \mathbb{Z}^n$ then we can use numerous known results to obtain the conditions of invertibility for $\pi_x(b)$ (see for example Antonevich [7], Gohberg, Feldman [1], Karapetyants, Samko [2], and Simonenko [1, 2]).

Any operator D is invertible iff the following conditions are satisfied:

(a) D is Fredholm;

(b) $\operatorname{ind} D = 0$;

(c) $\operatorname{Ker} D = 0$.

When considering operators of the form $\pi_x(b)$ it is reasonable to look through the conditions (a), (b), (c) because on the one hand it may well happen that $\pi_x(b)$ is Fredholm but not invertible (see 8.9) and on the other hand in a number of cases the conditions (a), (b), (c) for $\pi_x(b)$ can be calculated explicitly. The next theorem is a demonstration of such an approach.

Theorem 21.6 *Let* A *be a separable* C^**–algebra,* $C^*(A, T)$ *possesses the property* (*), $\dim H_x < \infty$ *for every* $x \in \hat{A}$, *and* $\operatorname{Tr} x(a)$ *be a continuous function for any* $a \in A$. *Let the homeomorphism* $t\colon \operatorname{Prim} A \to \operatorname{Prim} A$ (*induced by* T) *be such that for every* $p \in \operatorname{Prim} A$ *there exist a periodic point* p^+ *and a periodic point* p^- *of periods* $n^+ = n(p^+)$ *and* $n^- = n(p^-)$ *such that the trajectory of* $p^+(p^-)$ *is a limit set for the positive (negative) semiorbit* $t^k(p)$, $k = 1,2...$ $(k = -1,-2,...)$.

Then $b = \sum a_k T^k$ *is invertible iff the following conditions hold.*

(1) *For any periodic point* p (*and* $x \in \hat{A}$ *such that* $\operatorname{Ker} x = p$)

$$\det \Delta(p, \lambda) \neq 0 \qquad \textit{when} \qquad |\lambda| = 1 \tag{21.3}$$

where $\Delta(p, \lambda)$ *is a block matrix of dimension* $n(p) \times n(p)$ *with blocks*

$$d_{j,k} = \sum_{l} t^k(x)\left[\dot{a}_{j-k+ln}\right]\lambda^{j-k+ln} \qquad j,k = 1,\dots,p \quad .$$

(2) *For any* p

$$\nu(p^+) = \nu(p^-) \tag{21.4}$$

where

$$\nu(p^\pm) = \frac{1}{2\pi\ n(p^\pm)}\ [\arg\ \det\ \Delta(p^\pm,\lambda)]_{|\lambda|=1} \ .$$

(3) *The equation* $\pi_x(b)h = 0$ *has only the trivial solution for every* $x \in \hat{A}$.

Proof The operator $\pi_x(a)$ is an operator of multiplication (in $l^2(\mathbb{Z}, H_x)$) by the sequence $x(\hat{T}^z(a)) = t^z(x)[a]$. When $z \to \pm\infty$ in view of the conditions on the trajectory of x (or p = Ker x) the sequence $t^z(x)[a]$ tends to a periodic sequence $t^z(x_\pm)[a]$. Thus $\pi_x(b)$ can be viewed as a discrete convolution operator with coefficients tending to periodic ones (such operators have been thoroughly investigated).

Now:

(1) is a necessary and sufficient condition for π_x to be Fredholm;

(2) means that ind $\pi_x(b) = 0$;

(3) means that Ker $\pi_x(b) = 0$. □

Remarks 21.7

(1) If conditions (1) and (2) of the theorem are satisfied then (3) is a necessary and sufficient condition for b to be invertible. This statement is analogous to the Fredholm alternative for integral equations.

Observe that even when (1) and (2) hold the condition Ker $b = 0$ is not sufficient for b to be invertible. Thus only the consideration of objects such as $\pi_x(b)$ gives us an opportunity to obtain this analogy to the classical theorem.

(2) If one compares Theorem 21.6 with Theorem 19.32 then evidently:

(21.3) coincides with (19.47);

(21.4) coincides with (19.48).

Remark 21.8 If one applies Theorem 21.6 to the investigation of Example 20.5 then the necessary conditions (20.7) for b to be invertible follow immediately from Theorem 21.6. But the application of condition (3) of this theorem is not so elementary. If we assume that a_1 and a_2 of b are constants in neighbourhoods of $+\infty$ and $-\infty$ then the investigation of $\pi_x(\bar{b})$ can be reduced to the investigation of a system of linear algebraic equations which gives explicit conditions of invertibility in this example. Notice however that in this approach one has to investigate a matrix of dimension $4k + 2$, where k, defined in the Example, may happen to be very large while the hyperbolic approach reduces this problem to the investigation of matrices of second order.

In many situations, Theorems 21.2 and 21.3 can be used effectively as a theoretical instrument and we shall often use them later.

Let us present now some examples where the conditions of Theorem 21.6 are satisfied.

Example 21.9 In the space $L^2(\mathbb{R}^k)$ consider the algebra A of operators of multiplication by continuous functions $a(x)$ such that there exist uniform limits

$$\lim_{\rho \to +\infty} a(\rho, \varphi) = a(\varphi)$$

where (ρ, φ) are polar coordinates in $\mathbb{R}^k$ and $a(\varphi) \in C(S^{k-1})$. Clearly Prim A is $\mathbb{R}^k$ compactified by the infinitely distant sphere S_∞^{k-1}. For any non–zero $h \in \mathbb{R}^k$ let us define the operator

$$(Tf)(x) = f(x + h) \qquad f \in L^2(\mathbb{R}^k) \quad .$$

Evidently the homeomorphism t induced by T has the form

$$t(x) = \begin{cases} x + h , & x \in \mathbb{R}^k \\ x , & x \in S_\infty^{k-1} \end{cases} \quad .$$

Obviously t acts topologically freely; thus $C^*(A, T)$ possesses the property (*). For any $x \in \mathbb{R}^k$ one has

$$t^n(x) \xrightarrow[n\to\pm\infty]{} \pm \frac{h}{\|h\|} \in S_\infty^{k-1}$$

and all the points of S_∞^{k-1} are fixed, so $C^*(A, T)$ satisfies all the conditions of Theorem 21.6.

Remark The investigation of more general algebras of this sort will be presented in Section 24.

Example 21.10 In the space $L_n^2(\mathbb{R})$ of square integrable vector functions consider the algebra A of operators of multiplications by continuous matrix functions having limits at $+\infty$ and $-\infty$. Here $\hat{A}$ is equal to $\mathbb{R}$ compactified by $+\infty$ and $-\infty$. If we take

$$(Tf)(x) = f(x + h) \quad , \qquad f \in L_n^2(\mathbb{R}) \ , \quad h \neq 0 \ ,$$

then $C^*(A, T)$ clearly satisfies the conditions of Theorems 21.2 and 21.6.

Notice that for any $b \in C^*(A, T)$ the invertibility of $\pi_{+\infty}(b)$ and $\pi_{-\infty}(b)$ is a necessary condition for $\pi_x(b)$, $x \in \mathbb{R}$, to be Fredholm. Thus the invertibility of $\pi_{-\infty}(b)$ and $\pi_{+\infty}(b)$ follows from the invertibility of $\pi_x(b)$, $x \in \mathbb{R}$. Recall also that by Remark 21.4 it is enough to consider only one point x on each orbit. So we conclude that b is invertible iff all the operators $\pi_x(b)$, $x_0 \le x \le x_0 + h$, are invertible (here x_0 is an arbitrary point in $\mathbb{R}$). Observe that in this example the set of elements $\pi_x(b)$, $x_0 \le x \le x_0 + h$, depends on x continuously and consequently is a compact set. If all $\pi_x(b)$ are invertible then the set $\left\{\pi_x^{-1}(b)\right\}$, $x_0 \le x \le x_0 + h$, is again compact. Thus there exists a C such that $\left\|\pi_x^{-1}(b)\right\| < C$, $x_0 \le x \le x_0 + h$, and we see that in this particular example the main problem arising in Theorem 21.2 can be solved in a simple way.

Note however that in general the set $\pi_x(b)$, $x \in \hat{A}$, does not depend continuously on x. Even in this example the set $\pi_x(b)$, $x \in -\infty \cup \mathbb{R} \cup +\infty$, is not continuous at $+\infty$ and $-\infty$ and is not compact.

Hyperbolicity and the trajectorial approach

The trajectorial approach gives us an additional tool to investigate the hyperbolic properties of operators of the form aT (not using linear extensions).

Theorem 21.11 *Let* $C^*(A, T)$ *possesses the property* (*). *Then* aT *is hyperbolic iff all the operators* $\pi_x(aT)$, $x \in \hat{A}$, *are hyperbolic.*

Proof Evident in view of Theorem 21.2. □

21.12 Now we shall consider another notion which appears when considering hyperbolic extensions – the so-called exponential dichotomy.

With any element $a \in A$ let us link the cocycle $\mathcal{A}: \hat{A} \times \mathbb{Z}_+ \to L(H_x)$, $x \in \hat{A}$, defined by the formula

$$\mathcal{A}(x, n) = \hat{a}(n)(x) \qquad x \in \hat{A}\ , \quad n \in \mathbb{Z}_+ \tag{21.5}$$

where $a(n) \in A$ is the element defined by the equation

$$(aT)^n = T^n a(n) \ . \tag{21.6}$$

If a is invertible then (21.5) and (21.6) define the cocycle $\mathcal{A}(\cdot, n)$ for every $n \in \mathbb{Z}$.

Since

$$T^{n+k}a(n + k) = (aT)^{n+k} = (aT)^n(aT)^k = T^n a(n)\cdot T^k a(k)$$

$$= T^{n+k}[\hat{T}^k(a(n))\cdot a(k)]$$

we conclude that

$$\mathcal{A}(x, n + k) = \mathcal{A}(t^k(x), n)\cdot \mathcal{A}(x, k) \tag{21.7}$$

where $t: \hat{A} \to \hat{A}$ is the homeomorphism of $\hat{A}$ induced by $\hat{T}$.

21.13 An operator aT (where a is invertible) is called *exponentially* *dichotomic at a point* $x \in \hat{A}$ if there exists a projection $p(x) \in L(H_x)$ and constants $C, \gamma > 0$ such that

$$\left\| \mathcal{A}(x,n_1)p(x)\mathcal{A}^{-1}(x,n_2)\right\| \leq Ce^{-\gamma(n_1-n_2)} \qquad +\infty > n_1 \geq n_2 > -\infty$$

$$\left\| \mathcal{A}(x,n_1)(1-p(x))\mathcal{A}^{-1}(x,n_2)\right\| \leq Ce^{-\gamma(n_2-n_1)} \qquad +\infty > n_2 \geq n_1 > -\infty .$$

21.14 For any $x \in \hat{A}$ consider the space $l^2(\mathbb{Z}, H_x)$. A projection $p(x) \in L(H_x)$ defines the projection $P \in L(l^2(\mathbb{Z}, H_x))$, the diagonal operator (operator–valued function) $\{p(n)\}_{n\in\mathbb{Z}}$, where

$$p(0) = p(x) , \quad p(n) = \pi_x(aT)^n p(0)\pi_x(aT)^{-n} .$$

In view of the construction of P we have

$$\pi_x(aT)P = P\pi_x(aT) . \tag{21.8}$$

Theorem 21.15 *If* $C^*(A, T)$ *possesses the property* (*) *and* $a \in A$ *is an invertible element then* aT *is exponentially dichotomic in* $x \in \hat{A}$ *iff* $\pi_x(aT)$ *is hyperbolic.*

Proof Let aT be exponentially dichotomic in x and let $p(x)$ be the projection mentioned in **21.13**.

Then using (21.7) and (21.8) and **21.13** we have

$$r(\pi_x(aT)P) = \lim_{n\to\infty} \left\|\pi_x(aT)^n P^n\right\|^{\frac{1}{n}} = \lim_{n\to\infty} \left\|\pi_x(T^n a(n))P\right\|^{\frac{1}{n}}$$

$$= \lim_{n\to\infty} \sup_{k\in\mathbb{Z}} \left\|\pi_x(T^n)(\mathcal{A}(t^k x,n)p(k))\right\|^{\frac{1}{n}}$$

$$= \lim_{n\to\infty} \sup_{k\in\mathbb{Z}} \left\| \pi_x(T^n)(\mathcal{A}(t^k x,n)\pi_x(T^k)\mathcal{A}(x,k)p(0)\mathcal{A}^{-1}(x,k)\pi_x(T^k)\right\|^{\frac{1}{n}}$$

$$= \lim_{n\to\infty} \sup_{k\in\mathbb{Z}} \left\| \pi_x(T^{n+k})\mathcal{A}(x,\ n + k)p(0)\mathcal{A}^{-1}(x,k)\pi_x(T^k)\right\|^{\frac{1}{n}}$$

$$= \lim_{n\to\infty} \sup_{k\in\mathbb{Z}} \left\| \mathcal{A}(x,\ n + k)p(x)\mathcal{A}^{-1}(x,k)\right\|^{\frac{1}{n}} \leq e^{-\gamma} \quad .$$

In an analogous way one can prove the inequality

$$r\left([\pi_x(aT)\ (I - P)]^{-1}\right) \leq e^{-\gamma} \quad .$$

Thus $\pi_x(aT)$ is hyperbolic.

Now let $\pi_x(aT)$ be hyperbolic. An argument analogous to that used in Lemma 19.1 shows that in this case the spectral projection

$$P = \frac{1}{2\pi i} \int_{|\lambda|=1} (\lambda - \pi_x(aT))^{-1} d\lambda$$

belongs to $\pi_x(A)$. Thus P can be expressed as an operator–valued function

$$P = \{p(n)\}_{n\in\mathbb{Z}} \qquad p(n) \in L(H_x) \quad .$$

Since P is a spectral projection it commutes with $\pi_x(aT)$ which means that

$$p(n) = \pi_x(aT)^n\ p(0)\ \pi_x(aT)^{-n} \quad . \tag{21.9}$$

Let us set

$$p(x) = p(0) \quad .$$

We have (using (21.9))

$$\|\mathcal{A}(x,n)p(x)\mathcal{A}^{-1}(x,k)\| = \left\|\pi_x(T^n)\ \mathcal{A}(x,n)\ p(0)\ \mathcal{A}^{-1}(x,k)\pi_x(T^{-k})\right\|$$

$$= \left\|\pi_x(aT)^n\ p(0)\ \pi_x(aT)^{-k}\right\| = \left\|\pi_x(aT)^{n-k}\ p(k)\right\|$$

$$= \left\|\pi_x(T^{n-k})[\mathcal{A}(t^k x,\ n - k)\ p(k)]\right\| \le \left\|\pi_x[(aT)P]^{n-k}\right\|$$

$$\le C\ r(\pi_x(aT)P)^{n-k} \quad . \tag{21.10}$$

As P is the Riesz projection and $\pi_x(aT)$ is hyperbolic

$$r(\pi_x(aT)P) = e^{-\gamma} < 1$$

for some $\gamma > 0$.

Thus (21.10) means that aT is exponentially dichotomic in x. □

Corollary 21.16 *If* $C^*(A, T)$ *possesses the property* (*) *and* a *is an invertible element then* aT *is hyperbolic iff it is exponentially dichotomic at every* $x \in \hat{A}$.

Proof Use Theorems **21.11**, 21.15. □

Remark Since the hyperbolicity of aT is linked with the hyperbolicity of the corresponding linear extension β (see **19.12** and Theorem 19.13) the foregoing results **21.11–21.16** can be used in an obvious way to obtain conditions for the hyperbolicity of β.

22. The spectral radius of elements in $C^*(A, T_g)$

A spectral radius formulae for an operator $b = aT$ in the case of a commutative algebra was obtained in Section 4. In the case in which the operator $b \in C^*(A, T_g)$ is of the general form

$$b = \sum a_g T_g$$

there is no hope of obtaining a simple and effective formula for the spectral radius of b. Here we shall deduce some formulae for the spectral radius of b in terms of its coefficients $a_g \in A$.

22.1 Let $C^*(A, T_g)$ be a C^*–algebra which possesses the property (*) (see 12.2). For any element $b = \sum_{g\in M} a_g T_g$ where M is a finite subset of G we can define elements $N_1(b)$ and $N_2(b)$ by the equations

$$N_1(b) = \sum_{g\in M} |a_g|$$

$$N_2(b) = \sum_{g\in M} a_g a_g^* = N(bb^*)$$

where $|a_g| = \sqrt{a_g a_g^*}$ and $N(b)$ was defined in 12.2.

We denote by $\|b\|_p$, $1 \le p < \infty$, and $\|b\|_\infty$ the norms

$$\|b\|_p = \left[\sum_{g\in M} \|a_g\|^p\right]^{\frac{1}{p}}$$

$$\|b\|_\infty = \max_{g\in M} \|a_g\| \quad .$$

A group G is said to be *subexponential* if for any finite set $F \subset G$

$$\lim_{n\to\infty} |F^n|^{\frac{1}{n}} = 1 \quad . \tag{22.1}$$

Remark If G is subexponential then G is amenable. Evidently commutative groups and groups with a polynomial growth in the number of words of length n are subexponential. The question of whether there exist subexponential groups in which the number of words of length n grows faster than any degree of n is known as Milnor's problem. This problem was solved positively by R.I. Grigorchuk [1].

Theorem 22.2 *Let* $C^*(A, T_g)$ *be a* C^**–algebra which possesses the property* (*) *and let* G *be a subexponential group. Then for any element* b *of type*

$$b = \sum_{g \in M} a_g T_g \quad , \qquad |M| < \infty$$

the spectral radius r(b) *is given by the formula*

$$r(b) = \lim_{n\to\infty} \left\| N_1(b^n) \right\|_1^{\frac{1}{n}} = \lim_{n\to\infty} \left\| N_2(b^n) \right\|_2^{\frac{1}{2n}} = \lim_{n\to\infty} \|b^n\|_\infty^{\frac{1}{n}} = \lim_{n\to\infty} \|b^n\|_p^{\frac{1}{n}} .$$

Proof To begin with we prove the following lemma which is useful in its own right.

Lemma 22.3 *For any finite set* $d_1,\dots,d_k$ *of elements of the* C^**–algebra*

$$\frac{1}{k} \left\| \sum_{i=1}^{k} d_i d_i^* \right\| \le \left\| \sum_{i=1}^{k} |d_i| \right\|^2 \le k \left\| \sum_{i=1}^{k} d_i d_i^* \right\| . \tag{22.2}$$

Proof Let the C^*–algebra be realized as some algebra of operators acting on a Hilbert space H. Then for any $\xi \in H$ we have

$$\left\| \sum_{i=1}^{k} |d_i|\xi \right\|^2 \le \left(\sum_{i=1}^{k} \| |d_i|\xi \| \right)^2 \le k \left(\sum_{i=1}^{k} \| |d_i|\xi \|^2 \right)$$

$$= k \left\langle \sum_{i=1}^{k} d_i d_i^* \xi, \xi \right\rangle \le k \left\| \sum_{i=1}^{k} d_i d_i^* \right\| \cdot \|\xi\|^2$$

which implies the right–hand inequality in (22.2).

On the other hand the left hand inequality follows from the relations

$$\left\|\sum_{i=1}^{k} d_i d_i^*\right\| = \sup_{\|\xi\|=1} \sum_{i=1}^{k} \langle d_i d_i^* \xi,\xi\rangle$$

$$= \sup_{\|\xi\|=1} \sum_{i=1}^{k} \langle |d_i|\xi,\ |d_i|\xi\rangle \le \sum_{i=1}^{k} \sup_{\|\xi\|=1} \langle |d_i|\xi,\ |d_i|\xi\rangle$$

$$\le \sum_{i=1}^{k} \sup_{\|\xi\|=1} \left\langle \sum_{j=1}^{k} |d_j|\xi,\ \sum_{j=1}^{k} |d_j|\xi \right\rangle$$

$$= \sum_{i=1}^{k} \left\|\sum_{j=1}^{k} |d_j|\right\|^2 = k\left\|\sum_{j=1}^{k} |d_j|\right\|^2 . \qquad \square$$

Now we proceed to prove the theorem. First we prove that

$$r(b) = \lim_{n\to\infty} \left\|N_2(b^n)\right\|^{\frac{1}{2n}} .$$

For the element $b = \sum a_g T_g$ denote by $M(b)$ the set of elements $g \in G$ such that $a_g \neq 0$. By applying the right–hand inequality in (22.2) and the property (*) it follows that for any element b with finite $M(b)$

$$\|N(bb^*)\| \le \|b\|^2 = \left\|\sum_{g\in M(b)} a_g T_g\right\|^2 \le |M(b)|\ \|N(bb^*)\|$$

or by the definition of $N_2(b)$

$$\|N_2(b)\| \le \|b\|^2 \le |M(b)|\ \|N_2(b)\| . \tag{22.3}$$

Applying this inequality to b^n it follows that

$$\left\|N_2(b^n)\right\| \le \|b^n\|^2 \le |M(b^n)|\ \left\|N_2(b^n)\right\|$$

or

$$\left\|N_2(b^n)\right\|^{\frac{1}{2n}} \le \|b^n\|^{\frac{1}{n}} \le |M(b^n)|^{\frac{1}{2n}} \left\|N_2(b^n)\right\|^{\frac{1}{2n}} .$$

Therefore

$$\overline{\lim_{n\to\infty}} \left\|N_2(b^n)\right\|^{\frac{1}{2n}} \le r(b) \le \underline{\lim_{n\to\infty}} |M(b^n)|^{\frac{1}{2n}} \left\|N_2(b^n)\right\|^{\frac{1}{2n}} .$$

Observe now that $M(b^n) \subset [M(b)]^n$. This leads to

$$1 \le |M(b^n)|^{\frac{1}{2n}} \le |[M(b)]^n|^{\frac{1}{2n}} .$$

Using the subexponentiality of the group G we find that

$$\lim_{n\to\infty} |M(b^n)|^{\frac{1}{2n}} = 1$$

and consequently

$$r(b) = \lim_{n\to\infty} \left\|N_2(b^n)\right\|^{\frac{1}{2n}} .$$

Now we prove that

$$\lim_{n\to\infty} \left\|N_1(b^n)\right\|^{\frac{1}{n}} = \lim_{n\to\infty} \left\|N_2(b^n)\right\|^{\frac{1}{2n}} .$$

Applying Lemma 22.3 we have that

$$\frac{1}{|M(b^n)|} \left\|N_2(b^n)\right\| \le \left\|N_1(b^n)\right\|^2 \le |M(b^n)| \left\|N_2(b^n)\right\| .$$

But we have already observed that

$$\lim_{n\to\infty} |M(b^n)|^{\frac{1}{2n}} = 1$$

so we conclude that

$$\lim_{n\to\infty} \left\|N_1(b^n)\right\|^{\frac{1}{n}} = \lim_{n\to\infty} \left\|N_2(b^n)\right\|^{\frac{1}{2n}} .$$

To obtain the remaining equalities mentioned in the statement of the theorem notice that

$$\|b\|_\infty \leq \|N_1(b)\|$$

which is true since for every $g_0 \in M$ one has

$$\|a_{g_0}\| = \||a_{g_0}|\| = \sup_{\|\xi\|=1} \left\langle |a_{g_0}|\xi,\xi \right\rangle \leq \sup_{\|\xi\|=1} \left\langle \sum_M |a_{g_0}|\xi,\xi \right\rangle$$

$$= \|N_1(b)\|$$

so

$$\|b^n\| \leq \|b^n\|_1 \leq |M(b^n)| \ \|b^n\|_\infty < |M(b^n)| \ \left\|N_1(b^n)\right\|$$

and as

$$\lim_{n\to\infty} |M(b^n)|^{\frac{1}{n}} = 1$$

$$\lim_{n\to\infty} \|N_1(b^n)\|^{\frac{1}{n}} = r(b)$$

we deduce that

$$r(b) = \lim_{n\to\infty} \|b^n\|_1^{\frac{1}{n}} = \lim_{n\to\infty} \|b^n\|_\infty^{\frac{1}{n}} .$$

Finally recall that by the Jensen inequality (Beckenbach, Bellman [1], Ch. I, § 16)

$$\|b\|_1 \geq \|b\|_p \geq \|b\|_\infty$$

which in view of the previous formula finishes the proof. □

Now we pass to another formula for the spectral radius of b which is obtained with the help of the 'Fourier transform' of b.

22.4 If $C^*(A, T_g)$ possesses the property (*) then with any finite sum $b = \sum a_g T_g$ we can associate an operator–valued function on $\widehat{C^*(G)}$ (see 2.91) defined by

$$b = \sum a_g T_g \to b(\tau) = \sum a_g \otimes \tau(g) \quad , \qquad \tau \in \widehat{C^*(G)} \tag{22.4}$$

where $a_g \otimes \tau(g)$ is the C^*–tensor product (see 11.17) (we can assume that $a_g \otimes \tau(g)$ acts in $H \otimes H_\tau$ where H is a representation space for some faithful representation π: $A \to L(H)$).

Define the norm of the function $b(\tau)$ by

$$\|b(\tau)\|_{\widehat{C^*(G)}} = \sup_{\tau \in \widehat{C^*(G)}} \|b(\tau)\| \quad .$$

Theorem 22.5 *Let* $C^*(A, T_g)$ *possesses the property* (*) *and let* G *be a subexponential group. Then for any element*

$$b = \sum_M a_g T_g \quad , \qquad |M| < \infty$$

we have

$$r(b) = \lim_{n \to \infty} \|b^n(\tau)\|_{\widehat{C^*(G)}}^{\frac{1}{n}} \quad . \tag{22.5}$$

Proof We shall divide the proof into two steps.

Step I. Consider the mapping

$$b = \sum a_g T_g \to \bar{b} = \sum a_g \otimes g \in A \underset{I}{\times} G$$

where I denotes the representation of G by trivial automorphisms.

Lemma 22.6 $\|\bar{b}\| = \|b(\tau)\|_{\widehat{C^*(G)}}$.

Proof In view of Lemma 11.18 we have

$$A \underset{I}{\times} G \cong A \otimes C^*(G)$$

where the isomorphism is established via the natural regular representation of $A \underset{I}{\times} G$ (see the proof of Lemma 11.18).

But on the other hand

$$C^*(G) \cong \bigoplus_{\tau \in \widehat{C^*(G)}} \tau(C^*(G))$$

which means that

$$A \underset{I}{\times} G \overset{\varphi}{\cong} A \otimes \left[\bigoplus_{\tau \in \widehat{C^*(G)}} \tau(C^*(G)) \right] \overset{\psi}{\cong} \bigoplus_{\tau \in \widehat{C^*(G)}} [A \otimes \tau(C^*(G))]$$

and thus the mapping

$$\bar{b} \to [\psi \circ \varphi](\bar{b}) = \bigoplus_{\tau} b(\tau)$$

is an isomorphism so lemma is proved. □

Step II As $A \underset{I}{\times} G$ possesses the property (*) and

$$\|N_2(b)\| = \left\|\sum a_g a_g^*\right\| = \left\|\sum \bar{a}_g \bar{a}_g^*\right\| = \left\|N_2(\bar{b})\right\|$$

we have (applying (22.3))

$$\|N_2(b)\| \le \|\bar{b}\|^2 \le |M(b)|\ \|N_2(b)\|^2$$

and (using Lemma 22.6)

$$\|N_2(b)\| \le \|b(\tau)\|^2_{\widehat{C^*(G)}} \le |M(b)|\ \|N_2(b)\| \quad . \tag{22.6}$$

We have already proved that

$$r(b) = \lim_{n\to\infty} \left\|N_2(b^n)\right\|^{\frac{1}{2n}}$$

and by the same argument as in the proof of Theorem 22.2 we conclude that $\lim_{n\to\infty} |M(b^n)|^{\frac{1}{2n}} = 1$. Consequently (22.6) implies (passing to the limit)

$$r(b) = \lim_{n\to\infty} \|b^n(\tau)\|^{\frac{1}{n}}_{\widehat{C^*(G)}} \quad . \qquad \square$$

22.7 In the case when G is a commutative group all the irreducible representations of $C^*(G)$ are one–dimensional and are obtained from the characters $\chi \in \hat{G}$ (Theorem 2.93). The mapping (22.4) in this case has the form

$$b = \sum a_g T_g \to b(\chi) = \sum a_g \chi(g) \quad , \qquad x \in \hat{G} = \widehat{C^*(G)} \tag{22.7}$$

and so $b(\chi)$ here is an A–valued function on $\hat{G}$.

The formula (22.5) can now be rewritten as

$$r(b) = \lim_{n\to\infty} \|b^n(\chi)\|_{\hat{G}}^{\frac{1}{n}} \tag{22.8}$$

where $\|b(\chi)\|_{\hat{G}} = \sup_{\hat{G}} \|b(\chi)\|$.

In particular if $G = \mathbb{Z}^n$ then $\hat{\mathbb{Z}}^n = S^n$ and any character has the form

$$\lambda(z) = \lambda^z$$

where $z = (z_1,\dots,z_n) \in \mathbb{Z}^n$, $\lambda = (\lambda_1,\dots,\lambda_n) \in S^n$ and $\lambda^z = \lambda_1^{z_1}\cdot\ldots\cdot\lambda_n^{z_n}$. Thus the mapping (22.7) here is

$$b = \sum a_z T^z \to \sum a_z \lambda^z = b(\lambda) \quad , \qquad \lambda \in S^n$$

where $T^z = T_1^{z_1}\cdot\ldots\cdot T_n^{z_n}$.

In this case (22.8) is reduced to

$$r(b) = \lim_{n\to\infty} \|b^n(\lambda)\|_{S^n}^{\frac{1}{n}}$$

where

$$\|b(\lambda)\|_{S^n} = \max_{S^n} \|b(\lambda)\| \quad .$$

22.8 The formulae obtained in Theorems 22.2 and 22.5, though rather useful as a theoretical instrument (see Notes and Remarks) an in some cases lead to effective results (see Section 23), can hardly be used in the general situation for direct evaluation of the spectral radius. Here we present one non–trivial and rather peculiar example of the direct computation of the spectral radius with the help of Theorem 22.2.

Example In the space $L^2(\mathbb{R})$ let us consider an operator

$$b = a_1 T_{h_1} + a_2 T_{h_2}$$

where a_i $i = \overline{1,2}$ $(\in A)$, are operators of multiplication by measurable bounded functions and T_{h_i}, $i = 1,2$, act as

$$(T_{h_i} f)(x) = f(x + h_i) \qquad i = \overline{1,2} \quad , \quad h_i \in \mathbb{R} \quad .$$

We present here a formula for the spectral radius of b in the situation when h_i and a_i satisfy the following conditions:

(1) $h_1 > 0$, $h_2 < 0$;

(2) there exists a point t such that $a_i(x) = 0$, $i = \overline{1,2}$, if $x \bar{\in} [t + h_2, t + h_1)$;

(3) the functions $a_i(x)$, $i = \overline{1,2}$, are Riemann integrable.

If h_i *and* a_i, $i = \overline{1,2}$, *satisfy the conditions* (1)–(3) *then the spectral radius of* b *is evaluated by the formulae*

(i) *if* h_1/h_2 is an irrational number then

$$r(b) = \exp \frac{1}{h_1 - h_2} \left[\int_{t+h_2}^{t} \ln|a_1(x)|dx + \int_{t}^{t+h_1} \ln|a_2(x)|dx \right]$$

(ii) *if* $\dfrac{h_1}{h_1 - h_2} = \dfrac{p}{q}$ *where* $\dfrac{p}{q}$ is an irreducible fraction then

$$r(b) = \operatorname*{ess\ sup}_{\left[t,\ t + \frac{h_1}{p}\right)} \left[\prod_{j=0}^{p-1} a_2\left(x + \frac{h_1}{p} j\right) \prod_{s=1}^{q-p} a_1\left(x - \frac{h_1}{p} s\right) \right]^{\frac{1}{q}} .$$

Proof Without loss of generality one can assume that $t = 0$. We shall use the formula

$$r(b) = \lim_{n \to \infty} \left\| N_1(b^n) \right\|^{\frac{1}{n}} \tag{22.9}$$

(see Theorem 22.2; clearly $C^*(A, T_h)_{h\in\mathbb{R}}$ possesses the property (*)).

Direct computations show that

$$b^n = \sum_{m=0}^{n} a_{nm} T_{h_1}^m T_{h_2}^{n-m} \tag{22.10}$$

where

$$a_{nm}(x) = \sum_{\nu(k)=m} a_{k_1}(x)a_{k_2}(x + h_{k_1})a_{k_3}(x + h_{k_1} + h_{k_2})$$

$$\cdot \ldots \cdot a_{k_n}(x + h_{k_1} + \ldots + h_{k_{n-1}}) \tag{22.11}$$

and where $k = (k_1,\ldots,k_n)$ is a multi–index, $k_i \in \{1, 2\}$, and $\nu(k)$ is the number of numbers $j \in \overline{1,n}$ such that $k_j = 1$.

By condition (2) if $a_{nm}(x) \neq 0$ then all the points $x_0 = x$, $x_1 = x + h_{k_1},\ldots,x_{n-1} = x + h_{k_1} + \ldots + h_{k_{n-1}}$ belong to the interval $\Delta = [h_2, h_1)$. The key fact is that in this case the sequence $k_1,k_2,\ldots,k_{n-1}$ is defined uniquely.

Indeed for any $x \in \Delta$ only one point $x + h_1$ or $x + h_2$ belongs to Δ, thus the sequence x_j is defined by the rule

$$x_{j+1} = \begin{cases} x_j + h_1 , & x_j < 0 \\ x_j + h_2 , & x_j \geq 0 \end{cases} \tag{22.12}$$

that is

$$k_{j+1} = \begin{cases} 1 , & x_j < 0 \\ 2 , & x_j \geq 0 \end{cases} . \tag{22.13}$$

Obviously for every $x \in \Delta$ the sequence constructed via the rule (22.12) exists. Thus under condition (2) for every $x \in \Delta$ all the summands in (22.11) except one (depending on x) are zero, that is

$$N_1(b)^n(x) = \sum_{m=0}^{n} |a_{nm}(x)| = |a_{k_1}(x)a_{k_2}(x_1)\cdot\ldots\cdot a_{k_n}(x_{n-1})| \tag{22.14}$$

where the sequence $k_1,k_2,\ldots,k_{n-1}$ depends on x.

Note, by the way, that for the multi–index $k(x) = (k_1,\ldots,k_n)$ the number $\nu(k)$ may be different for different points x, which means that the number of non–zero summands in (22.10) may be more than one.

Let us establish a bijective mapping of $\Delta = [h_2, h_1)$ onto the unit circle S^1 by the formula

$$x \to \tilde{x} = e^{i\frac{2\pi x}{h_1 - h_2}} .$$

One can easily verify that under this mapping the mapping $\tilde{x}_j \to \tilde{x}_{j+1}$ is given by

$$\tilde{x}_{j+1} = e^{i\omega}\,\tilde{x}_j \tag{22.15}$$

where $\omega = \dfrac{2\pi h_1}{h_1 - h_2}$.

But (22.15) is simply the rotation of the circle by an angle ω. If $\frac{\omega}{2\pi}$ is irrational (that is, h_1/h_2 is irrational) then in view of condition (3) and (22.14) we have

$$\lim_{n\to\infty} \left\|N_1(b^n)\right\|^{\frac{1}{n}} = \lim_{n\to\infty} \sup_{\tilde{x}\in S^1} \left[\prod_{j=0}^{n-1} |f(\tilde{x}\cdot e^{ij\omega})|\right]^{\frac{1}{n}}$$

$$= \exp \frac{1}{2\pi} \int_{S^1} \ln|f(\tilde{x})|\,d\tilde{x}$$

where

$$f(\tilde{x}) = \begin{cases} a_1(x) , & x \in [h_2, 0) \\ a_2(x) , & x \in [0, h_1) \end{cases}$$

which implies (i).

If $\frac{\omega}{2\pi} = \frac{p}{q}$ where $\frac{p}{q}$ is an irreducible fraction then

$$\lim_{n\to\infty} \sup_{\tilde{x}\in S^1} \left[\prod_{j=0}^{n-1} |f(\tilde{x}\cdot e^{ij\omega})|\right]^{\frac{1}{n}} = \sup_{\tilde{x}\in S^1} \left[\prod_{j=0}^{q-1} |f(\tilde{x}\cdot e^{ij\omega})|\right]^{\frac{1}{q}}$$

which by routine computation implies (ii). □

23. Estimates for the spectral radius of elements in $C^*(A, T_g)$

In the previous section we obtained some formulae for the spectral radius of elements in $C^*(A, T_g)$. Although these formulae definitely have some theoretical and aesthetic value (for example, by means of them isomorphism theorems (such as Corollary 12.17) for subexponential groups can be proved (see Brenner [2, 4])), if we really want to calculate the numerical value of the spectral radius we can hardly use them.

In this section we deduce some rather effective and simple to work with estimates for the spectral radius of elements in $C^*(A, T_g)$. In some cases (see Corollaries 23.12, 23.14) these estimates lead to explicit formulae.

23.1 Let us fix any finite set $M \subset G$. By G_M we denote the subgroup generated by M. Consider any point $x_0 \in \text{Prim } A$. We say that $P(x_0) \subset \text{Prim } A$ is a *limit set* for M and x_0 if $P(x_0)$ is the minimal t_g–invariant ($g \in M$) set such that:

(i) $P(x_0)$ contains all the points which are not separated from $P(x_0)$ (a point x is not separated from $P(x_0)$ iff for any neighbourhood $V(x)$ of x and for any neighbourhood $U(P(x_0))$ of $P(x_0)$ we have $U(P(x_0)) \cap V(x) \neq \emptyset$);

(ii) for any neighbourhood $U(P(x_0))$ there exists a number $K(U) \in \mathbb{N}$ such that for every x belonging to the G_M*–orbit* of x_0 (that is,

$x \in \{t_g(x_0)\}_{g \in G_M})$

$$\{M^n(x)\} \subset U$$

for all $n \in \mathbb{N}$ excluding perhaps a finite set of numbers

$$n_1, n_2, \ldots, n_{k_x} \qquad k_x < K(U)$$

where $\{M^n(x)\} = \{t_g(x)\}_{g \in M^n}$.

Remarks

(1) The limit set $P(x_0)$ is evidently closed. If Prim A is Hausdorff then condition (i) means simply the closedness of $P(x_0)$. In general (i) means more than closedness. For example, if one considers the example in Remark 19.29(1) he can notice that if, say, $\{1 - 0\} \in P(x_0)$ then $\{1\}$ and $\{1 + 0\}$ also belong to $P(x_0)$.

(2) If P_1 and P_2 both satisfy conditions (i) and (ii) then $[P_1 \cap P_2] \neq \emptyset$. Indeed, if $P_1 \cap P_2 = \emptyset$ then P_1 and P_2 are separated from each other (since they are both quasicompact) so taking any non–intersecting neighbourhoods $U_1(P_1)$ and $U_2(P_2)$ one arrives at a contradiction with (ii). Now obviously the set $[P_1 \cap P_2]$ also satisfies (i) and (ii). Thus (since the whole of Prim A satisfies (i) and (ii)) in view of Zorn's lemma we conclude that the set $P(x_0)$ with all the desired properties does exist.

(3) If G_M is commutative then the minimality of $P(x_0)$ and (i), (ii) implies the t_g–invariance of $P(x_0)$.

Indeed since $P(x_0)$ possesses (i) and (ii) it follows that $t_{g_0}(P(x_0))$, $g_0 \in G_M$, also possesses (i) and (ii). For (i) this is evident, and to verify (ii) observe that any x in a G_M–orbit of x_0 has the form $x = t_{g_0}^{-1}(x_1)$ (for some x_1 in the G_M–orbit of x_0, and any neighbourhood $U(t_{g_0}(P(x_0)))$ has the form $t_{g_0}[U_1(P(x_0))]$ for some neighbourhood $U_1(P(x_0))$.

Now the minimality of $P(x_0)$ implies

$$t_{g_0}(P(x_0)) \supset P(x_0)$$

for any $g_0 \in G_M$. And since

$$t_{g_0}(P(x_0)) \supset P(x_0) \Leftrightarrow P(x_0) \supset t_{g_0^{-1}}(P(x_0))$$

we conclude that

$$t_{g_0}(P(x_0)) = P(x_0) \quad .$$

(4) If M consists of a single element g_0 then $P(x_0) = P_+(x_0) \cup P_-(x_0)$ where $P_\pm(x_0)$ is the set of all the limit points of the sequence $\left\{t_{g_0}^{\pm n}(x_0)\right\}$, $n = 1,2,\ldots$.

Lemma 23.2 *Let* P *be any set satisfying* (i) *in* 23.1, *and let*

$$\sup_{\mathrm{P}} \breve{a}(x) = \mathrm{R} \quad .$$

Then for any $\varepsilon > 0$ *there exists* U(P) *such that*

$$\sup_{\mathrm{U(P)}} \breve{a}(x) < \mathrm{R} + \varepsilon$$

(*recall that* $\breve{a}(x) = \|[a/x]\|$ $x \in$ Prim A).

Proof Consider the set

$$W = \{x\colon \breve{a}(x) \geq R + \varepsilon\} \quad .$$

W is quasicompact (Theorem 2.87). Suppose that

$$W \cap U(P) \neq \emptyset \qquad \text{for any} \quad U(P) \quad .$$

The sets $[W \cap U(P)]$ form a filter. But since W is quasicompact, any filter has a limit point, that is, there exists an $x_0 \in W$ such that

$$V(x_0) \cap [W \cap U(P)] \neq \emptyset$$

for any neighbourhood $V(x_0)$ and any $U(P)$. Thus x_0 is not separated from P and it follows that $x_0 \in P$. So $W \cap P \neq \emptyset$ and we arrived at a contradiction. □

Remark 23.3 The lemma implies that for any finite number of elements $a_1,\ldots,a_n$ and any $\varepsilon > 0$ there exists a $U(P)$ such that

$$\sup_{U(P)} \overbrace{\left[\sum_{i=1}^{n} a_i\right]}(x) < \sup_{P} \overbrace{\left[\sum_{i=1}^{n} a_i\right]}(x) + \varepsilon$$

(just put $a = \sum a_i$). But it may well happen that there does not exist a $U(P)$ such that

$$\sup_{U(P)} \left[\sum \breve{a}_i(x)\right] < \sup_{P} \left[\sum \breve{a}_i(x)\right] + \varepsilon \quad . \tag{23.1}$$

Example Consider the C^*–algebra A of continuous 2×2 matrix functions $a(t)$ on $[0, 1]$ such that $a(1)$ is diagonal. In this case Prim $A = ([0, 1) \cup \{1^+\} \cup \{1^-\})$ where the topology on $[0, 1)$ is the standard one while basic neighbourhoods for $\{1^+\}$ and $\{1^-\}$ are

$$\{1^+\} \cup (1 - \eta,\ 1) \qquad \text{for} \quad \{1^+\}$$
$$\{1^-\} \cup \{1 - \delta,\ 1) \qquad \text{for} \quad \{1^-\} \quad .$$

Let $P = \{1^+\} \cup \{1^-\}$. Take any a_1 and a_2 such that

$$a_1(1) = \begin{bmatrix} 0 & 0 \\ 0 & 1 \end{bmatrix} \qquad \text{and} \qquad a_2(1) = \begin{bmatrix} 1 & 0 \\ 0 & 0 \end{bmatrix} \quad .$$

Then $\breve{a}_1(1^+) = 0$, $\breve{a}_1(1^-) = 1$, $\breve{a}_2(1^+) = 1$, $\breve{a}_2(1^-) = 0$, and consequently

$$\sup_P \ [\breve{a}_1(x) + \breve{a}_2(x)] = 1 \quad .$$

On the other hand since $\breve{a}_1(x) \approx 1$ and $\breve{a}_2(x) \approx 1$ when $x \in [0,\ 1)$ and is near $\{1\}$ we conclude that in any neighbourhood $U(P)$

$$\sup_{U(P)} [\breve{a}_1(x) + \breve{a}_2(x)] \geq \frac{3}{2} \quad .$$

Unfortunately in our forthcoming considerations we shall need sums of the form

$$\sum \breve{a}_i(x)$$

and some estimates like (23.1). So how can we cope with this situation? One can see that if in the foregoing example we consider the points $\{1^+\}$ and $\{1^-\}$ as one point (which is really natural in view of the definition of A) then $\|a_1(1)\| = \|a_2(1)\| = 1$ and clearly there exists a neighbourhood $u(1)$ such that

$$\sup_{u(1)} [\breve{a}_1(x) + \breve{a}_2(x)] < \|a_1(1)\| + \|a_2(1)\| + \varepsilon$$

so everything will be alright.

The procedure of gluing together points such as $\{1^+\}$ and $\{1^-\}$ is called Hausdorffization and we shall now describe it.

23.4 Let x_1 and x_2 be two points in Prim A. We define the equivalence relation $x_1 \sim x_2$ iff

$$f(x_1) = f(x_2) \qquad \text{for every} \quad f \in C(\text{Prim } A)$$

and denote by $\tilde{x}$ the equivalence class of x under this relation.

By $\breve{a}(\tilde{x})$ we denote $\left\| \left[{}^{a}/_{J_{\tilde{x}}} \right] \right\|$ where $J_{\tilde{x}} = \bigcap_{x \in \tilde{x}} x$.

Remark

(1) If Prim A is Hausdorff then all the equivalence classes consists of single points.

(2) Observe that if P satisfies (i) in 23.1 then $x \in P$ implies $\tilde{x} \subset P$.

Lemma 23.5 *Let* P *satisfy condition* (i) *in* 23.1. *Then for any finite set* $a_1,\ldots,a_n$ *of elements in* A *and any* $\varepsilon > 0$ *there exists* U(P) *such that*

$$\sup_{U(P)} \sum_{i=1}^{n} \breve{a}_i(x) < \sup_{\tilde{x} \subset P} \sum_{i=1}^{n} \breve{a}_i(\tilde{x}) + \varepsilon \quad .$$

Proof Let

$$R = \sup_{\tilde{x} \subset P} \sum_{i=1}^{n} \breve{a}_i(\tilde{x}) \quad . \tag{23.2}$$

Consider the set

$$W = \left\{ x: \sum \breve{a}_i(x) \geq R + \varepsilon \right\} \quad . \tag{23.3}$$

Suppose that $W \cap U(P) \neq \emptyset$ for every $U(P)$. Since neighbourhoods of P are generated by neighbourhoods of points in P there exists $x_0 \in P$ such that

$$W \cap U(x_0) \neq \emptyset \qquad \text{for every} \quad U(x_0) \quad .$$

We shall denote by R_1

$$R_1 = \underset{W,x_0}{\text{e s s sup}}\ \breve{a}_1(x)$$

that is, the maximal number such that for any $\delta > 0$, and any $U(x_0)$ there exists an $x \in [U(x_0) \cap W]$ such that

$$\breve{a}_1(x) \geq R_1 - \delta$$

(thus there exists a $U_1(x_0)$ such that

$$\breve{a}_1(x) < R_1 + \delta \qquad \text{for every} \quad x \in [W \cap U_1(x_0)] \text{).}$$

Fix $\delta < \frac{\varepsilon}{2n}$ and consider the set

$$V_1 = \{x\colon \breve{a}_1(x) \geq R_1 - \delta\} \quad .$$

We have that

$$V_1 \cap W \cap U(x_0) \neq \emptyset \qquad \text{for every} \quad U(x_0) \quad .$$

Next consider the set $W_1 = V_1 \cap W$ and take

$$R_2 = \underset{W_1,x_0}{\text{e s s sup}}\ \breve{a}_2(x) \quad .$$

Let

$$V_2 = \{x\colon \breve{a}_2(x) \geq R_2 - \delta\} \quad .$$

We have

$$V_2 \cap V_1 \cap W \cap U(x_0) \neq \emptyset \qquad \text{for every} \quad U(x_0) \quad .$$

Now take $W_2 = [V_2 \cap V_1 \cap W]$ and $R_3 = \operatorname*{ess\,sup}_{W_2, x_0} \breve{a}_3(x)$. Continuing this procedure we obtain the sets $V_1,...,V_n$ such that

$$V_1 \cap V_2 \cap ... \cap V_n \cap U(x_0) \neq \emptyset \qquad \text{for every} \quad U(x_0)$$

and corresponding numbers $K_1,...,K_n$ such that there exists a $U_0(x_0)$ with the property

$$R_i + \delta \geq \breve{a}_i(x) \geq R_i - \delta \tag{23.4}$$

for $x \in [V_1 \cap V_2 \cap ... \cap V_n \cap U_0(x_0)] = V_0$.

The neighbourhoods $U(x_0)$ generate the filter

$$V_1 \cap V_2 \cap ... \cap V_n \cap U(x_0) \quad .$$

Since all $V_1,...,V_n$ are quasicompact this filter has limit points $x_1 \in V_1,...x_n \in V_n$. In view of the construction, all these points are not separated from x_0 and thus $x_i \in \tilde{x}_0$, $i = \overline{1,n}$, so

$$\breve{a}_i(\tilde{x}_0) \geq R_i - \delta \quad . \tag{23.5}$$

Now (23.4) implies

$$\sum_{i=1}^{n} R_i + n\delta \geq \sum_{i=1}^{n} \breve{a}_i(x) \qquad \text{for} \quad x \in V_0$$

and at the same time (since $V_0 \subset W$) by (23.3)

$$\sum_{i=1}^{n} \breve{a}_i(x) \geq R + \varepsilon \quad .$$

Thus

$$R + \varepsilon \leq \sum_{i=1}^{n} R_i + n\delta$$

and in view of (23.5)

$$\sum_{i=1}^{n} \breve{a}_i(\tilde{x}_0) \geq \sum_{i=1}^{n} R_i - n\delta \geq R + \varepsilon - 2n\delta > R \quad .$$

We thus arrive at a contradiction with (23.2). □

23.6 The triangle procedure. Now we describe a method of obtaining an estimate for $\left\|N_1(b^n)\right\|$ (see 22.1) the idea of which is due to V.V. Brenner.

Along with the algebra A consider the C^*–algebra $\breve{A}$ generated by functions $\breve{a}(x)$, $x \in \text{Prim } A$, $a \in A$, with the sup norm.

Remark If Prim A is Hausdorff then $\breve{A} = C(\text{Prim } A)$ (Dixmier [1], 3.3.9, and the Dauns–Hofmann theorem 2.90).

Let $\breve{A} \times_{\hat{T}_g} G$ be the cross–product associated with the automorphisms

$\hat{T}_g(\breve{a}) = (\widetilde{T_g a T_g^*})$.

Let $M \subset G$ be a finite set. If an operator b has the form

$$b = \sum_{g \in M} a_g T_g \tag{23.6}$$

then we denote by $\breve{b}$ the element in $\breve{A} \times_{\hat{T}_g} G$ of the form

$$\breve{b} = \sum_{g \in M} \breve{a}_g \otimes g \quad . \tag{23.7}$$

Clearly b^n can be written as

$$b^n = \sum_{g \in M^n} (a_n)_g T_g \tag{23.8}$$

where

$$(a_n)_g = \sum_{\substack{g_1 \cdot g_2 \cdot \ldots \cdot g_n = g \\ g_i \in M}} a_{g_1} \cdot \hat{T}_{g_1}(a_{g_2}) \cdot \ldots \cdot \hat{T}_{g_1 \cdot \ldots \cdot g_{n-1}}(a_{g_n}) \tag{23.9}$$

and $(\breve{b})^n$ as

$$(\breve{b})^n = \sum_{g \in M^n} (\breve{a}_n)_g \otimes g \tag{23.10}$$

where

$$(\breve{a}_n)_g = \sum_{\substack{g_1 \cdot g_2 \cdot \ldots \cdot g_n = g \\ g_i \in M}} \breve{a}_{g_1} \cdot \hat{T}_{g_1}(\breve{a}_{g_2}) \cdot \ldots \cdot \hat{T}_{g_1 \cdot \ldots \cdot g_{n-1}}(\breve{a}_{g_n}) \ . \tag{23.11}$$

Let

$$N_1(\breve{b}) = \sum_{g \in M} \breve{a}_g \ .$$

Evidently (in view of (23.9) and (23.11))

$$\begin{aligned} \widecheck{N_1(b^n)}(x) &= \left[\sum_{g \in M^n} \widecheck{|(a_n)_g|} \right](x) \le \sum_{g \in M^n} \widecheck{|(a_n)_g|}(x) \\ &\le \sum_{g \in M^n} (\breve{a}_n)_g(x) = N_1[(\breve{b})^n](x) \ . \end{aligned} \tag{23.12}$$

Now we deduce an upper estimate for $N_1[(\breve{b})^n](x)$ and thus an upper estimate for $\widecheck{N_1(b^n)}(x)$. In order to describe the idea of how the estimate is made let us consider first an example.

Example Let $b = a_g T_g + a_h T_h$ (that is, $M = \{g, h\}$). Then

$$N_1[(\breve{b})^3](x) = \Big[\breve{a}_g\hat{T}_g(\breve{a}_g)(\hat{T}_{g^2}(\breve{a}_g + \breve{a}_h)) + \breve{a}_g\cdot\hat{T}_g(\breve{a}_h)(\hat{T}_{gh}(\breve{a}_g + \breve{a}_h))$$

$$+ \breve{a}_h\hat{T}_h(\breve{a}_g)(\hat{T}_{hg}(\breve{a}_g + \breve{a}_h)) + \breve{a}_h\hat{T}_h(\breve{a}_h)(\hat{T}_{h^2}(\breve{a}_g + \breve{a}_h))\Big](x) .$$

Note that

$$N_1[(\breve{b})^2](x) = \Big[\breve{a}_g\hat{T}_g(\breve{a}_g) + \breve{a}_g\hat{T}_g(\breve{a}_h) + \breve{a}_h\hat{T}_h(\breve{a}_g) + \breve{a}_h\hat{T}_h(\breve{a}_h)\Big](x) .$$

Denote by $m_2(b)(x)$ the number

$$m_2(b)(x) = \max_{s\in M^2} \hat{T}_s(\breve{a}_g + \breve{a}_h)(x) .$$

Now the foregoing implies

$$N_1[(\breve{b})^3](x) \le N_1[(\breve{b})^2](x)\cdot m_2(b)(x) .$$

Continuing this argument for $N_1[(\breve{b})^2](x)$ we have

$$N_1[(\breve{b})^3](x) \le m_0(b)(x)\cdot m_1(b)(x)\cdot m_2(b)(x)$$

where

$$m_1(b)(x) = \max_{s\in M} \hat{T}_s(\breve{a}_g + \breve{a}_h)(x)$$

$$m_0(b)(x) = (\breve{a}_g + \breve{a}_h)(x) .$$

In general when b has the form $b = \sum_{g\in M} a_g T_g$ an argument analogous to that considered in the example, along with (23.12), gives us the following estimate

$$\widetilde{N_1(b^n)}(x) \le N_1[(\breve{b})^n](x) \le \prod_{i=0}^{n-1} m_i(b)(x) \tag{23.13}$$

where

$$m_i(b)(x) = \max_{s\in M^i} \hat{T}_s\left(\sum_{g\in M} \breve{a}_g\right)(x)$$

and $M^0 = \{e\}$.

We call this method of obtaining (23.13) the *triangle procedure* because in fact it consists of a sequence of steps where first one obtains an estimate on M^{n-1}, then on M^{n-2},..., and the sequence

$$M^{n-1}, M^{n-2}, \ldots, M, \ \{e\}$$

forms a 'triangle'.

23.7 Let $M \subset G$ be a finite set. The subset $P \subset \text{Prim } A$ defined by

$$P = \bigcup_{\text{Prim } A} P(x)$$

where $P(x)$ is described in **23.1** we call the *limit set* for M.

Our main result in this section is the following.

Theorem 23.8 *Let* $\mathrm{M} \subset \mathrm{G}$ *be a finite set such that* $\mathrm{C}^*(\mathrm{A}, \mathrm{T}_g)_{g\in \mathrm{G}_\mathrm{M}}$ *possesses the property* (*) *and* G_M *is a subexponential group. Then*

$$\mathrm{r}\left(\sum_{g\in \mathrm{M}} \mathrm{a}_g \mathrm{T}_g\right) \leq \sup_{\tilde{\mathrm{x}}\subset \mathrm{P}} \sum_{g\in \mathrm{M}} \breve{\mathrm{a}}_g(\tilde{\mathrm{x}}) \tag{23.14}$$

where G_M *is the group generated by* M, P *is the limit set for* M, *and* $\tilde{\mathrm{x}}$ *is the equivalence class of* x (see 23.4).

Proof Let $b = \sum_{g\in M} a_g T_g$. Then by Theorem 21.2

$$r(b) = \max_{\hat{A}} r(\pi_y(b)) \tag{23.15}$$

so in fact it is enough to verify (23.14) for any fixed $y \in \hat{A}$ and $\pi_y(b)$.

By Theorem 22.2

$$r(\pi_y(b)) = \lim_{n\to\infty} \left\| N_1(\pi_y(b^n)) \right\|^{\frac{1}{n}} .$$

Fix any $x_0 \in \hat{A}$. We shall denote by $G_M x_0$ the G_M–orbit of Ker x_0 on Prim A. Obviously (see (23.8), (23.9))

$$\left\| N_1(\pi_{x_0}(b^n)) \right\| = \sup_{G_M x_0} \left[\widetilde{\sum_{g\in M^n} |(a_n)_g|} \right](x) . \qquad (23.16)$$

Now the estimate we need will be obtained by means of the triangle procedure described above, together with Lemma 23.5.

Fix any $\varepsilon > 0$ and consider the neighbourhood $U(P(x_0))$ such that

$$\sup_{U(P(x_0))} \sum_{g\in M} \breve{a}_g(x) < \sup_{\tilde{x}\subset P(x_0)} \sum_{g\in M} \breve{a}_g(\tilde{x}) + \varepsilon \quad ;$$

we can do this by Lemma 23.5.

Let $K(u)$ be the number mentioned in **23.1** (ii). By (23.13) and (23.16) we have

$$\left\| N_1(\pi_{x_0}(b^n)) \right\| = \sup_{G_M x_0} \prod_{i=0}^{n-1} m_i(b)(x) \le R^{K(u)}(R_{P(x_0)} + \varepsilon)^{n-K(u)}$$

where

$$R = \sup_{G_M x_0} \sum_{g\in M} \breve{a}_g(x) \ ; \qquad R_{P(x_0)} = \sup_{\tilde{x}\subset P(x_0)} \sum_{g\in M} \breve{a}_g(\tilde{x}) .$$

Consequently

$$r(\pi_{x_0}(b)) = \lim_{n\to\infty} \left\| N_1(\pi_{x_0}(b^n)) \right\|^{\frac{1}{n}} \le R_{P(x_0)} + \varepsilon .$$

As ε is arbitrary, in view of (23.15), we have finished. □

In fact this theorem leads to some additional effective estimates. Note that (23.14) can be written as

$$r(b) \leq \sup_{\tilde{x} \subset P} N_1(\breve{b})(\tilde{x}) \qquad (23.17)$$

where

$$N_1(\breve{b})(\tilde{x}) = \sum_{g \in M} \breve{a}_g(\tilde{x}) \quad .$$

Corollary 23.9 *Let the conditions of the theorem be satisfied. Then*

$$r(b) \leq \lim_{n \to \infty} \left[\sup_{\tilde{x} \subset P} N_1(\breve{b^n})(\tilde{x}) \right]^{\frac{1}{n}} \quad . \qquad (23.18)$$

Proof Note that

$$[r(b^n)]^{\frac{1}{n}} = r(b) \qquad (23.19)$$

and the limit set for M^n is a subset of the limit set for M. Now (23.18) follows from (23.19) and (23.17). □

23.10 If V is a set in $\hat{A}$ (Prim A) invariant under all t_g, $g \in G_M$, then by $b|_V$ we denote the object corresponding to the 'restriction' of b onto V, that is, the corresponding element in $A|_V \underset{\hat{T}}{\times} G_M$.

Since

$$N_1(\breve{b^n})(\tilde{x}) \leq N_1[(\breve{b})^n](\tilde{x})$$

and by Theorem 22.2

$$r(\breve{b})|_P = \lim_{n \to \infty} \left[\sup_{\tilde{x} \subset P} N_1[(\breve{b})^n](\tilde{x}) \right]^{\frac{1}{n}}$$

the previous corollary leads to

Corollary 23.11 $r(b) \leq r(\breve{b})|_P$.

Corollary 23.12 *If* A *is a commutative* C^**–algebra then*

$$r(b) = r(b)|_P = \max_{x \in P} r(\pi_x(b)) \quad . \tag{23.20}$$

Proof Since $b|_P$ is a representation of $C^*(A, T_g)_{g\in G_M}$ we always have

$$r(b) \geq r(b)|_P \quad .$$

Since A is commutative

$$N_1(\breve{b^n}) = N_1(b^n) \quad .$$

In view of (23.18) this implies

$$r(b) \leq r(b)|_P \quad .$$

Thus

$$r(b) = r(b)|_P \quad .$$

The right–hand equality in (23.20) follows from Theorem 21.2. □

Remark 23.13 In particular, if P consists only of fixed points then $A|_P \underset{\hat{T}}{\times} G \cong A|_P \otimes C^*(G)$ (here the right–hand term is the standard tensor product (see 11.17)) and Corollary 23.12 gives us a method of obtaining an explicit formula for the spectral radius (see also the example in Remark 23.15).

Corollary 23.14 *Consider a* C^**–algebra* A *and a unitary operator* T *such that* $TAT^* = A$. *Then*

$$r(aT) = r(aT)|_P = \max_P r(\pi_x(aT)) \quad .$$

Proof Direct computation (using the formula $r(b) = \lim_{n\to\infty} \|b^n\|^{\frac{1}{n}})$ shows that

$$r(aT) = r(a \otimes 1)$$

where $a \otimes 1$ is an element in $A \times_{\hat{T}^z} \mathbb{Z}$.

To $A \times_{\hat{T}^z} \mathbb{Z}$ we can apply the results of Sections 22 and 23 (since $A \times_{\hat{T}^z} \mathbb{Z}$ possesses the property (*)). As $b|_P$ is a representation of $A \times_{\hat{T}^z} \mathbb{Z}$ we have

$$r(aT) \geq r(a \otimes 1)|_P = \max_P r(\pi_x(aT)) \tag{23.21}$$

where the right–hand equality follows from Theorem 21.2. On the other hand obviously

$$\left\|N_1((a \otimes 1)^n)|_P\right\| = \sup_{\tilde{x} \subset P} N_1(\widecheck{(aT)^n})(\tilde{x})$$

and thus Theorem 22.2 and (23.18) imply

$$r(a \otimes 1) \leq r(a \otimes 1)|_P$$

which along with (23.21) completes the proof. □

Remark 23.15 If $C^*(A, T_g)$ possesses the property (*) then for any t_g–invariant closed set P and any $b \in C^*(A, T_g)$ we have

$$r(b) \geq r(b)|_P = \max_P r(\pi_x(b)) \quad .$$

Here we used Theorem 21.2 and the fact that $b|_P$ is a representation of $C^*(A, T_g)$. Thus (23.18) gives us in fact upper and lower estimates

$$r(b)|_P = \max_P r(\pi_x(b)) \le r(b) \le \lim_{n\to\infty} \left[\sup_{\tilde{x}\subset P} N_1(\widetilde{b^n})(\tilde{x})\right]^{\frac{1}{n}} . \qquad (23.22)$$

In some cases the left–hand estimate can be written in explicit terms.

Example Let $G_M \cong \mathbb{Z}^n$ and let the limit set P consists only of fixed points of t_g, $g \in M$. Then the representation $b|_P$ can be written as

$$A|_P \times_I \mathbb{Z}^n \cong A|_P \otimes C^*(\mathbb{Z}^n)$$

where I is the representation of $\mathbb{Z}^n$ by trivial automorphisms (Lemma 11.18).

Thus

$$b|_P = \sum a_z T^z|_P \to \sum a_z \lambda^z|_P$$

where $\lambda = (\lambda_1,...,\lambda_n) \in S^n$, and S^n is the character group for $\mathbb{Z}^n$. It follows that

$$r(b)|_P = \max_{P\times S^n} r\left(\sum \pi(a_z)\lambda^z\right) . \qquad (23.23)$$

In particular if A is a commutative C^*–algebra we have

$$r(b)|_P = \max_{P\times S^n} \left|\sum a_z(x)\lambda^z\right| . \qquad (23.24)$$

When G_M is not commutative then instead of $\hat{G}_M$ we have to consider $\widehat{C^*(G_M)}$ (see 2.91).

23.16 Elements $g_1,...,g_n \in G$ are said to be *independent* (with respect to A) if

(1) $C^*(A, T_g)_{g\in G_M}$, $M = \{g_1,...,g_n\}$, possesses the property (*)

(2) for any $h \in G_M$ having non–zero 'degree'

(that is, having the form

$$h = g_{i_1}^{m_{i_1}} \cdot \ldots \cdot g_{i_k}^{m_{i_k}} , \qquad g_{i_s} \in M$$

where

$$\Big| \sum_{g_{i_s} = g_1} m_{i_s} \Big| + \Big| \sum_{g_{i_s} = g_2} m_{i_s} \Big| + \ldots + \Big| \sum_{g_{i_s} = g_n} m_{i_s} \Big| \neq 0 \quad)$$

we have $h \neq e$.

Remarks

(1) If $g_1, \ldots, g_n$ are mutually commuting elements then clearly they are independent iff $C^*(A, T_z)$ possesses the property (*). Here by T_z we denote the elements of the form

$$T_z = T_{g_1}^{z_1} \cdot \ldots \cdot T_{g_n}^{z_n} , \qquad z = (z_1, \ldots, z_n) \in \mathbb{Z}^n \quad .$$

(2) If $x \in \hat{A}$ is a fixed point for all t_{g_i}, $i = \overline{1,n}$, and g_i are independent then the mappings

$$A \ni a \to x(a)$$
$$T_{g_i} \to \lambda_i 1 \qquad |\lambda_i| = 1$$

induce irreducible representations of $C^*(A, T_g)_{g \in G_M}$.

(3) If g_i are independent and mutually commuting then the above–mentioned representations are all the irreducible representations corresponding to x.

Lemma 23.17 *If* A *is a commutative* C**–algebra,* $g_1, \ldots, g_n$ *are independent and the limit set for* $\{g_1, \ldots, g_n\}$ *consists only of fixed points of* t_{g_i}, $i = \overline{1,n}$, *then*

$$r\left(\sum_{i=1}^{n} a_i T_{g_i}\right) = \max_P \left(\sum_{i=1}^{n} |a_i(x)|\right) .$$

Proof By Theorem 23.8 $r\left(\sum_{i=1}^{n} a_i T_{g_i}\right) \le \max_P \left(\sum_{i=1}^{n} |a_i(x)|\right)$.

Now having in mind the example in Remark 23.15 and item (2) of the foregoing Remarks we conclude

$$r\left(\sum_{i=1}^{n} a_i T_{g_i}\right) \ge \max_P \left(\max_{(\lambda_1,\dots,\lambda_n)\in S^n} \left|\sum a_i(x)\lambda_i\right|\right) .$$

Now choose $\lambda_i = \arg a_i(x)$, $i = \overline{1,n}$, and we are done. □

24. Invertibility of elements in $C^*(A,T_g)$. Convex and rationally independent systems of shifts on $\mathbb{R}^k$.

The previous sections give us a tool for obtaining some necessary and sufficient conditions for the invertibility of elements in $C^*(A, T_g)$ in an explicit form, and we now proceed to do this.

24.1 Let $M = \{g_1,\dots,g_n\}$ be a finite set of elements of G. We shall denote by P_j, $j = \overline{1,n}$, the limit set for $M_j = \left\{g_i g_j^{-1}\right\}_{\substack{i\neq j \\ i=\overline{1,n}}}$.

Theorem 24.2 *Let* $b = \sum_{i=1}^{n} a_i T_{g_i}$. *If there exists a number* $j_0 \in \{1,2,\dots,n\}$ *such that:*

(1) $G_{M_{j_0}}$ *is subexponential and* $C^*(A, T_g)_{g\in M_{j_0}}$ *possesses the property* (*);

(2) a_{j_0} *is invertible;*

(3) *there exists an* $\varepsilon > 0$ *such that for every* $\tilde{x} \subset P_{j_0}$

$$\left[\check{a}_{j_0}^{-1}(\tilde{x})\right]^{-1} > \left(\sum_{i\neq j_0} \breve{a}_i(\tilde{x})\right) + \varepsilon \tag{24.1}$$

then b *is invertible.*

Proof Let the conditions (1), (2) and (3) be satisfied. We have

$$a_{j_0}^{-1}\, bT_{g_{j_0}}^{-1} = \left(I + \sum_{i\neq j_0} a_{j_0}^{-1} a_i T_{g_i} T_{g_{j_0}}^{-1}\right) = I + b' \quad .$$

Thus b is invertible if and only if $(-1) \,\overline{\in}\, \sigma(b')$. By Theorem 23.8

$$r(b') \leq \sup_{\tilde{x}\subset P_{j_0}} \sum_{i\neq j_0} \left(\widetilde{a_{j_0}^{-1} a_i}\right)(\tilde{x})$$

$$\leq \sup_{\tilde{x}\subset P_{j_0}} \check{a}_{j_0}^{-1}(\tilde{x})\left(\sum_{i\neq j_0} \breve{a}_i(\tilde{x})\right) < 1$$

where the last inequality is true in view of (24.1). Thus $(-1) \,\overline{\in}\, \sigma(b')$ and consequently b is invertible. □

Remarks

(1) Clearly if $\breve{a}(\tilde{x})$ is continuous on P_{j_0} then we can omit ε in (24.1) since P_{j_0} is quasicompact.

Thus, for example, if Prim A is Hausdorff then instead of (24.1) we simply demand

$$\left[\check{a}_{j_0}^{-1}(x)\right]^{-1} > \sum_{i\neq j_0} \breve{a}_i(x) \quad , \qquad x \in P_{j_0} \tag{24.2}$$

and if A is commutative then simply

$$\left|a_{j_0}(x)\right| > \sum_{i \neq j_0} |a_i(x)| \quad , \qquad x \in P_{j_0} \quad . \tag{24.3}$$

(2) In general, however, $\breve{a}(\tilde{x})$ need not be continuous.

Example Consider the algebra A which was described in the example in Remark 19.29 (1). One can easily find two independent (see 23.16) homeomorphisms of Prim A such that the corresponding limit set will be the whole of Prim A.

Observe that $\breve{a}(\tilde{x})$ in this example is not continuous. Indeed take for example $a \in A$ such that

$$a_{11}(1) = a_{11}(1 + 0) = 1 \ , \qquad a_{22}(1) = 2 \ , \qquad a_{22}(1 + 0) = 3 \ .$$

Since $\{\tilde{1}\} = \{1\} \cup \{1 - 0\} \cup \{1 + 0\}$ we have

$$\breve{a}(\{\tilde{1}\}) = 3 \quad .$$

But in any neighbourhood of $\{\tilde{1}\}$ there are points ($\tilde{x} < 1$) where $\breve{a}(\tilde{x}) < \frac{5}{2}$.

(3) The conditions of the theorem are not necessary.

Example Consider the space $L^2(\mathbb{R})$ and an operator

$$b = (c_1 + c_2T)(c_3 + c_4T) = a_0 + a_1T + a_2T^2$$

where c_i, $i = \overline{1,n}$, are the operators of multiplication by continuous functions $c_i(x)$ having limits $c_i(\infty)$ at infinity, and

$$(Tf)(x) = f(x + h) \quad , \qquad f \in L^2(\mathbb{R}) \quad , \qquad h \in \mathbb{R}\backslash\{0\} \quad .$$

Consequently

$$a_0(x) = c_1(x) \cdot c_3(x);$$

$$a_1(x) = c_2(x)\cdot c_3(x + h) + c_1(x)\cdot c_4(x)$$
$$a_2(x) = c_2(x)c_4(x + h) \quad .$$

Now choose for example $c_2(x) \equiv -c_3(x) \equiv 1$, let $c_1(x) \equiv c_4(x)$ and let c_1 satisfy the conditions:

(1) $c_1(x_0) = 1$

(2) $c_1(x_1) = 0$

(3) $c_1(\infty) = \frac{1}{2}$.

Then all a_0, a_1, a_2 are not invertible but both of the operators $(c_1 + c_2T)$ and $(c_3 + c_4T)$ are invertible (see 18.2) and thus b is invertible, while condition (2) of the theorem is not satisfied.

If we choose $c_2(x) \equiv -c_3(x) \equiv 1$ and $c_1(x) \equiv c_4(x) = \frac{1}{2}$ then b is again invertible, a_0, a_1, a_2 are invertible, but condition (3) is not satisfied.

Now we proceed to the necessary conditions for invertibility.

Lemma 24.3 *Let* $\mathrm{M} = \{g_1,\ldots,g_n\}$ *be a set such that for some* $j_0 \in \{1,\ldots,n\}$ *the elements in* M_{j_0} *are independent. Let* P *be any set in* $\hat{\mathrm{A}}$ *which consists of fixed points for all* t_g, $g \in \mathrm{M}_{j_0}$. *If* $b = \sum_{i=1}^{n} a_i T_{g_i}$ *is invertible then all the operators*

$$\sum_{i=1}^{n} \pi(a_i)\lambda_i \qquad \pi \in P \ , \quad |\lambda_i| = 1 \ , \quad i = \overline{1,n}$$

are invertible.

Proof b is invertible iff $bT_{g_{j_0}}^{-1}$ is invertible and the operators

$$\sum_{i=1}^{n} \pi(a_i)\lambda_i \ , \qquad |\lambda_i| = 1 \ , \quad i = \overline{1,n} \ ,$$

are all invertible iff the operators

$$(\pi(a_{j_0}) + \sum_{i \neq j_0} \pi(a_i)\lambda_i) \ , \qquad |\lambda_i| = 1 \ , \quad i \neq j_0 \tag{24.4}$$

are all invertible.

Now observe that in view of item (2) of the Remarks in **23.16**, the mappings $A \ni a \to \pi(a)$, $T_{g_i g_{j_0}^{-1}} \to \lambda_i$, $i \neq j_0$, generate a representation of $C^*(A, T_g)_{g \in M_{j_0}}$ and consequently since $bT_{g_{j_0}}^{-1}$ is invertible all the operators in (24.4) are invertible as well. □

Theorem 24.4 *Let* A *be a commutative algebra and* M, $\mathrm{M}_{\mathrm{j}_0}$; P, b *be as described in the previous lemma. Then for any connected component* $\Delta \subset$ P *there exists a number* j_Δ *such that*

$$\left| a_{j_\Delta}(x) \right| > \sum_{i \neq j_\Delta} |a_i(x)| \ , \qquad x \in \Delta \ .$$

Proof By the previous lemma

$$\sum_{i=1}^{n} a_i(x)\lambda_i \neq 0 \ , \qquad x \in P \ , \quad |\lambda_i| = 1 \ , \quad i = \overline{1,n} \ .$$

Fix any connected component $\Delta \subset P$ and any $x_0 \in \Delta$. Then we have

$$\sum_{i=1}^{n} a_i(x_0)\lambda_i \neq 0 \ , \qquad |\lambda_i| = 1 \ , \quad i = \overline{1,n} \ . \tag{24.5}$$

Choose j_Δ from the condition

$$\left|a_{j_\Delta}(x_0)\right| = \max_{i=\overline{1,n}} |a_i(x_0)| \quad . \tag{24.6}$$

(24.6) together with (24.5) implies

$$\left|a_{j_\Delta}(x_0)\right| > \sum_{i\neq j_\Delta} |a_i(x_0)| \quad . \tag{24.7}$$

Indeed suppose that (24.7) is not true, that is,

$$\left|a_{j_\Delta}(x_0)\right| \leq \sum_{i\neq j_\Delta} |a_i(x_0)| \quad .$$

Thus for any $i_0 \in \{1,\ldots,n\}$ we have

$$\left|a_{i_0}(x_0)\right| \leq \sum_{i\neq i_0} |a_i(x_0)| \tag{24.8}$$

and (24.8) means that there exists an n–polygon in a plane which has edges of length $|a_i(x_0)|$, $i = \overline{1,n}$. Analytically this means that there exist complex numbers $\lambda_1,\ldots,\lambda_n$, $|\lambda_i| = 1$, such that

$$\sum_{i=1}^{n} a_i(x_0)\lambda_i = 0$$

which contradicts (24.5). Consequently (24.7) is true.

Since $|a_i(x)|$, $i = \overline{1,n}$, are continuous and Δ is connected it follows that (24.7) is true for any $x \in \Delta$ and in particular the number j_Δ is defined uniquely. □

Theorem 24.5 *Let* A *be a commutative algebra and* M = $\{g_1,\ldots,g_n\}$ *be a* *set such that*:

(i) *for every* $j \in \{1,\dots,n\}$ *the elements in* M_j *are independent and* G_{M_j} *is subexponential*;

(ii) *all* P_j, $j = \overline{1,n}$, (see **24.1**) *coincide* (= P).

If P *is connected and consists of fixed points for* t_g, $g \in M_j$, $j = \overline{1,n}$, *then*

$$b = \sum_{i=1}^{n} a_i T_{g_i}$$

is invertible iff there exists a number j_0 *such that*

(1) a_{j_0} *is invertible*

(2) $\left|a_{j_0}(x)\right| > \sum_{i \neq j_0} |a_i(x)|$, $x \in P$.

Proof By Theorem 24.2 and the following Remark (1), conditions (1) and (2) are sufficient for b to be invertible. By Theorem 24.4 condition (2) is necessary. So we have to prove that (1) is also necessary.

Suppose on the contrary that b is invertible while a_{j_0} is not. Then there exists a point x_0 such that $a_{j_0}(x_0) = 0$. Consider $\pi_{x_0}(bT_{g_{j_0}}^{-1})$ (see **21.1** with $G = G_{M_{j_0}}$). Since b is invertible $\pi_{x_0}(bT_{g_{j_0}}^{-1})$ is invertible as well and consequently there exists an ε–neighbourhood of $\pi_{x_0}(bT_{g_{j_0}}^{-1})$ (in $L(l^2(G))$) such that all the operators in this neighbourhood are invertible. Now we shall find a non–invertible operator in this neighbourhood and thus shall arrive at a contradiction.

Consider the function $a(g) \in l^\infty(G)$ defined by

$$a(g) = \begin{cases} 0 \ , & g = e \\ \frac{\varepsilon}{3} & \text{when} \quad g \neq e \ \text{ and } \ \left| a_{j_0}(t_g(x_0)) \right| < \frac{\varepsilon}{2} \\ a_{j_0}(t_g(x_0)) & \text{when} \quad \left| a_{j_0}(t_g(x_0)) \right| \geq \frac{\varepsilon}{2} \ . \end{cases}$$

Let a be the operator on $l^2(G)$ of multiplication by this function.

Consider the operator

$$b_1 = a + \sum_{i \neq j_0} \pi_{x_0}(a_i T_{g_i g_{j_0}^{-1}}) \ .$$

In view of the construction of a we have

$$\left\| \pi_{x_0}(bT_{g_{j_0}}^{-1}) - b_1 \right\| = \left\| \pi_{x_0}(a_{j_0}) - a \right\| \leq \frac{\varepsilon}{3} + \frac{\varepsilon}{2} < \varepsilon \ .$$

Now observe that b_1 has a non–zero kernel. Indeed let d be the operator defined by

$$d = \sum_{i \neq j_0} d_i \ \pi_{x_0}(T_{g_i g_{j_0}^{-1}})$$

where d_i are the operators of multiplication by the functions

$$d_i(g) = \begin{cases} \dfrac{a_i(t_g(x_0))}{a(g)} \ , & g \neq e \\ 0 \ , & g = e \ . \end{cases}$$

Since condition (2) is fulfilled, Lemma 23.17 and the construction of a imply

$$r(d) < 1$$

and thus $(I + d)$ is invertible.

Let ξ_0 be a non-zero vector defined by

$$\xi_0 = (I + d)^{-1}\delta_e$$

where $\{\delta_g\}_{g\in G}$ is the standard basis in $l^2(G)$. Since $r(d) < 1$ we obtain

$$(I + d)^{-1} = \sum_{n=0}^{+\infty} (-d)^n$$

and routine calculations show that

$$b_1[(I + d)^{-1}\delta_e] = [a(I + d)][(I + d)^{-1}\delta_e] \quad .$$

Thus we have

$$b_1\xi_0 = a(I + d)\left[(I + d)^{-1}\delta_e\right] = a\delta_e = 0$$

which finishes the proof. □

With the help of this theorem we can describe the spectrum of some operators which have many summands (cf., for example, Theorem 13.10).

Let P be a closed connected set in Prim A and A be a commutative algebra. Denote

$$r_1(b) = \max_P \sum_{j=1}^{n} |a_j(x)|$$

$$r_2(b) = \max_{k=\overline{1,n}} \min_P \left[|a_k(x)| - \sum_{j\neq k} |a_j(x)|\right] \quad .$$

If $r_2(b) > 0$ there exists a unique number k_0 such that

$$\left|a_{k_0}(x)\right| > \sum_{j\neq k_0} |a_j(x)| \quad , \qquad x \in P \quad .$$

Corollary 24.6 *Let* A *be a commutative algebra and* M = $\{g_0 = e, g_1,\dots,g_n\}$ *be a set such that:*

(i) *the elements of* $M_0 = \{g_1,\dots,g_n\}$ *are independent and* G_{M_0} *is subexponential;*

(ii) *all* P_j, $j = \overline{0,n}$, (see **24.1**) *coincide* (= P).

If P *is connected and consists of fixed points for* t_{g_i}, $i = \overline{1,n}$, *then the spectrum of the operator* $b = \sum_{i=1}^{n} a_i T_{g_i}$ *is:*

(a) *the annulus*

$$r_2(b) \le |\lambda| \le r_1(b)$$

if $r_2(b) > 0$ *and* a_{k_0} *is invertible;*

(b) *the disk*

$$|\lambda| \le r_1(b)$$

in any other case.

Proof By Lemma 23.17 the spectral radius of b is $r_1(b)$. Routine calculations show that the elements of every M_j, $j = \overline{0,n}$, are independent and G_{M_j} are subexponential.

If $r_2(b) \le |\lambda| \le r_1(b)$ then the operator $(\lambda I - b)$ does not satisfy condition (2) of Theorem 24.5 and consequently the annulus $r_2(b) \le |\lambda| \le r_1(b)$ belongs to the spectrum.

If $r_2(b) > 0$ and $|\lambda| < r_2(b)$ then condition (2) of the theorem for $(\lambda I - b)$ is satisfied with $j_0 = k_0$ and condition (1) is satisfied iff a_{k_0} is invertible. □

Now we shall consider an example where the results obtained above can be used effectively.

Convex and rationally independent systems of shifts on $\mathbb{R}^k$.
Consider the space $L^2(\mathbb{R}^k)$. Let A be the algebra of operators of multiplication by functions $a(x) \in L^\infty(\mathbb{R}^k)$ such that there exist uniform limits

$$\lim a(\rho, \varphi) = a(\varphi)$$

where (ρ, φ) are polar coordinates in $\mathbb{R}^k$ and $a(\varphi) \in C(S^{k-1})$.

For a fixed vector $h \in \mathbb{R}^k$ we define the operator

$$(T_h f)(x) = f(x + h) \quad , \qquad f \in L^2(\mathbb{R}^k) \quad .$$

Consider the operators of the form

$$b = \sum_{i=1}^{n} a_i T_{h_i} \quad , \qquad a_i \in A \quad . \tag{24.9}$$

We shall not determine Prim A explicitly, but notice that evidently the infinitely distant sphere S_∞^{k-1} belongs to Prim A.

24.7 We shall call a system $M = \{h_1,\ldots,h_n\}$ of vectors *convex* if h_i, $i = \overline{1,n}$, are vertices of a convex polygon in $\mathbb{R}^n$.

Lemma 24.8 *If* $\mathrm{M} = (\mathrm{h}_1,\ldots,\mathrm{h}_\mathrm{n}\}$ *is a convex system then all the* P_j, $\mathrm{j} = \overline{1,\mathrm{n}}$, (see **24.1**) *coincide with* S_∞^{k-1}.

Proof Since for any $h \in \mathbb{R}^k$ we have

$$\lim_{\rho \to +\infty} [\hat{T}_h(a)](\rho, \varphi) = a(\varphi)$$

the infinitely, distant sphere consists of fixed points for every t_h and thus belongs to any limit set.

On the other hand M is a convex system iff for every M_j, $j = \overline{1,n}$, there exists a hyperplane F_j in $\mathbb{R}^k$ such that 0 lies on one side of F_j while all $h \in M_j$ lie on the other. This implies

$$\min_{h \in M_j^m} \|h\| \xrightarrow[m\to\infty]{} \infty$$

and thus $P_j \subset S_\infty^{k-1}$. □

24.9 We shall say that a system $M = \{h_1,...,h_n\}$ is *rationally independent* if the system $M_1 = \{h_2 - h_1,...,h_n - h_1\}$ of vectors is linearly independent with respect to the field of rational numbers.

Obviously here instead of the field of rational numbers one can take the ring of integers.

Lemma 24.10 *If* M *is a rationally independent system then all* M_j, $j = \overline{1,n}$, *are independent sets (with respect of* A) (see **23.16**).

Proof First of all let us notice that one can use any M_j, $j = \overline{1,n}$, in definition 24.9. In fact, suppose on the contrary that for some $j_0 \in \overline{2,n}$ there exist integers $k_1,...,k_{j_0-1},k_{j_0+1},...,k_n$ not all zero such that

$$k_1(h_1 - h_{j_0}) +...+ k_{j_0-1}(h_{j_0-1} - h_{j_0}) + k_{j_0+1}(h_{j_0+1} - h_{j_0})$$

$$+...+ k_n(h_n - h_{j_0}) = 0 \quad .$$

Consider the following two possibilities:

(1) $k_2 = k_3 = ... = k_n = 0$. Thus $k_1 \neq 0$ and $-k_1(h_{j_0} - h_1) = 0$ which means that M is not rationally independent.

(2) Among the numbers $k_2,k_3,...,k_n$ there are some that are non–zero. In this case we have

$$k_1h_1 + ... + k_nh_n - \left(\sum_{i\neq j_0} k_i\right)h_{j_0} = 0 \quad .$$

Denoting $\left(-\sum_{i\neq j_0} k_i\right)$ by k_{j_0} we deduce (by routine computation) that

$$k_2(h_2 - h_1) + \dots + k_{j_0}\left(h_{j_0} - h_1\right) + \dots + k_n(h_n - h_1) = 0$$

and it follows again that M is not rationally independent.

Now since the elements in M_1 are mutually commuted in order to finish the proof we have to show that $C^*(A, T_z)$ possesses the property (*) where

$$T_z = T_{z_1g_1+\dots+z_{n-1}g_{n-1}} \qquad z = (z_1,\dots,z_{n-1})\in\mathbb{Z}^{n-1},\ M_1 = (g_1,\dots,g_{n-1}) \ .$$

As $g_1,\dots,g_n$ are linearly independent with respect to $\mathbb{Z}$ we conclude that for any finite set $F \subset \mathbb{Z}^{n-1}$ there exists an ε–ball $B_\varepsilon \subset \mathbb{R}^k$ such that for any point $x \in \mathbb{R}^k$ and any non–zero $z = (z_1,\dots,z_{n-1}) \in F$ we have

$$(x + B_\varepsilon) \cap (x + B_\varepsilon + z_1g_1 + \dots + z_{n-1}g_{n-1}) = \emptyset \ . \tag{24.10}$$

For any operator $b = \sum_F a_zT_z$ and any $\delta > 0$ consider the set

$$W_\delta = \left\{x \in \mathbb{R}^k\colon\ |a_0(x)| \geq \sup_{\mathbb{R}^k} |a_0(x)| - \delta\right\} .$$

Let x be some point such that $\mu(x + B_\varepsilon \cap W_\delta) > 0$ (where μ is the standard Lebesgue measure on $\mathbb{R}^k$).

Let f be any vector in $L^2(\mathbb{R}^k)$ which is concentrated on $(x + B_\varepsilon \cap W_\delta)$, i.e.,

$$\int_{\mathbb{R}^k\setminus(x+B_\varepsilon\cap W_\delta)} |f^2(x)|\,dx = 0 \ . \tag{24.11}$$

In view of (24.10) and (24.11) one has

$$(a_zT_z f) \perp (a_0f) \ , \qquad z \neq 0\ ,\quad z \in F$$

and by the definition of W_δ

$$\|a_0 f\| \geq (\|a_0\| - \delta)\|f\| \quad .$$

It follows now that

$$\|bf\| \geq (\|a_0\| - \delta)\|f\|$$

and since δ is arbitrary we conclude that $C^*(A, T_z)$ possesses the property (*). □

Now we can easily obtain the conditions for invertibility of operators with convex and rationally independent systems of shifts.

In fact, Lemmas 24.8 and 24.10, and Theorem 24.5 imply

Theorem 24.11 *Let the system* $\mathrm{M} = \{\mathrm{h}_1,\ldots,\mathrm{h}_n\}$ *be convex and rationally independent. The operator* b *of the form* (24.9) *is invertible iff there exists a number* j_0 *and* $\delta > 0$ *such that*

(1) $\left|\mathrm{a}_{\mathrm{j}_0}(\mathrm{x})\right| > \delta$ *almost everywhere*

(2) $\left|\mathrm{a}_{\mathrm{j}_0}(\varphi)\right| > \sum\limits_{j \neq j_0} |a_j(\varphi)| \ , \quad \varphi \in S^{k-1} \ .$

Corollary 24.6 gives us a corresponding simple description of the spectrum of b.

25. Spectrum of weighted shifts and the Lyapunov exponents.

If A is not a commutative algebra then the description of the spectrum $\sigma(aT)$ unfortunately cannot be presented in such an efficient manner as the one given for the commutative case in Sections 4 and 7, and Theorem 13.10. Nevertheless in the case when a is a continuous operator–valued function over a compact space X having compact values Ju. D. Latushkin and A.M. Stepin have brought to light the deep interrelation between the structure of $\sigma(aT)$ and the Lyapunov exponents of the corresponding cocycle.

In essence this section contains a presentation of their ideas.

25.1 Here we confine ourselves to the consideration of the spectrum of aT where $a \in A \cong C(X, L(H))$ is the algebra of continuous operator–valued functions over a metrizable compact space X (H is a separable Hilbert space) and $(TaT^*)(x) = a(\alpha(x))$ where α is a homeomorphism of X. One of the main tools in this consideration will be the multiplicative ergodic theorem (Theorem 25.3) which is true under the assumption that all the values $a(x)$, $x \in X$, are compact operators. Thus we shall obtain a description of the spectrum under this assumption:

$$a(x) \text{ is a continuous operator–valued function having compact values } (\in K(H)) \quad . \tag{25.1}$$

Multiplicative ergodic theorem.

25.2 (cf. **21.12**) Let us define the cocycle $\mathcal{A}: X \times \mathbb{Z}_+ \to L(H)$ for α and a via the formulae

$$\begin{aligned} \mathcal{A}(x, n) &= a(\alpha^n(x)) \cdot \ldots \cdot a(\alpha(x)) \quad , \qquad n = 1,2,\ldots, \\ \mathcal{A}(x, 0) &= I \quad . \end{aligned} \tag{25.2}$$

Observe that

$$(aT)^n = T^n \mathcal{A}(\cdot, n) \tag{25.3}$$

$$\mathcal{A}(x, n+k) = \mathcal{A}(\alpha^n x, k)\, \mathcal{A}(x, n) \quad . \tag{25.4}$$

Now we recall the statement of the multiplicative ergodic theorem in a form convenient for our purposes.

Theorem 25.3 (Ruelle [1]) (Multiplicative ergodic theorem) *Let* $a(x)$ *satisfy the assumption* (25.1). *For any ergodic measure* μ *with respect to* α *there exists an* α*–invariant set* $X_\mu \subset X$, $\mu(X_\mu) = 1$, *such that for any* $x \in X_\mu$ *there exists the limit*

$$\mathcal{A}_x = \lim_{n\to\infty} [\mathcal{A}^*(x, n)\, \mathcal{A}(x, n)]^{\frac{1}{2n}} \quad . \tag{25.5}$$

Let $e^{\lambda^1} > e^{\lambda^2} > \ldots > e^{\lambda^s} > \ldots$ *be a sequence of eigenvalues of the non–negative compact operator* $\mathcal{A}_x$ *(where we set* $s = \infty$ *if the sequence is infinite),* U_x^j *be the invariant subspaces corresponding to* e^{λ^j}, $j = 1,2,\ldots$ $(< s + 1)$, *and let us denote*

$$V_x^0 = \{0\} , \qquad V_x^1 = H , \qquad V_x^{j+1} = H \ominus \left(U_x^1 + \ldots + U_x^j\right) ,$$

$$j = 1,2,\ldots \; (< s + 1) , \qquad V_x^{s+1} = \operatorname{Ker} \mathcal{A}_x .$$

Then for every $x \in X_\mu$ *there exists the* Lyapunov *(characteristic)* exponent

$$\lambda(x, v) = \lim_{n\to\infty} \frac{1}{n} \ln\|\mathcal{A}(x, n)v\| , \qquad v \in H \tag{25.6}$$

and

$$\lambda(x, v) = \lambda^j , \qquad v \in \left[V_x^j \ominus V_x^{j+1}\right] \qquad j = 1,2,\ldots \; (< s + 1)$$

$$\lambda(x, v) = -\infty \qquad v \in V_x^{s+1} .$$

In particular

$$\lambda^1 = \lim_{n\to\infty} \frac{1}{n} \ln\|\mathcal{A}(x, n)\| . \tag{25.7}$$

25.4 Along with the cocycle $\mathcal{A}(x, n)$ it is reasonable to consider the cocycle

$$\tilde{\mathcal{A}}(x, n) = a^*(\alpha^{1-n}(x))\cdot\ldots\cdot a^*(\alpha^{-1}(x))\cdot a^*(x) , \qquad n = 1,2,\ldots$$

which arises from the operator $(aT)^* = T^* a^*$ since

$$(aT)^{*n} = T^{*n} \tilde{\mathcal{A}}(\cdot, n) .$$

By Theorem 25.3 for every $x \in X_\mu$ there exists the limit

$$\tilde{\mathscr{A}}_x = \lim_{n\to\infty} \left[\tilde{\mathscr{A}}^*_{(x,n)} \; \tilde{\mathscr{A}}_{(x,n)}\right]^{\frac{1}{2n}} \quad . \tag{25.8}$$

It is shown by Ruelle [1] that the set of eigenvalues of $\tilde{\mathscr{A}}_x$ is the same as for $\mathscr{A}_x$: $e^{\lambda^1} > e^{\lambda^2} > \ldots > e^{\lambda^s} > \ldots$ and if by U_x^{-j} we denote the invariant subspace for $\tilde{\mathscr{A}}_x$ corresponding to e^{λ_j} then $\dim U_x^j = \dim U_x^{-j}$ (notice that since $\mathscr{A}_x$ is compact $\dim U_x^i < \infty$).

Consider the subspaces

$$V_x^{-j} = U_x^{-1} + \ldots + U_x^{-j} \quad , \qquad j = 1,2,\ldots \ (< s+1) \quad .$$

In view of the definition of V_x^j and the foregoing observation we have

$$\operatorname{codim} V_x^{j+1} = \dim V_x^{-j} \quad .$$

Moreover Ruelle [1] also proved that:

(1) $V_x^{j+1} \cap V_x^{-j} = \{0\}$ μ–almost everywhere; (25.9)

(2) $V_x^{j+1} + V_x^{-j} = H$ μ–almost everywhere. (25.10)

If we denote by W_x^j the subspace

$$W_x^j = V_x^j \cap V_x^{-j} \quad , \qquad j = 1,2,\ldots, \ (< s+1)$$

then (25.9) and (25.10) imply

$$V_x^j = V_x^{j+1} \oplus W_x^j \tag{25.11}$$

$$V_x^{-j} = W_x^1 \oplus \ldots \oplus W_x^j \quad . \tag{25.12}$$

Finally we note that the numbers $\lambda(x, v)$, dim U^j_x, dim W^j_x, and s are invariant with respect to α (on X_μ). If we want to underline their connection with the measure μ we shall write λ^j_μ, dim U^j_μ... .

The spectral radius of aT.

25.5 If A is a commutative algebra (that is, $A \cong C(X)$) and $a(x) \neq 0$ then in view of the Birkhoff–Hinchin ergodic theorem (Theorem 2.19) for every $x \in X_\mu$ we have

$$\lim_{n\to\infty} \frac{1}{n} \ln\|\mathcal{A}(x, n)\| = \lim_{n\to\infty} \frac{1}{n} \sum_{i=1}^{n} \ln|a(\alpha^i(x))|$$

$$= \int_X \ln|a(x)|\,d\mu$$

which means that in this case all the Lyapunov exponents (for the measure μ, a and α) coincide and are equal to

$$\lambda^1_\mu = \int_X \ln|a(x)|\,d\mu \quad .$$

So the statement of Theorem 4.4 in this situation has the form

$$\ln\ r(aT) = \max_{\mu\in\Lambda} \left\{\lambda^1_\mu\right\} \quad .$$

In this form it can be generalized to a formulae of the spectral radius of a weighted shift operator with an operator–valued coefficient:

Theorem 25.6 (Latushkin, Stepin [2]) *Let* a *satisfy the assumption* (25.1). *Then*

$$\ln\ r(aT) = \sup_{\mu\in\Lambda} \left\{\lambda^1_\mu\right\} \tag{25.13}$$

where Λ *is the set of all the ergodic measures on* X *with respect to* α.

Proof In view of (25.3) and since T is unitary we have

$$r(aT) = \lim_{n\to\infty} \max_{x\in X} \|\mathcal{A}(x, n)\|^{\frac{1}{n}} \tag{25.14}$$

and consequently for any $\mu \in \Lambda$ (25.7) implies (for any $x \in X_\mu$)

$$e^{\lambda_\mu^1} = \lim_{n\to\infty} \|\mathcal{A}(x, n)\|^{\frac{1}{n}} \le \lim_{n\to\infty} \max_{x\in X} \|\mathcal{A}(x, n)\|^{\frac{1}{n}} = r(aT) \quad .$$

Thus to prove the theorem it is enough to show that for any $\varepsilon > 0$ there exists a $\mu \in \Lambda$ such that

$$\lambda_\mu^1 \ge \ln r(aT) - \varepsilon \quad . \tag{25.15}$$

Without loss of generality we can assume that $r(aT) > 0$ (if $r(aT) = 0$ then in view of the foregoing observation (25.13) is true).

Let B and S be the unit ball and the unit sphere in H with the weak topology and let $X_B = X \times B$, $X_S = X \times S$ with the product topology,

$$\mathring{X}_B = \{(x,v) \in X_B\colon a(\alpha(x))v \neq 0\}, \ \tilde{a}(x,v) = \ln\|a(\alpha(x))v\|, \ (x,v) \in \mathring{X}_B \ .$$

Let us define the 'projectivization' $\tilde{\alpha}\colon X_B \to X_B$ of α via the formula

$$\tilde{\alpha}(x, v) = \begin{cases} \left(\alpha(x), \dfrac{a(\alpha(x))v}{\|a(\alpha(x)v)\|}\right), & (x, v) \in \mathring{X}_B \\ (\alpha(x), 0) , & (x, v) \bar{\in} \mathring{X}_B \ . \end{cases} \tag{25.16}$$

Note that since $a(x) \in K(H)$, $x \in X$, the function $\tilde{a}$ is continuous on $\mathring{X}_B$.

The mapping $\tilde{\alpha}$ is a Borel endomorphism of X_B as a composition of mappings

$$(x,\ v) \to (\alpha(x),\ a(\alpha(x))v)$$

$$(x,\ v) \to \left(x,\ \frac{v}{\|v\|}\right)\ , \qquad v \neq 0$$

$$(x,\ 0) \to (x,\ 0)\quad .$$

The image $\tilde{\alpha}(X_B) = \tilde{\alpha}(\mathring{X}_B) \cup (X \times \{0\})$ is also a Borel set in X_B since

$$\tilde{\alpha}(\mathring{X}_B) = X_S \cap \{(x,\ v):\ x \in X,\ v \in \operatorname{Im}\ a(\alpha(x))\}$$

$$\operatorname{Im}\ a(x) = \bigcup_{k=0}^{\infty} \operatorname{Im}\left(a(x)\,|_{\{v:\|v\|<k\}}\right)$$

and the operator $a(x)$ maps the closed ball $\{v:\ \|v\| \leq k\}$ into a strongly closed set (see Halmos [1], Corollary to Problem 130).

The important property of $\tilde{a}$ for our purposes is

$$\tilde{a}(x,\ \lambda v) = \ln|\lambda| + \tilde{a}(x,\ v)\ , \qquad \lambda \neq 0\ , \quad (x,\ v) \in \mathring{X}_B\ . \tag{25.17}$$

Observe that if $\tilde{\alpha}^j(x,\ v) \in \mathring{X}_B$, $j = 0,\ldots,n-1$, then

$$\tilde{\alpha}^n(x,\ v) = \left(\alpha^n x\ ,\ \frac{\mathcal{A}(x,\ n)v}{\|\mathcal{A}(x,\ n)v\|}\right)$$

and for such $(x,\ v)$ (25.17) implies

$$\|\mathcal{A}(x,\ n)v\|^{\frac{1}{n}} = \exp \frac{1}{n} \sum_{j=0}^{n-1} \tilde{a}(\tilde{\alpha}^j(x,\ v))\quad . \tag{25.18}$$

A measure μ satisfying (25.15) will be defined below as the projection on X of some Borelian probability measure ν on X_B which we shall construct now.

Since $\mathcal{A}(x,\ k)$ are compact there exist $(x_k,\ v_k) \in X_s$ such that

$$\|\mathcal{A}(x_k, k)v_k\| = \max_{x \in X} \|\mathcal{A}(x, k)\| \quad . \tag{25.19}$$

Recall that we are considering the case $r(aT) > 0$, thus

$$\tilde{\alpha}^j(x_k, v_k) \in \overset{\circ}{X}_B \quad , \qquad j = 0,\ldots,k-1 \quad .$$

Let us define the Borelian probability measures ν_k on X_B by

$$\nu_k(f) = \frac{1}{k} \sum_{j=0}^{k-1} f(\tilde{\alpha}^j(x_k, v_k)) \quad , \qquad f \in C(X_B) \ . \tag{25.20}$$

By the Banach–Alaoglu theorem there exists a subsequence ν_{n_k} and a probability measure ν such that

$$\nu(f) = \lim_{k\to\infty} \nu_{n_k}(f) \quad , \qquad f \in C(X_B) \quad . \tag{25.21}$$

We mention that $\tilde{a} \,\overline{\in}\, C(X_B)$. Nevertheless $\nu_k(\tilde{a})$ are well defined by (25.20) and in view of the choice of (x_k, v_k) and by (25.14)

$$\lim_{k\to\infty} \nu_{n_k}(\tilde{a}) = \ln r(aT) \quad . \tag{25.22}$$

Let us show that:

(a) $\nu(\overset{\circ}{X}_B) = 1$;

(*b*) $\tilde{a}$ is ν–integrable on X_B ;

(c) $\nu(\tilde{a}) \geq \ln r(aT)$.

To obtain this for any $N = 1,2,\ldots$ define the sets

$$F_N = \{(x, v) \colon \tilde{a}(x, v) \leq -N\} \supset X_B \setminus \overset{\circ}{X}_B$$

and consider the continuous functions on X_B:

$$\tilde{a}_N(x, v) = \begin{cases} \tilde{a}(x, v) , & (x, v) \in X_B \setminus F_N \\ -N , & (x, v) \in F_N . \end{cases}$$

Clearly

$$\tilde{a}(x,v) \le \tilde{a}_N(x,v) \le \max_{x \in X} \ln\|a(x)\| \overset{\text{def}}{=\!=} M , \qquad (x,v) \in X_B . \tag{25.23}$$

The left–hand inequality in (25.23) implies

$$\nu_{n_k}(\tilde{a}) \le \nu_{n_k}(\tilde{a}_N)$$

and passing to the limit when $k \to \infty$ we obtain, in view of the continuity of $\tilde{a}_N$ and (25.21), (25.22), that

$$\nu(\tilde{a}_N) = \lim_{k\to\infty} \nu_{n_k}(\tilde{a}_N) \ge \lim_{k\to\infty} \nu_{n_k}(\tilde{a}) = \ln\, r(aT) . \tag{25.24}$$

By the right–hand inequality in (25.23)

$$\nu(\tilde{a}_N) \le -N\, \nu(F_N) + M$$

and this, in view of (25.24), implies

$$\nu(F_N) \xrightarrow[N\to\infty]{} 0 .$$

Thus

$$\nu(X_B \setminus \overset{\circ}{X}_B) = 0$$

and (a) is proved.

To prove (b) note that on $\overset{\circ}{X}_B$ (that is, ν–almost everywhere) we have $\tilde{a}_N \searrow \tilde{a}$, where the convergence is pointwise. Thus $\tilde{a}$ is ν–integrable.

Finally the foregoing observation and (25.24) leads to

$$\nu(\tilde{a}) = \lim_{N\to\infty} \nu(\tilde{a}_N) \geq \ln r(aT) \tag{25.25}$$

so (c) is also true.

Now we shall prove that ν is $\tilde{\alpha}$–invariant. To do this it is enough to show that

$$\nu(h\circ\tilde{\alpha}) = \nu(h) \quad , \qquad h \in C(X_B) \quad . \tag{25.26}$$

It has already been mentioned that $\tilde{\alpha}$ is continuous on $\overset{\circ}{X}_B$ and it follows that

$$h\circ\tilde{\alpha} \in C_b(\overset{\circ}{X}_B) \quad , \qquad h \in C(X_B) \tag{25.27}$$

where $C_b(\overset{\circ}{X}_B)$ is the set of all the functions continuous on $\overset{\circ}{X}_B$ and bounded on X_B.

Let us show first that

$$\nu(f) = \lim_{k\to\infty} \nu_{n_k}(f) \quad , \qquad f \in C_b(\overset{\circ}{X}_B) \quad . \tag{25.28}$$

Indeed, for any $\delta > 0$ there exist open sets $\overline{Y}_1 \subset Y_2 \subset X_B$ such that

$$\begin{aligned} &X_B \backslash \overset{\circ}{X}_B && \subset Y_1 \\ &Y_1 \cap (X_B \backslash Y_2) && = \emptyset \\ &\nu(Y_2) && < \delta \end{aligned}$$

In view of the separability of H, X_B is metrizable (see Halmos [2], the solution of Problem 18) and thus is absolutely normal (Kuratowski [1], §21, IV).

In view of the Tietze theorem (Kuratowski [1], §14, IV) one can choose $f_1 \in C(X_B)$ in such a way that

$$f_1(x, \nu) = f(x, \nu) \quad , \qquad (x, \nu) \overline{\in} Y_1$$

and

$$|f_1(x, v)| \leq 2 \sup_{X_B} |f(x, v)| \quad . \tag{25.29}$$

There also exists a non–negative $f_2 \in C(X_B)$ with the properties

$$f_2(x, v) = 0 \quad , \qquad (x, v) \bar{\in} Y_2$$
$$f_2(x, v) = \Phi \quad , \qquad (x, v) \in Y_1$$
$$\|f_2\| \leq \Phi$$

where

$$\Phi = \sup \{ |f(x, v) - f_1(x, v)| : (x, v) \in Y_1 \} \leq 3\|f\|$$

(see (25.29)).

Now we have

$$\left|\nu(f) - \nu_{n_k}(f)\right| \leq |\nu(f - f_1)| + \left|\nu(f_1) - \nu_{n_k}(f_1)\right| + \nu_{n_k}(|f - f_1|)$$

$$\leq \Phi\nu(Y_1) + \left|\nu(f_1) - \nu_{n_k}(f_1)\right| + \left|\nu_{n_k}(f_2) - \nu(f_2)\right| + \Phi\nu(Y_2)$$

$$\leq \left|\nu(f_1) - \nu_{n_k}(f_1)\right| + \left|\nu_{n_k}(f_2) - \nu(f_2)\right| + 6\|f\|\cdot\delta$$

which in view of the arbitrariness of δ and (25.21) proves (25.28).

Now (25.26) follows from the equalities

$$\nu(h\circ\tilde{\alpha}) - \nu(h) = \lim_{k\to\infty} \left[\nu_{n_k}(h\circ\tilde{\alpha}) - \nu_k(h)\right]$$

$$= \lim_{k\to\infty} \frac{1}{n_k} \left[h(\tilde{\alpha}^{n_k}(x_{n_k}, v_{n_k}) - h(x_{n_k}, v_{n_k}))\right]$$

$$= 0 \quad .$$

As ν is $\tilde{\alpha}$–invariant and $\tilde{\alpha}(X_B)$ is Borel we conclude

$$\nu(\tilde{\alpha}(X_B)) = 1$$

which means that $\tilde{\alpha}$ is a ν mod 0–endomorphism of X_B.

Consequently the Birkhoff–Hinchin ergodic theorem (see Theorem 2.19) implies the existence of a set $Y \subset X_B$, $\nu(Y) = 1$, such that for every $(x, v) \in Y$ there exists

$$\lim_{k\to\infty} \frac{1}{k} \sum_{j=0}^{k-1} \tilde{a}(\tilde{\alpha}^j(x, v)) = \bar{a}(x, v) \tag{25.30}$$

and

$$\nu(\bar{a}) = \nu(\tilde{a}) \geq \ln r(aT) \quad . \tag{25.31}$$

By the way this argument implies (in view of (25.18)) the existence of a point $(x_0, v_0) \in X_B$ for which

$$\ln r(aT) \leq \nu(\tilde{a}) = \lim_{k\to\infty} \frac{1}{k} \ln\|\mathcal{A}(x_0, k)v_0\|$$

$$\leq \overline{\lim_{k\to\infty}} \frac{1}{k} \ln\|\mathcal{A}(x_0, k)\| \quad . \tag{25.32}$$

Now we are ready to construct the desired measure μ (for (25.15)). For any $\varepsilon > 0$ and the set Y_ε defined by

$$Y_\varepsilon = \{(x, v) \in Y\colon \bar{a}(x, v) \geq \nu(\tilde{a}) - \varepsilon\}$$

we have

$$\nu(Y_\varepsilon) > 0 \quad .$$

Let $p_1(x, v) = x$ and $p_2(x, v) = v$. By μ and $\tilde{\mu}$ we denote the projections of the measure ν on X and B respectively. By the Fubini theorem

$$\nu(Y_\varepsilon) = \int_{p_2(Y_\varepsilon)} \mu(\{x \in X\colon (x, v) \in Y_\varepsilon\})d\tilde{\mu} > 0 \quad .$$

So there exists a vector $\bar{v} \in p_2(Y_\varepsilon)$ such that

$$\mu(X_\varepsilon) > 0 \tag{25.33}$$

where

$$X_\varepsilon = \{x \in X: (x, \bar{v}) \in Y_\varepsilon\} \quad .$$

μ is clearly α–invariant by the construction of $\tilde{\alpha}$ and the $\tilde{\alpha}$–invariance of ν.

Passing if necessary to the decomposition of X into ergodic components we can assume that μ is ergodic.

By the multiplicative ergodic theorem and (25.33) there exists a set X_μ and a point $\bar{x} \in X_\mu \cap X_\varepsilon$ with the property

$$\lim_{k\to\infty} \frac{1}{k} \ln\|\mathscr{A}(\bar{x}, k)\| = \lambda_\mu^1 \quad .$$

Thus for the point $(\bar{x}, \bar{v}) \in Y_\varepsilon$ we have (using (25.18), (25.30), (25.31))

$$\ln r(aT) - \varepsilon \le \nu(\tilde{a}) - \varepsilon \le \bar{a}(\bar{x}, \bar{v}) = \lim_{k\to\infty} \frac{1}{k} \sum_{j=0}^{k-1} \tilde{a}(\tilde{\alpha}^j(\bar{x}, \bar{v}))$$

$$\le \lim_{k\to\infty} \frac{1}{k} \ln\|\mathscr{A}(\bar{x}, k)\| = \lambda_\mu^1 \quad .$$

The proof is complete. □

Remark While proving the theorem we have shown that

$$r(aT) = \overline{\lim_{k\to\infty}} \|\mathscr{A}(x_0, k)\|^{\frac{1}{k}}$$

for some $x_0 \in X$ (see (25.32)).

In fact this can be shown also by means of the idea used in the proof of Corollary 23.14.

25.7 The Lyapunov exponents are in fact a rather effective tool when considering the hyperbolicity of an operator (see Section 19).

We recall that an operator b is said to be hyperbolic if its spectrum does not meet the unit circle.

In the remaining part of this section we shall assume that $\alpha: X \to X$ acts topologically freely, that is, for every $n = 1,2,\ldots$, the set

$$X_n = \{x \in X: \alpha^n(x) = x\}$$

has an empty interior.

Though Prim A in our case is much greater than X (see Dixmier [1], 4.7.22) it is evident (since $a(x)$ changes continuously on X) that the topologically free action of α on X implies the topologically free action of $\hat{T}$ on Prim A (see **12.13**). Thus the results of Section 12 imply:

Theorem 25.8 *If* A *and* T *are as defined in* 25.1 *and* α: X $\to$ X *acts topologically freely then*

$$C^*(A, T) \cong A \times_{\hat{T}} \mathbb{Z} \quad .$$

This theorem and Corollary 13.6 imply in turn

Corollary 25.9 *If* A *and* T *are as considered in Theorem* 25.8 *then* $\sigma(aT)$ *is invariant under rotations.*

Using Lemma 19.1 we have

Lemma 25.10 *If* A *and* T *are as considered in Theorem* 25.8 *and* aT *is hyperbolic then the Riesz projection*

$$P = \frac{1}{2\pi i} \int_{|\lambda|=1} (\lambda - aT)^{-1} d\lambda$$

belongs to A.

25.11 Bearing in mind the difference between α in 25.1 and t in 19.5 (just as we did in Section 19) we can associate with aT the linear extension β of α^{-1}: $X \to X$ on $H(X) = \bigoplus_{x \in X} H_x$ (where $H_x = H$) defined by the formula (whereby (x, h) we denote the vector in H_x equal to $h \in H$)

$$\beta(x, h) = (\alpha^{-1}(x), a(\alpha^{-1}(x))h) \quad , \qquad x \in X \ , \quad h \in H_x$$

(cf. **19.12**) then Lemma 25.10 leads to

Theorem 25.12 aT *is hyperbolic if the corresponding linear extension is hyperbolic* (cf. *Theorem* 19.13).

Here the hyperbolicity of β is defined exactly in the same way as it was done in definition **19.4**.

25.13 The ideas used in Section 19 give us an opportunity to describe the stable and unstable subspaces H^s and H^u in terms of the so–called Lyapunov numbers which are introduced in the following way (cf. **19.8**).

For any $h \neq 0$ and any $x \in X$

$$\overline{\lambda}^{+}(x, h) = \overline{\lim_{n\to\infty}} \frac{1}{n} \ln\| \mathcal{A}(x, n)h\|$$

$$\underline{\lambda}^{+}(x, h) = \underline{\lim_{n\to\infty}} \frac{1}{n} \ln\| \mathcal{A}(x, n)h\|$$

$$\overline{\lambda}^{-}(x, h) = \overline{\lim_{n\to\infty}} - \frac{1}{n} \ln\|h_n\|$$

$$\underline{\lambda}^{-}(x, h) = \underline{\lim_{n\to\infty}} - \frac{1}{n} \ln\|h_n\|$$

where $\{h_n\}_{n=1}^{\infty}$ is a sequence such that $h_0 = h$ and $a(\alpha^{-n+1}(x))h_n = h_{n-1}$, $n = 1,2,...,$ (we set $\underline{\lambda}^{-} = \overline{\lambda}^{-} = -\infty$ if there is no such sequence). Clearly $\underline{\lambda}^{-}$ and $\overline{\lambda}^{-}$ in general depend on the sequence $\{h_n\}_{n=1}^{\infty}$ with the property described.

We denote by H_x^s and H_x^u the stable and unstable fibres over $x \in X$, that is,

$$H_x^s = P(x)H_x$$

$$H_x^u = (I - P)(x)H_x$$

(cf. **19.20** (4)).

Let us set

$$\sum = \{t \in \mathbb{R}\colon e^t \in |\sigma(aT)|\} \quad .$$

In view of Corollary 25.9 for aT considered in it we have

$$\sum = \{t \in \mathbb{R}\colon e^t \in \sigma(aT)\} \quad .$$

Now Lemma 19.9 (having in mind **25.11** and Theorem 25.12) can be reformulated as

Lemma 25.14 *Let* aT *be a hyperbolic operator. Then:*

(a) $h \in H_x^s$ *iff* $\overline{\lambda}^{+}(x, h)$ *and* $\underline{\lambda}^{+}(x, h)$ *are less than* 0;

(b) *if* $h \in H_x^s$ *then* $\overline{\lambda}^{-}(x, h) < 0$ *for any sequence* $\{h_n\}$ *mentioned in* 25.13;

(c) $h \in H_x^u$ *iff there exists a sequence* $\{h_n\}$ *mentioned in* **25.13** *such that* $\underline{\lambda}^{-}(x, h) > 0$;

(d) *if* $h \in H_x^u$ *then* $\underline{\lambda}^{+}(x, h) > 0$.

Moreover $\overline{\lambda}^{+}(x, h)$ *and* $\underline{\lambda}^{+}(x, h)$ *both lie in the same connected component of* $\sum$ *and also* $\underline{\lambda}^{-}(x, h)$ *and* $\overline{\lambda}^{-}(x, h)$ *for any particular* $\{h_n\}$ *defined in* **25.13** *both lie in the same connected component of* $\sum$.

25.15 (cf. **19.10**) For any two positive numbers $t_1 \neq t_2$ not belonging to the spectrum $\sigma(aT)$ let Δ_{t_1,t_2} be the part of the spectrum lying in the ring $\{t\colon t_1 < |t| < t_2\}$ and P_{t_1,t_2} be the corresponding Riesz projection

$$H_x^{t_1,t_2} = \operatorname{Im} P_{t_1,t_2}(x) \quad .$$

For any positive t not lying in $\sigma(aT)$ by Δ_t we denote the part of $\sigma(aT)$ lying in the circle $\{z\colon |z| < t\}$, by P_t the corresponding Riesz projection, and $H_x^t = \operatorname{Im} P_t(x)$.

To any connected component $\Delta \subset \sigma(aT)$ there corresponds the bundle H_x^Δ defined by

$$H_x^\Delta = \begin{cases} \bigcap\limits_{\substack{t_1,t_2 \,\bar{\in}\, \sigma(aT) \\ (t_1,t_2) \supset \Delta}} H_x^{t_1,t_2} \,, & \text{if } \Delta \not\ni 0 \\ \bigcap\limits_{\substack{t \,\bar{\in}\, \sigma(aT) \\ [0,\, t) \supset \Delta}} H_x^t \,, & \text{if } \Delta \ni 0 \quad . \end{cases}$$

Now Lemma 25.14 leads to

Lemma 25.16 (cf. Lemma 19.11)

(1) $H_x^{t_1,t_2}$ *coincides with the closed linear span of vectors* $h \neq 0$ *such that* $\underline{\lambda}^+(x, h)$ *and* $\overline{\lambda}^+(x, h)$ *both belong to the segment* $(\ln t_1, \ln t_2)$, $\overline{\lambda}^-(x, h) < \ln t_2$ *for every* $\{h_n\}$ *defined in* 25.13 *and* $\underline{\lambda}^-(x, h) > \ln t_1$ *for some* $\{h_n\}$.

(2) H_x^t *coincides with the closed linear span of vectors* $h \neq 0$ *such that* $\underline{\lambda}^\pm(x, h), \overline{\lambda}^\pm(x, h) < \ln t$.

(3) H_x^Δ *coincides with the closed linear span of vectors* $h \neq 0$ *such that*:

(i) $\underline{\lambda}^\pm(x, h), \overline{\lambda}^\pm(x, h) \in \ln \Delta$ *if* $\Delta \ni 0$ (*where* $\ln 0 = -\infty$);

(ii) $\underline{\lambda}^{+}(x, h), \overset{+}{\overline{\lambda}}(x, h) \in \ln \Delta, \overset{-}{\overline{\lambda}}(x, h) \le \ln t_2$ *for every* $\{h_n\}$ *defined in* 25.13 *and* $\underline{\lambda}^{-}(x, h) \ge \ln t_1$ *for some* $\{h_n\}$;

when $\Delta = [t_1, t_2], t_1 > 0$.

25.17 Evidently if $X = X_1 \cup X_2$ where X_1 and X_2 are closed non–intersecting α–invariant sets then

$$\sigma(aT) = \sigma(aT|_{X_1}) \cup \sigma(aT|_{X_2}) .$$

Thus the principal case when considering the spectrum of aT is the case of an α–irreducible space X (see 18.6).

Theorem 25.18 *Let* A *and* T *be as considered in Theorem* 25.8. *Then* $|\sigma(aT)|$ *consists of a set of segments (corresponding to the rings of* σ(aT))

$$|\sigma(aT)| = \bigcup_{k=1}^{N} \left[r_k^-, r_k^+\right] \tag{25.34}$$

and

(1) $\dim H_X^{\Delta_k} < \infty$ *if* $\Delta_k \not\ni 0$ *(here* $\Delta_k = \left\{t: r_k^- \le |t| \le r_k^+\right\}$;

(2) *if* $\dim H = \infty$ *then* $\sigma(aT) \ni 0$;

(3) *if* X *is* α*–irreducible then* $N \le \dim H$;

(4) *if* $N = \infty$ *and* X *is* α*–irreducible then* $\lim_{k\to\infty} r_k^{\pm} = 0$; *in this case* $\bigcup_{k=1}^{N} \left[r_k^-, r_k^+\right] \overset{\text{def}}{=\!=} \left[r_1^-, r_1^+\right] \cup \left[r_2^-, r_2^+\right] \cup \dots \cup \left[r_\infty^-, r_\infty^+\right]$ *where* $r_\infty^-, r_\infty^+ = 0$ *and we can assume that* $r_k^- > r_{k+1}^+$.

Proof Since $\sigma(aT)$ is invariant under rotations every segment $\left[r_k^-, r_k^+\right]$ in $|\sigma(aT)|$ corresponds to the whole of the ring $\Delta_k = \left\{t: r_k^- \le |t| \le r_k^+\right\}$ $\subset \sigma(aT)$.

If $\Delta_k \bar{\in} 0$ then as $a(x)$ is compact $\dim H_x^{\Delta_k} < \infty$ due to the fact that $\beta: H_x^{\Delta_k} \to H_{\alpha^{-1}(x)}^{\Delta_k}$ is invertible (β is defined in 25.11). Thus (1) is proved.

If $\dim H = \infty$ then aT is not invertible (β is not invertible as $a(x)$ is compact). Thus $0 \in \sigma(aT)$ and (2) is true.

To prove (4) it is enough to show that for any $t > 0$, $t \bar{\in} \sigma(aT)$, there exist not more than a finite number of segments $\left[r_k^-, r_k^+\right]$ such that $r_k^{\pm} > t$. So let us prove this.

Suppose on the contrary that we have infinitely many segments $\left[r_k^-, r_k^+\right]$ with $r_k^{\pm} > t$. From what we have already said it follows that

$$\dim[\operatorname{Im}(I - P_t)(x)] = N_0 < \infty$$

(P_t is defined in 25.15).

Remark We write here N_0 (not depending on x) since $\dim[\operatorname{Im}(I - P_t)(x)]$ is a continuous function (as $P_t(x)$ is continuous with respect to x) and is α–invariant. Thus since X is α–irreducible $\dim[\operatorname{Im}(I - P(x)] \equiv \text{const}$.

Now consider the numbers

$$s_0 > r(aT) > s_1 > s_2 > \dots > s_{N_{0+1}} = t$$

such that $s_i \bar{\in} \sigma(aT)$ and $|\sigma(aT)| \cap [s_{i+1}, s_i] \neq \emptyset$. We have already shown that

$$0 < \dim\left[\operatorname{Im} P_{s_i, s_{i+1}}(x)\right] = N^i < \infty$$

(in view of the foregoing remark N^i does not depend on x).

As

$$(I - P)(x) = \bigoplus_{i=0}^{N_0} P_{s_i, s_{i+1}}(x)$$

we conclude

$$N_0 = \overset{N_0}{\underset{i=0}{\oplus}} N^i$$

and arrive at a contradiction. Thus (4) is proved. Obvious changes in the foregoing argument prove (3) and thus the whole of lemma. □

25.19 Now let us fix some ergodic measure μ on X. If aT is hyperbolic then in view of Lemma 25.14 there exists a number $j_\mu < \infty$ such that

$$\lambda^{j_\mu+1} < 0 < \lambda^{j_\mu} .$$

Set $j_\mu = 0$ if $\lambda^1_\mu = 0$. Here λ^j_μ are the Lyapunov exponents (see Theorem 25.3).

The next lemma shows that the fibres $V_{x,\mu}$ corresponding to the negative and positive Lyapunov exponents coincide with the 'traces' of H^s_x and H^u_x on X_μ.

Lemma 25.20 *If* aT *is hyperbolic and* j_μ *is that defined in* 25.19 *then*

$$\left.\begin{aligned} V_x^{j_\mu+1} &= H_x^s \\ \overset{j_\mu}{\underset{j=1}{\oplus}} W_x^j &= H_x^u \end{aligned}\right\} \text{ on } X_\mu$$

where V^j_x *and* X_μ *are those mentioned in* Theorem 25.3 *and* W^j_x *are defined in* 25.4.

Proof The first equality is evident in view of Theorem 25.3 and Lemma 25.14.

To prove the second one it is enough to show (since dim $H^u_x < \infty$ and in view of (25.9), (25.12) that

$$V_x^{-j_\mu} \subset H_x^u \quad . \tag{25.35}$$

To see this recall (Ruelle [1], p.261) that for any non–zero $h \in V_x^{-j_\mu}$ there exists the unique sequence $\{h_n\}$ mentioned in **25.13** such that

$$\underline{\lambda}^-(x, h) = \overline{\lambda}^-(x, h) \geq \lambda^{j_\mu} > 0$$

which in view of Lemma 25.14 proves (25.35) and thus finishes the proof. □

Remark 25.21 If X is α–irreducible then we have already shown (see the proof of Theorem 25.18) that $\dim H_x^u \equiv \text{const} < \infty$ on X. Thus in this case $\dim\left[\bigoplus_{j=1}^{j_\mu} W_x^j\right]$ does not depend on μ.

25.22 Lemma 25.20 gives us a tool to link the spectral description obtained in Theorem 25.18 with the Lyapunov exponents.

If X is α–irreducible then by Theorem 25.18 (3), (4) one can find numbers $t_0 > t_1 > \ldots$ such that

$$t_0 > r_1^+ \geq r_1^- > t_1 > r_2^+ \geq r_2^- > \ldots \geq 0 \quad . \tag{25.36}$$

Theorem 25.23 *If* X *is α–irreducible and* A, T *are as mentioned in Theorem* 25.8 *then for any particular ergodic measure* μ *there exist numbers*

$$\mathrm{j}_1 = 1 < \mathrm{j}_2 < \ldots < \mathrm{j}_{N+1} = \mathrm{s}_\mu + 1 \quad , \qquad N \leq \infty$$

such that

$$\lambda_\mu^{\mathrm{j}_k} < \mathrm{t}_{k-1} < \lambda_\mu^{\mathrm{j}_{k-1}} \qquad \mathrm{k} = 2,3,\ldots$$

and

(1) $H_x^{t_{k-1}} = V_x^{j_k}$ *on* X_μ

(2) $H_x^{t_k, t_{k-1}} = \sum_{j=j_k}^{j_{k+1}-1} W_x^j$ *on* X_μ

(3) $\lambda_\mu^j \in \left[\ln r_k^-, \ln r_k^+\right]$ *for* $j_k \le j < j_{k+1}$.

Here t_k *are the numbers mentioned in* (25.36) and s_μ in Theorem 25.3.

Proof Since for every t_k the operator $t_k^{-1}aT$ is hyperbolic (1) and (2) are evident in view of Lemma 25.20, and (3) follows from Lemma 25.16. □

Theorem 25.24 *Under the conditions of Theorem* 25.23 *we have*

$$\ln r_k^- = \inf \left\{\lambda_\mu^j : j_k \le j < j_{k+1}, \mu \in \Lambda\right\} \tag{25.37}$$

$$\ln r_k^+ = \sup \left\{\lambda_\mu^j : j_k \le j < j_{k+1}, \mu \in \Lambda\right\} . \tag{25.38}$$

Proof (25.38) follows by applying Theorem 25.6 to the operator $(P_{t_k, t_{k-1}} a)T = aTP_{t_k, t_{k-1}}$. On the subspace Im $P_{t_k, t_{k-1}}$ this operator is invertible. Applying Theorem 25.6 to its inverse we obtain (25.37). □

26. Wiener–type theorems. Fullness.

Considering the conditions for invertibility of an element d it is often important to know what sort of element d^{-1} can be (if it exists). For example the well–known theorem of N. Wiener [1] asserts that if a function can be expressed as an absolutely convergent Fourier series then its inverse (if it exists) can also be expressed as an absolutely convergent series.

Using algebraic terminology one would say in this case that the algebra of absolutely convergent Fourier series is full (see **2.58**).

In this section we present the corresponding results for some subalgebras of $C^*(A, T_g)$. These results will also be used in the next section and in the second part of the book.

Throughout this section G will be a commutative group.

26.1 Let B be a Banach algebra. We shall denote by $S(\hat{G}, B)$ a (non–closed) subalgebra of $C(\hat{G}, B)$ ($C(\hat{G}, B)$ is the algebra of all continuous functions $f\colon \hat{G} \to B$ with the sup norm) which consists of functions of the form

$$f(\chi) = \sum_{g \in G} \chi(g) b_g \qquad \chi \in \hat{G}, \quad b_g \in B$$

where $\sum_{g \in G} \|b_g\| < \infty$ (in particular, no more than a countable number of coefficients b_g are non–zero).

26.2 Recall that a subalgebra $\mathscr{B}_1$ of an algebra $\mathscr{B}$ is called *full* if any element $b \in \mathscr{B}_1$ which is invertible in $\mathscr{B}$ is invertible in $\mathscr{B}_1$ as well.

Fullness is clearly equivalent to the fact that

$$\sigma_{\mathscr{B}_1}(b) = \sigma_{\mathscr{B}}(b)$$

where $\sigma_{\mathscr{B}}(b))(\sigma_{\mathscr{B}_1}(b))$ is the spectrum of b in $\mathscr{B}(\mathscr{B}_1)$.

Evidently if $\mathscr{B}_1$ is full then the closure of $\mathscr{B}_1$ is also full.

An obvious but useful result is

Lemma 26.3 *The commutant of any set $\mathscr{D} \subset \mathscr{B}$ is a full subalgebra in $\mathscr{B}$.*

Very important for us is the following Bochner–Phillips–Allan theorem which is actually a Banach algebra analogue to the Wiener theorem.

Theorem 26.4 (Bochner, Phillips [1]) *Let* G *be a discrete group then* $S(\hat{G}, B)$ *is a full subalgebra of* $C(\hat{G}, B)$.

Now we describe some full subalgebras in algebras associated with automorphisms.

Recall that in this book we deal with a group G (independently of its structure (see for example 26.9)) as with a discrete group.

26.5 We denote by S the (non–closed) subalgebra of $A \underset{\mathcal{T}}{\times} G$ of elements of the form

$$\sum_{g \in G} a_g \otimes g \;, \qquad \text{where} \qquad \sum_{g \in G} \|a_g\| < \infty \;.$$

Theorem 26.6 S *is a full subalgebra of* $A \underset{\mathcal{T}}{\times} G$.

Proof Let $b = \sum a_g \otimes g$, $\sum \|a_g\| < \infty$, be an invertible element in $A \underset{\mathcal{T}}{\times} G$. Then by Proposition 13.1

$$b(\chi) = \sum \chi(g) a_g \otimes g$$

is also invertible for every $\chi \in \hat{G}$.

We can consider $b(\chi)$ as an element of $S(\hat{G}, A \underset{\mathcal{T}}{\times} G)$ and by Theorem 26.4

$$[b(\chi)]^{-1} = \sum \chi(g)\; V_g$$

where $V_g \in A \underset{\mathcal{T}}{\times} G$ and $\sum \|V_g\| < \infty$.

By Hewitt, Ross [1], (23.19) for any fixed $g_0 \in G$

$$\int_{\hat{G}} [b(\chi)]^{-1} \chi(g_0^{-1}) d\chi = \sum_{G} V_g \int_{\hat{G}} \chi(g \cdot g_0^{-1}) d\chi = V_{g_0} \;. \tag{26.1}$$

Consider the C^*–subalgebra $\mathcal{B} \subset C(\hat{G}, G \underset{\mathcal{T}}{\times} A)$

$$\mathcal{B} = \overline{\left\{\sum \chi(g) a_g \otimes g \;, \quad \sum \|a_g\| < \infty\right\}}$$

where ——— denotes the uniform closure.

We have (see **12.3**)

$$\left\|\sum \chi(g)a_g \otimes g\right\|_{C(G,A \underset{\mathcal{T}}{\times} G)} = \left\|\sum a_g \otimes g\right\|_{A \underset{\mathcal{T}}{\times} G} \geq \|a_g\| \quad .$$

Thus the mapping

$$N_{g_0}\left(\sum \chi(g)a_g \otimes g\right) = a_{g_0} \otimes g_0$$

is defined correctly and can be extended onto the whole of $\mathcal{B}$. Obviously for every $b(\chi) \in \mathcal{B}$

$$\int_{\hat{G}} b(\chi)\ \chi(g_0^{-1})d\chi = N_{g_0}(b(\chi)) \quad . \tag{26.2}$$

Observe that since $\mathcal{B}$ is a C^*–algebra it is full (Theorem 2.78). For any $b \in S$ we have $b(\chi) \in \mathcal{B}$ and thus $[b(\chi)]^{-1} \in \mathcal{B}$. Now comparison of (26.1) and (26.2) finishes the proof. □

Remark 26.7 While proving the theorem we have also proved that if $b = \sum a_g \otimes g$ belongs to S and is invertible then

$$b(\chi) = \sum \chi(g)a_g \otimes g$$

is invertible for every $\chi \in \hat{G}$ and

$$[b(\chi)]^{-1} = \sum \chi(g)\ \bar{a}_g \otimes g$$

where

$$\sum \bar{a}_g \otimes g = b^{-1} \in S \quad .$$

As an immediate corollary of Theorem 26.6 we obtain

Theorem 26.8 *If* G *is a commutative group and* $C^*(A, T_g)$ *possesses the property* (*) (*or* G *acts topologically freely on* A) *then the subalgebra of elements of the form*

$$\sum a_g T_g \qquad \textit{where} \qquad \sum \|a_g\| < \infty$$

is full.

Proof Use Theorems 12.8 and 12.14. □

There are also some other complete subalgebras in $A \underset{\hat{T}}{\times} G$ and consequently in $C^*(A, T_g)$.

26.9 Let G be any commutative group isomorphic to $\mathbb{R}^m \times \mathbb{Z}^p \times K$ where K is a compact group (in fact any locally compact and compactly generated commutative group is such a group (see Hewitt, Ross [1], (9.8))). For any element $g \in G$ we define

$$|g| = |\lambda_1| + \dots + |\lambda_m| + |z_1| + \dots + |z_p| \tag{26.3}$$

where $g = (\lambda, z, k)$, $\lambda = (\lambda_1, \dots, \lambda_m) \in \mathbb{R}^m$, $z = (z_1, \dots, z_p) \in \mathbb{Z}^p$, $k \in K$. We denote by E the subalgebra of $A \underset{\mathcal{T}}{\times} G$ which has *exponentially decreasing coefficients*, that is, $E \subset S$ (see **26.5**) and for any $b \in E$ of the form

$$b = \sum a_g \otimes g$$

there exists a $\gamma > 0$ such that

$$\sum e^{\gamma|g|} \|a_g\| < \infty \quad . \tag{26.4}$$

If G has a finite set of generators then it is isomorphic to $\mathbb{Z}^p \times F$ where F is a finite group (see Hewitt, Ross [1], (A.27)) and consequently in this case (26.3) turns into

$$|g| = |z_1| + \dots + |z_p|$$

and (26.4) is equivalent to the condition that there exist $c < \infty$ and $\nu > 0$ such that

$$\|a_g\| < C\, e^{-\nu|g|} \quad . \tag{26.5}$$

Observe also that in this case (26.5) implies $b \in S$ (just use Lemma 11.16 and Theorem 15.2).

Theorem 26.10 *Let* G ≈ $\mathbb{R}^m \times \mathbb{Z}^p \times K$. *Then* E *is a full subalgebra in* $A \underset{\mathcal{T}}{\times} G$.

Proof Let $b \in E$ be invertible and have the form $b = \sum a_g \otimes g$. Consider the function $b(\xi)$, $\xi = (\xi_1,\ldots,\xi_{m+p}) \in \mathbb{C}^{m+p}$, defined by

$$b(\xi) = \sum e^{\langle \xi, g\rangle} a_g \otimes g$$

where $\langle \xi, g\rangle = \langle \xi_1\lambda_1 +\ldots+ \xi_m\lambda_m + \xi_{m+1}z_1 +\ldots+ \xi_{m+p}z_p\rangle$, $g = (\lambda,z,k)$, $\lambda = (\lambda_1,\ldots,\lambda_m) \in \mathbb{R}^m$, $z = (z_1,\ldots,z_p) \in \mathbb{Z}^p$, $k \in K$.

Since $b \in E$ by (26.4) $b(\xi)$ is an analytic function on the set $\{\xi\colon |\mathrm{Re}\, \xi_i| < \gamma,\ i = \overline{1,m+p}\}$.

(26.4) also implies the existence of $\varepsilon > 0$ such that

$$\|b(\xi) - b\|_1 \le \frac{1}{2\|b^{-1}\|_1} \qquad \text{when } |\mathrm{Re}\, \xi_i| < \varepsilon\ ,\ i = \overline{1,m+p} \tag{26.6}$$

where $\|b\|_1 = \sum \|a_g\|$.

Since S is full (see Theorem 26.6) (26.6) implies that $[b(\xi)]^{-1} \in S$ and (see **2.58** and Lemma 2.7)

$$\|[b(\xi)]^{-1}\|_1 \le 2\|b^{-1}\|_1 \quad . \tag{26.7}$$

As $b(\xi)$ is analytic when $|\mathrm{Re}\, \xi_i| < \gamma$, $i = \overline{1,m+p}$, it follows that $[b(\xi)]^{-1}$ is analytic when $|\mathrm{Re}\, \xi_i| < \min(\varepsilon, \gamma)$, $i = \overline{1,m+p}$.

But if $\mathrm{Re}\, \xi_i = 0$, $i = \overline{1,m+p}$, then $e^{\langle \xi, \cdot\rangle}$ is a character and we have

$$[b(\xi)]^{-1} = \sum e^{\langle \xi, g \rangle} \, \bar{a}_g \otimes g$$

where $\sum \bar{a}_g \otimes g = b^{-1}$.

By the uniqueness of the analytic extension we have

$$[b(\xi)]^{-1} = \sum e^{\langle \xi, g \rangle} \, \bar{a}_g \otimes g \qquad \text{when } |\mathrm{Re}\ \xi_i| < \min(\varepsilon, \gamma), \ i = \overline{1, m+p} \ .$$

Now (26.7) means that $b \in E$. □

26.11 Considering an algebra $C^*(A, T_g)$ where $G \approx \mathbb{R}^m \times \mathbb{Z}^p \times K$ we say that an element $b \in C^*(A, T_g)$ has exponentially decreasing coefficients if it has the form

$$b = \sum a_g T_g \qquad \text{where } \sum \|a_g\| < \infty$$

and there exists a $\gamma > 0$ such that

$$\sum e^{\gamma |g|} \, \|a_g\| < \infty$$

(see 26.9).

Just as in Theorem 26.8 we conclude that the following statement is true.

Theorem 26.12 *If* $G \approx \mathbb{R}^m \times \mathbb{Z}^p \times K$ *and* $C^*(A, T_g)$ *possess the property* (*) *(or* G *acts topologically freely on* A*) then the elements which have exponentially decreasing coefficients form a full subalgebra in* $C^*(A, T_g)$.

Remark Observe that any element of the form

$$b = \sum_{g \in M} a_g T_g \qquad |M| < \infty$$

may be viewed as an element with exponentially decreasing coefficients.

27. The approximation approach.

In Section 6 we observed that though invertibility is not a stable property under small perturbations of an operator in the *strong* topology, nevertheless in some cases if we change the operators generating automorphisms in such a way that automorphisms are changed slightly then the spectrum $\sigma(aT)$ also changes slightly. Here we continue to investigate this effect in the general situation and obtain results proving that our guess (see Remark 6.5 (2)) was correct.

The group G here will be assumed to be the same as in Section 26.

An algebra A will be assumed to be separable though we shall not use this until the proof of sufficiency in Theorem 27.7.

27.1 We recall that the strong topology in the set Aut A of automorphisms of A is one for which the basis of neighbourhoods of $u \in \text{Aut } A$ is formed by sets $O(u,a_1,\dots,a_k,\varepsilon)$, $a_i \in A$, $i = \overline{1,k}$, $\varepsilon > 0$, where

$$O(u,a_1,\dots,a_k,\varepsilon) = \{v \in \text{Aut } A\colon \|v(a_i) - u(a_i)\| < \varepsilon,\ i = \overline{1,k}\}\ . \qquad (27.1)$$

Examples

(1) Let $A = C(M)$ where M is a metrizable compact space with a metric ρ. Then any automorphism $u\colon A \to A$ has the form

$$(ua)(x) = a(\alpha(x)) \quad , \qquad x \in M$$

where $\alpha\colon M \to M$ is a homeomorphism.

We define the distance between two automorphisms u_1 and u_2 as

$$d(u_1, u_2) = \sup_M \rho(\alpha_1(x), \alpha_2(x))$$

where α_i are the homeomorphisms corresponding to u_i, $i = 1,2$. Evidently $u_n \to u$ in the above–mentioned topology iff $d(u_n, u) \to 0$.

(2) Let G be a compact group and $A = C(G)$. Let U_g be the representation of G by the automorphisms of A:

$$(U_{g_0}a)(g) = a(gg_0^{-1}) \quad .$$

In this case we have $U_{g_i} \to U_{g_0} \Leftrightarrow g_i \to g_0$.

27.2 Let $\mathcal{U}$: $G \to \text{Aut } A$ be a representation of a group G by automorphisms of A. The basis of neighbourhoods of $\mathcal{U}$ is formed by the sets $O(\mathcal{U}, g_1,\ldots,g_n, a_1,\ldots,a_k, \varepsilon)$ of representations $\mathcal{V}$: $G \to \text{Aut } A$ such that

$$\mathcal{V}(g_i) \in O(\mathcal{U}(g_i), a_1,\ldots,a_k, \varepsilon) \quad , \qquad i = \overline{1,n} \tag{27.2}$$

(see **27.1**).

Obviously if G is finitely generated and $g_1,\ldots,g_m$ are its generators then the sets $O(\mathcal{U}, g_1,\ldots,g_m, a_1,\ldots,a_k, \varepsilon)$ form a basis of neighbourhoods of $\mathcal{U}$.

By $O(\mathcal{U})$ we denote an arbitrary neighbourhood of $\mathcal{U}$.

If a unitary representation V: $g \to V_g$, $g \in G$, is such that $V_gAV_g^* = A$ then we shall also write $V \in O(\mathcal{U})$ if $\hat{V} \in O(\mathcal{U})$ (recall that $\hat{V}$ is the representation of G by the automorphisms $\hat{V}_g(a) = V_gaV_g^*$, $a \in A$).

27.3 If $b = \sum a_g \otimes g \in S$ (see **26.5**) then we denote $\sum \|a_g\|$ by $\|b\|_1$.

27.4 Let $\mathcal{V}$: $G \to \text{Aut } A$ be a representation of a group G by automorphisms of A. By $\mu_{\mathcal{V}}$ we denote the mapping from $l^1(G, A)$ into $A \underset{\mathcal{V}}{\times} G$ defined by

$$\mu_{\mathcal{V}}(\{a_g\}_{g\in G}) = \sum a_g \otimes g \in A \underset{\mathcal{V}}{\times} G \quad . \tag{27.3}$$

The norm in $l^1(G, A)$ we shall also denote by $\| \quad \|_1$ when it will not be confusing (cf. **27.3**).

If $b = \sum a_g \otimes g \in S\left(\subset A \underset{\mathcal{T}}{\times} G, \text{ see } \mathbf{26.5}\right)$ then we denote by $\eta_{\mathcal{V}}(b)$ the element in $A \underset{\mathcal{V}}{\times} G$:

$$\eta_{\mathcal{V}}(b) = \mu_{\mathcal{V}}\{a_g\}_{g\in G}$$

(see (27.3)).

27.5 If $b = \sum a_g \otimes g \in S$ and $V\colon g \to V_g$ is a unitary representation of G then by b_V we denote the operator

$$b_V = \sum a_g V_g \quad .$$

Our first technical step in obtaining the principal results of this section (Theorems 27.7, 27.9, 27.11) is the following

Lemma 27.6 *Let* $\mathrm{b}_1, \mathrm{b}_2 \in \mathrm{S}$ (*see* 26.5). *For any* $\varepsilon > 0$ *there exists a neighbourhood* $\mathrm{O}(\mathcal{T})$ *such that for all* $\mathcal{V} \in \mathrm{O}(\mathcal{T})$ *we have*

$$\left\|\mu_{\mathcal{V}}^{-1}[\eta_{\mathcal{V}}(\mathrm{b}_1)\cdot\eta_{\mathcal{V}}(\mathrm{b}_2)] - \mu_{\mathcal{T}}^{-1}[\mathrm{b}_1\cdot\mathrm{b}_2]\right\|_1 < \varepsilon \quad .$$

Proof Let $b_i = \sum a_g^i \otimes g$, $i = 1,2$. Fix $\delta > 0$ and choose $M \subset G$ such that $|M| < \infty$ and $\left\|b_i^M - b_i\right\|_1 < \delta$ where $b_i^M = \sum\limits_M a_g^i \otimes g$, $i = 1,2,$. We have

$$\left\|\mu_{\mathcal{V}}^{-1}[\eta_{\mathcal{V}}(b_1)\cdot\eta_{\mathcal{V}}(b_2)] - \mu_{\mathcal{T}}^{-1}[b_1\cdot b_2]\right\|_1$$

$$\le \left\|\mu_{\mathcal{V}}^{-1}\left[\eta_{\mathcal{V}}(b_1)\cdot\eta_{\mathcal{V}}(b_2) - \eta_{\mathcal{V}}\left(b_1^M\right)\cdot\eta_{\mathcal{V}}(b_2)\right]\right\|_1 + \left\|\mu_{\mathcal{V}}^{-1}\left[\eta_{\mathcal{V}}\left(b_1^M\right)\cdot\eta_{\mathcal{V}}(b_2)\right.\right.$$

$$\left.\left. - \eta_{\mathcal{V}}\left(b_1^M\right)\cdot\eta_{\mathcal{V}}\left(b_2^M\right)\right]\right\|_1 + \left\|\mu_{\mathcal{V}}^{-1}\left[\eta_{\mathcal{V}}\left(b_1^M\right)\cdot\eta_{\mathcal{V}}\left(b_2^M\right)\right] - \mu_{\mathcal{T}}^{-1}\left[b_1^M\cdot b_2^M\right]\right\|_1$$

$$+ \left\|\mu_{\mathcal{T}}^{-1}\left[b_1^M\cdot b_2^M - b_1^M\cdot b_2\right]\right\|_1 + \left\|\mu_{\mathcal{T}}^{-1}\left[b_1^M\cdot b_2 - b_1\cdot b_2\right]\right\|_1$$

$$\le \|b_2\|_1 \left\|b_1 - b_1^M\right\|_1 + \left\|b_1^M\right\|_1 \left\|b_2 - b_2^M\right\|_1$$

$$+ \left\| \mu_{\mathcal{V}}^{-1} \left[\eta_{\mathcal{V}}\left(b_1^M\right) \cdot \eta_{\mathcal{V}}\left(b_2^M\right) \right. - \mu_{\mathcal{G}}^{-1} \left[b_1^M \cdot b_2^M \right] \right\|_1$$

$$+ \left\| b_1^M \right\|_1 \left\| b_2^M - b_2 \right\|_1 + \|b_2\|_1 \left\| b_1^M - b_1 \right\|_1$$

$$\leq C \cdot \delta + \left\| \mu_{\mathcal{V}}^{-1} \left[\eta_{\mathcal{V}}\left(b_1^M\right) \cdot \eta_{\mathcal{V}}\left(b_2^M\right) \right. - \mu_{\mathcal{G}}^{-1} \left[b_1^M \cdot b_2^M \right] \right\|_1 \tag{27.4}$$

where

$$C = 4 \max(\|b_1\|_1, \|b_2\|_1) \quad .$$

Consider now

$$K = \left\| \mu_{\mathcal{V}}^{-1} \left[\eta_{\mathcal{V}}\left(b_1^M\right) \cdot \eta_{\mathcal{V}}\left(b_2^M\right) \right. - \mu_{\mathcal{G}}^{-1} \left[b_1^M \cdot b_2^M \right] \right\|_1 \quad . \tag{27.5}$$

Observe that

$$\eta_{\mathcal{V}}\left(b_1^M\right) \cdot \eta_{\mathcal{V}}\left(b_2^M\right) = \sum_{h \in M} a_h^1 \otimes h \cdot \sum_{g \in M} a_g^2 \otimes g$$

$$= \sum_h \sum_g a_h^1 \cdot \mathcal{V}_h\left(a_g^2\right) \otimes hg$$

$$= \sum_{s \in M^2} \left[\sum_{h \in M} a_h^1 \; \mathcal{V}_h\left(a_{sh^{-1}}^2\right) \right] \otimes s$$

and thus (27.5) can be rewritten as

$$K = \sum_{s \in M^2} \left\| \sum_{h \in M} \left(a_h^1 \; \mathcal{V}_h\left(a_{sh^{-1}}^2\right) - a_h^1 \; \mathcal{G}_h\left(a_{sh^{-1}}^2\right) \right) \right\|$$

$$\leq \|b_1\|_1 \sum_{s \in M^2} \sum_{h \in M} \left\| \mathcal{V}_h\left(a_{sh^{-1}}^2\right) - \mathcal{G}_h\left(a_{sh^{-1}}^2\right) \right\| \quad . \tag{27.6}$$

Choose $\delta < \frac{\varepsilon}{C + 1}$ and $O(\mathcal{G})$ such that (27.6) $< \delta$ for any $\mathcal{V} \in O(\mathcal{G})$.

Now the statement of the lemma follows from (27.4). □

We proceed to one of the main results of this section.

Theorem 27.7 *An element* b ∈ S (*see* 26.5) *is invertible iff there exists a neighbourhood* $O(\mathcal{T})$ *and a constant* K > 0 *such that for every* $\mathcal{V} \in O(\mathcal{T})$ $\eta_{\mathcal{V}}(b)$ (*see* 27.4) *is invertible and* $\left\|[\eta_{\mathcal{V}}(b)]^{-1}\right\| < K$.

Proof

Necessity If b is invertible then $b^{-1} \in S$ by Theorem 26.6, and Lemma 27.6 implies the existence of $O(\mathcal{T})$ with the property

$$\left\|\eta_{\mathcal{V}}(b)\cdot\eta_{\mathcal{V}}(b^{-1})-e\right\| < \tfrac{1}{2} \quad \text{and} \quad \left\|\eta_{\mathcal{V}}(b^{-1})\eta_{\mathcal{V}}(b)-e\right\| < \tfrac{1}{2}\ , \quad \mathcal{V} \in O(\mathcal{T})$$

(since $\|\cdot\|_1 > \|\cdot\|$). Here e is the identity of $A \underset{\mathcal{V}}{\times} G$.

This means (by 2.58 and Lemma 2.7) that both $F_1 = \eta_{\mathcal{V}}(b)\cdot\eta_{\mathcal{V}}(b^{-1})$ and $F_2 = \eta_{\mathcal{V}}(b^{-1})\eta_{\mathcal{V}}(b)$ are invertible and thus $\eta_{\mathcal{V}}(b)$ is invertible and

$$[\eta_{\mathcal{V}}(b)]^{-1} = \eta_{\mathcal{V}}(b^{-1})F_1^{-1} = F_2^{-1}\ \eta_{\mathcal{V}}(b^{-1}) \quad .$$

Moreover Lemma 2.7 implies $\left\|F_1^{-1} - e\right\| \le \dfrac{\|F_1 - e\|}{1 - \|F_1 - e\|} < 1$.

Thus

$$\left\|F_1^{-1}\right\| < 2$$

and consequently

$$\left\|[\eta_{\mathcal{V}}(b)]^{-1}\right\| = \left\|\eta_{\mathcal{V}}(b^{-1})\cdot F_1^{-1}\right\| \le 2\left\|\eta_{\mathcal{V}}(b^{-1})\right\| \le 2\|b^{-1}\|_1$$

which finishes the proof of necessity.

Sufficiency If $O(\mathcal{T})$ and K are as mentioned in the theorem then there exists a sequence $\mathcal{V}_\lambda \to \mathcal{T}$ such that all $\eta_{\mathcal{V}_\lambda}(b)$ are invertible and

$\left\| \left[\eta_{\mathcal{V}_\lambda}(b) \right]^{-1} \right\| < K.$

Consider the regular representation of $A \underset{\mathcal{V}}{\times} G$, that is (recall **11.11** and Remark (β) after Theorem 11.12), a representation in $l^2(G, H)$ (H is the space where A acts) defined by

$$\pi_{\mathcal{V}}(a)(g \otimes h) = g \otimes \mathcal{V}_g(a)h$$

$$\pi_{\mathcal{V}}(I \otimes g_0)(g \otimes h) = \left(g \cdot g_0^{-1} \right) \otimes h \quad .$$

The condition $\mathcal{V}_\lambda \to \mathcal{G}$ implies $\pi_{\mathcal{V}_\lambda}(a) \to \pi_{\mathcal{G}}(a)$ strongly and consequently

$$\pi_{\mathcal{V}_\lambda}\left(\eta_{\mathcal{V}_\lambda}(b) \right) \to \pi_{\mathcal{G}}(b)$$

$$\left[\pi_{\mathcal{V}_\lambda}\left(\eta_{\mathcal{V}_\lambda}(b) \right) \right]^* \to [\pi_{\mathcal{G}}(b)]^*$$

strongly and (since $\pi_{\mathcal{V}}$ is a faithful representation of $A \underset{\mathcal{V}}{\times} G$)

$$\left\| \pi_{\mathcal{V}_\lambda}\left(\eta_{\mathcal{V}_\lambda}(b) \right)^{-1} \right\| = \left\| \left[\pi_{\mathcal{V}_\lambda}\left(\eta_{\mathcal{V}_\lambda}(b) \right)^{-1} \right]^* \right\| = \left\| \eta_{\mathcal{V}_\lambda}(b)^{-1} \right\| < K \ .$$

To finish the proof we use the following simple but useful observation:

Lemma 27.8 *If a sequence of invertible operators* A_n *on a Banach space* E *is such that*

$$\mathrm{A}_n \xrightarrow{\text{strongly}} \mathrm{A}$$

$$\mathrm{A}_n^* \xrightarrow{\text{strongly}} \mathrm{A}^*$$

and there exists a constant $\mathrm{K} > 0$ *such that*

$$\left\| \mathrm{A}_n^{-1} \right\| = \left\| \mathrm{A}_n^{*-1} \right\| < \mathrm{K}$$

then A *is invertible.*

Proof Indeed on the contrary let A be not invertible. Then there exists a sequence $\xi_i \in E$, $\|\xi_i\| = 1$, such that $\|A\xi_i\| \to 0$ or such a sequence exists for A^*. Consider the first situation (the second can be treated in an analogous way). Fix any i_0 such that

$$\|A\xi_{i_0}\| < \frac{1}{K} .$$

Now for sufficiently large n we have

$$\|A_n\xi_{i_0}\| < \frac{1}{K}$$

which implies $\|A_n^{-1}\| > K$ and we arrive at a contradiction. □

This lemma, along with the foregoing argument, shows that $\pi_{\mathcal{J}}(b)$ and consequently b are invertible. The proof of the theorem is complete. □

Consider now an algebra $C^*(A, T_g)$. The foregoing theorem leads to the next statement.

Theorem 27.9 *Let* $C^*(A, T_g)$ *be an algebra which possesses the property* (*) (*or* $\{T_g\}_{g\in G}$ *acts topologically freely on* A). *If an operator* b *having the form*

$$b = \sum a_g T_g \qquad \textit{where} \quad \sum \|a_g\| < \infty$$

is invertible then:

(1) *there exists* $O(\hat{T})$ *and a constant* $K > 0$ *such that for every unitary representation* $V: g \to V_g$, $g \in G$, *which belongs to* $O(\hat{T})$ (see **27.2**) b_V (see **27.5**) *is invertible and* $\|b_V^{-1}\| < K$;

(2) *for any sequence* $V_\lambda: g \to V_{\lambda_g}$, $g \in G$, *of unitary representations such that* $V_{\lambda_g} A V^*_{\lambda_g} = A$ *and* $\hat{V}_\lambda \to \hat{T}$ *there exists* $K > 0$ *such that for all but finitely many* λ, b_{V_λ} *are invertible and* $\left\| b^{-1}_{V_\lambda} \right\| < K$.

If $V_{\lambda_g} \to T_g$ *strongly or if* $C^*(A, V_{\lambda_g})$ *possess the property* (*) *then* (2) *implies the invertibility of* b.

Proof Since $C^*(A, T_g)$ possesses the property (*) we have (Theorem 12.8)

$$C^*(A, T_g) \cong A \underset{\hat{T}}{\times} G \quad .$$

As $A \underset{\hat{V}}{\times} G \ni \sum a_g \otimes g \to \sum a_g V_g$ is a representation of $A \underset{\hat{V}}{\times} G$ the necessity of conditions (1) and (2) for the invertibility of b follows from Theorem 27.7.

If $C^*(A, V_{\lambda_g})$ possess the property (*) then

$$C^*(A, V_{\lambda_g}) \cong A \underset{\hat{V}_\lambda}{\times} G$$

and the sufficiency of (2) for the invertibility of b follows from the proof of Theorem 27.7.

Suppose finally that (2) is satisfied and $V_{\lambda_g} \to T_g$ strongly. Then (by Kadison [1]) $V^*_{\lambda_g} \to T^*_g$ strongly and consequently

$$b_{V_\lambda} \to b \quad ; \qquad b^*_{V_\lambda} \to b^*$$

strongly and

$$\left\|b_{V_\lambda}^{-1}\right\| = \left\|b_{V_\lambda}^{-1*}\right\| < K \quad .$$

Thus (in view of Lemma 27.8) b is invertible. □

Remarks

(1) If $C^*(A, T_g)$ does not possess the property (*) then conditions (1) and (2) of the theorem are not necessary for the invertibility of b.

Example Consider the unit circle S^1 and the space $L^2_\mu(S^1)$ where μ is the normalized Lebesgue measure. Let T be the operator of the form

$$(Tf)(x) = f(\pi(x)) \qquad f \in L^2_\mu(S^1)$$

where $\pi(x)$ is the rotation of the circle by the angle π (thus $T^2 = I$). Consider the algebra $C^*(T) = C^*(CI, T)$. Obviously since $T^2 = I$ this algebra does not possess the property (*).

Consider now the operators V_φ, $\varphi \in [0, 2\pi]$, defined by

$$(V_\varphi f)(x) = f(\varphi(x)) \qquad f \in L^2_\mu(S^1)$$

where $\varphi(x)$ is the rotation of S^1 by the angle φ. Let $b = \gamma I - T \in C^*(T)$ and $b_{V_\varphi} = \gamma I - V_\varphi \in C^*(V_\varphi) = C^*(CI, V_\varphi)$. For any φ $\hat{V}_\varphi = \hat{T}$ is the trivial automorphism of $A = CI$. The operator b is invertible iff $\gamma^2 \neq 1$. On the other hand if $\frac{\varphi}{\pi}$ is an irrational number then b_{V_φ} is not invertible when $|\gamma| = 1$ (see the Remark in 8.5).

(2) The condition $\hat{V}_\lambda \to \hat{T}$ does not imply in general the strong convergence of V_{λ_g} to T_g.

Example In the space $L^2_\mu(S^1)$ (see the previous example) consider the algebra A of operators of multiplication by continuous functions $a \in C(S^1)$.

Let $T = V_{\varphi_0}$ where $\frac{\varphi_0}{\pi}$ is an irrational number (V_φ were introduced in the previous example). Then T acts topologically freely on A. Define $V_\lambda = e^{i\lambda x}T$. We have $\hat{V}_\lambda = \hat{T}$. But V_λ does not tend strongly to T if $\lambda \not\to 0$.

27.10 It is well known that the spectrum is upper semicontinuous in the uniform operator topology: for an operator F and any $\varepsilon > 0$ there exists a $\delta > 0$ such that the inequality $\|F - E\| < \delta$ implies that $\sigma(E)$ belongs to the ε–neighbourhood $\sigma_\varepsilon(F)$ of $\sigma(F)$ ($\sigma_\varepsilon(F) = \{\lambda \in \mathbb{C}: \rho(\lambda, \sigma(F)) < \varepsilon\}$ where $\rho(\lambda, M) = \inf_{z\in M} |\lambda - z|$) (see 2.12).

On the other hand in general the spectrum is not upper semicontinuous in the strong topology. The foregoing results however can be used to establish the fact that while the convergence $\mathcal{V} \to \mathcal{T}$ implies only the 'strong' convergence $\eta_{\mathcal{V}}(b) \to b$ (see the proof of Theorem 27.7) the spectrum $\sigma(\eta_{\mathcal{V}}(b))$ is upper semicontinuous.

Theorem 27.11 *Let* b $\in$ S (see 26.5). *For any* $\varepsilon > 0$ *there exists a neighbourhood* $O(\mathcal{T})$ *such that for every* $\mathcal{V} \in O(\mathcal{T})$, $\sigma(\eta_{\mathcal{V}}(b)) \subset \sigma_\varepsilon(b)$.

Proof The sets $\sigma(b)$ and $\sigma_\varepsilon(b)$ are clearly contained in the ball $R_\varepsilon = \{z \in \mathbb{C}: |z| \le \|b\|_1 + 1 + \varepsilon\}$. Let λ_0 be any point in $\overline{[R_\varepsilon \backslash \sigma_\varepsilon(b)]}$. Denote

$$P_1(\lambda_0) = \|\lambda_0 I - b\|_1 , \qquad P_2(\lambda_0) = \left\|(\lambda_0 I - b)^{-1}\right\|_1 .$$

Fix any $\delta < \min(\varepsilon, \frac{1}{4})$. Choose any neighbourhood $U(\lambda_0)$ with the property:

$$\lambda \in U(\lambda_0) \Rightarrow \begin{cases} |\lambda - \lambda_0| < \delta , \\ |\lambda - \lambda_0| P_i(\lambda_0) < \delta , \qquad i = 1,2, \\ \dfrac{|\lambda - \lambda_0| P_2^2(\lambda_0)}{1 - P_2(\lambda_0)|\lambda - \lambda_0|} < \delta . \end{cases} \tag{27.7}$$

Now cover $\overline{[R_\varepsilon \setminus \sigma_\varepsilon(b)]}$ with such neighbourhoods and then choose any finite subcover $U(\lambda_1),\ldots,U(\lambda_n)$. Observe that by the choice of δ this subcover does not intersect with $\sigma(b)$.

For every $k = \overline{1,n}$ by Lemma 27.6 there exists $O_k(\mathcal{T})$ such that for any $\mathcal{V} \in O_k(\mathcal{T})$

$$\left\|(\lambda_k I - \eta_{\mathcal{V}}(b))\eta_{\mathcal{V}}\left[(\lambda_k I - b)^{-1}\right] - I\right\|_1 < \delta$$

$$\left\|\eta_{\mathcal{V}}\left[(\lambda_k I - b)^{-1}\right](\lambda_k I - \eta_{\mathcal{V}}(b)) - I\right\|_1 < \delta \quad . \tag{27.8}$$

Now take $O(\mathcal{T}) = \bigcap_{k=1}^{n} O_k(\mathcal{T})$. This is the neighbourhood we are looking for.

Indeed, for any $\lambda \in U(\lambda_k)$, $k = \overline{1,n}$, we have

$$\left\|(\lambda I - \eta_{\mathcal{V}}(b))\eta_{\mathcal{V}}\left[(\lambda I - b)^{-1}\right] - I\right\|_1$$

$$= \left\|((\lambda - \lambda_k)I + \lambda_k I - \eta_{\mathcal{V}}(b))\cdot\eta_{\mathcal{V}}\left[(\lambda I - b)^{-1} - (\lambda_k I - b)^{-1} + (\lambda_k I - b)^{-1}\right] - I\right\|_1$$

$$\leq |\lambda - \lambda_k| \ \|(\lambda I - b)^{-1}\|_1 + \|\lambda_k I - b\|_1 \left\|\left[(\lambda I - b)^{-1} - (\lambda_k I - b)^{-1}\right]\right\|_1$$

$$+ \left\|(\lambda_k I - \eta_{\mathcal{V}}(b))\cdot\eta_{\mathcal{V}}\left[(\lambda_k I - b)^{-1}\right] - I\right\|_1 \quad . \tag{27.9}$$

Observe that by Lemma 2.7 and 2.58

$$\left\|(\lambda I - b)^{-1} - (\lambda_k I - b)^{-1}\right\|_1 \leq \frac{|\lambda - \lambda_k| \ P_2^2(\lambda_k)}{1 - P_2(\lambda_k)|\lambda - \lambda_k|} < \delta$$

where the right–hand inequality follows from (27.7). Thus

$$\left\|(\lambda I-b)^{-1}\right\|_1 \le \left\|(\lambda_k I-b)^{-1}\right\|_1 + \delta = P_2(\lambda_k) + \delta \quad .$$

Consequently in view of (27.7), (27.8) we conclude

$$(27.9) < \delta(P_2(\lambda_k) + \delta) + P_1(\lambda_k)\delta + \delta \le 3\delta - \delta^2 < 1 \ . \qquad (27.10)$$

Since $\| \cdot \|_1 \ge \| \cdot \|$ (27.10) implies the invertibility of $(\lambda I - \eta_{\mathcal{V}}(b))\cdot\eta_{\mathcal{V}}[(\lambda I - b)^{-1}]$. The same argument shows that $\eta_{\mathcal{V}}[(\lambda I - b)^{-1}]\ (\lambda I - \eta_{\mathcal{V}}(b))$ is invertible as well and so $(\lambda I - \eta_{\mathcal{V}}(b))$ is invertible which finishes the proof. □

Corollary 27.12 *Let* $C^*(A, T_g)$ *be an algebra which possesses the property* (*) *(or* $\{T_g\}_{g\in G}$ *acts topologically freely on* A). *If an operator* b *has the form*

$$b = \sum a_g T_g \qquad \textit{where} \qquad \sum \|a_g\| < \infty$$

then for any $\varepsilon > 0$ *there exists a neighbourhood* $O(\hat{T})$ *such that for every unitary representation* $V\colon g \to V_g$, $g \in G$, *which belongs to* $O(\hat{T})$ (see **27.2**) *and* b_V *defined in* **27.5**, *we have*

$$\sigma(b_V) \subset \sigma_\varepsilon(b) \quad .$$

Proof By Theorem 12.8 or Corollary 12.17

$$C^*(A, T_g) \cong A \underset{\hat{T}}{\times} G \quad ,$$

and $A \underset{\hat{V}}{\times} G \ni \sum a_g \otimes g \to \sum a_g V_g$ is a representation of $A \underset{\hat{V}}{\times} G$. Now the result follows from the theorem. □

Notes and remarks

Section 17. The results of this section constitute a refinement and development of the corresponding assertions obtained in a number of papers for particular weighted shift operators: see Parrot [1], Kravchenko [1], Karapetyants [1], Karapetyants, Samko [1, 2], Antonevich [4, 8], Antonevich, Ryvkin [1], Antonevich, Vol'pert [1], Karlovich, Kravchenko [1], and Lebedev [5, 6].

Here we follow Lebedev [1, 2, 7] and Antonevich, Lebedev [1, 3].

Section 18. A number of examples presented here were investigated (using other methods) by different authors. Particular cases of **18.4**(a),(d) were investigated by Kravchenko [1]; (b) by Karapetyants [1], Antonevich, Ryvkin [1]; (c) by Lebedev [8]. **18.5–18.7** is due to Antonevich [4, 8, 14].

Section 19. As we mentioned at the beginning of this section its ideas deeply intersect with the ideas arising in the theory of differential equations and dynamical systems. The amount of work on the theory of hyperbolic dynamical systems (Anosov systems) is enormous; we refer here for example to the survey by Pesin [1]. Linear extensions of dynamical systems are treated systematically in Bronstein [1], Samoilenko [1], and Johnson, Palmer, Sell [1].

Theorem 19.13 is a C^*–algebra analogue of the well–known Mather theorem [1] (see also Bronstein [1], p.99). The main idea of the proof here (the use of the isomorphism theorem after establishing a result similar to Lemma 19.1) belongs to Antonevich [6, 11] where the definition of the hyperbolic extension and relation in the case when $A \cong$ HOM E (E is a vector bundle) was introduced, and corresponding versions of Theorems 19.21, 19.30, and 19.32 were obtained.

In connection with Theorem 19.13 we would like to mention also the results of Karlovich and Kravchenko [3, 4] (see also Karlovich, Kravchenko, Litvinchuk [1]).

The important work concerning the investigation of linear extensions interrelated to C^*–algebras of continuous operator–valued functions was made by Latushkin, Stepin [1, 2] who generalized the corresponding finite dimensional results of Sacker, Sell [1, 2] (see also the surveys by Johnson, Palmer, Sell [1,] and Sell [1]). The results of this section may be viewed as general C^*–algebra statements of this sort; they were obtained by Lebedev

[12, 13, 16, 17] and the exposition here follows these papers. Lemma 19.16 is due to Radbel' [1].

A criterion for an extension to be hyperbolic in terms of quasihyperbolicity and dimensional invariants has been obtained by Latushkin, Stepin ([2], Theorem 4.11).

Section 20. The method of reducing the investigation of operators of the form $\sum a_i T^i$ to the investigation of a two–term operator is due to Antonevich [11]. Here we presented the cross–product variant of this method. Example 20.5 is also due to Antonevich [11].

Section 21. The trajectorial approach as a method of obtaining invertibility conditions for particular classes of functional equations generated by weighted composition operators was used by a number of authors: Antonevich [11, 14], Hevelev [1, 2], Karlovich [1], Karlovich, Kravchenko [2], Lebedev [9, 10, 12, 13], Myasnikov, Sazonov [1], Semenjuta [1], and Soldatov [1]. We emphasize again that the necessity of all $\pi_x(b)$ to be invertible for the invertibility of b is evident and the problem was to obtain sufficiency. The main idea of the proof of Theorem 21.2 arose in the paper by Lebedev [9] in the case of a commutative algebra A and $G = \mathbb{Z}$. The exposition presented here follows Lebedev [13]. Since then Batkhin [3, 4] has generalized the assertion of Theorem 21.2 up to the case when A is some 'densely separably generated' (but perhaps non–separable) C^*–algebra and G is a discrete amenable (but perhaps non–countable) group acting topologically freely. Theorem 21.6 is a straightforward C^*–algebraic reinterpretation of the corresponding results of Antonevich [11]. The close connection of hyperbolicity and exponential dichotomy is treated extensively in the theory of differential equations. Perron (see Massera, Sheffer [1]) brought to light the correlation between exponential dichotomy and the solvability of non–homogeneous equations.

Results **21.12–21.16** are simply general C^*–algebra reformulations of the corresponding results of Latushkin, Stepin [2] (Section 3.4) it is worth mentioning that various ideas interrelated to the trajectorial decomposition are of great value in the theory of *–dynamical systems (see for example Vershik [1], and Vershik, Fedorov [1]).

Section 22. The main ideas concerning the evaluation of the spectral radius formulae presented here arise in the work by Brenner [1–4] where the corresponding formulae were obtained in the case of a commutative algebra A and $G = \mathbb{Z}^n$ and it was observed that they can be generalized to the case of a subexponential group. Using these formulae V.V. Brenner has found an original proof of the isomorphism theorem (and this method was generalized further by Antonevich [11, 14]). The spectral radius formulae obtained here are due to Lebedev [13, 16]. The example in 22.8 is due to Antonevich and Makhnist. It is worth mentioning that perhaps the first spectral radius formulae of the type considered here appeared in the classical paper by Beurling [1].

Very interesting applications of the spectral radius formulae of sums of shift operators with constant coefficients to the problems of combinatorial group theory and probability theory have been found by Grigorchuk [1, 2].

Section 23. Some spectral radius estimates for operators in $L^p[0, 1]$, $1 \leq p \leq \infty$, and $C[0, 1]$ dealing with the formula

$$f(t) \to \sum_{k=1}^{m} a_k(t)f(\varphi_k(t))$$

where φ_k: $[0, 1] \to [0, 1]$ are transformations of the 'delay' type were obtained by Berezanskij [1, 2] and Kurbatov [3, 4].

The idea of using the 'triangle procedure' (see **23.6**) for obtaining spectral radius estimates is due to Brenner [1–3] who used it in the case of a commutative algebra A and $G = \mathbb{Z}^n$ but the idea of how to use it in the more general case existed in the essence of his argument. Thus the results of this section are in fact a general C^*–algebra generalization of his method obtained by Lebedev [16].

Section 24. The criteria of invertibility obtained in **24.1–24.6** take their origin in the work by Brenner [2, 3] and Antonevich [8, 12, 14]. In their present form they are due to Lebedev [16].

Convex and rationally independent systems of shifts were investigated by Antonevich [8, 12, 14] for the case when the weights have limits at infinity

$$\lim_{x\to\infty} a(x) = a(\infty) \quad .$$

The investigation of the case presented here was made by Lebedev [16].

Section 25. Multiplicative ergodic theorem is a rather intensively treated subject; see for example Gol'shejd, Margulis [1], Kornfel'd, Sinaj [1], Krengel [1], and Walters [1]. The most important standard reference in this connection is Oseledets [1]. The infinite dimensional situation considered in the text was investigated in the paper by Ruelle [1] (see also Thieullen [1]).

Theorem 25.6 and the proof presented here is due to Latushkin, Stepin [1, 2]. The results of **25.7–25.24** are also in essence due to Latushkin and Stepin. We presented them here in a form which goes well with the whole of the theory previously developed in this book (especially using Section 19). In connection with the use of the multiplicative ergodic theorem presented here we would like to mention the survey by Johnson, Palmer, Sell [1], where for example it is proved that the dynamical spectrum boundary lies in the Lyapunov spectrum.

To obtain the idea of using the Lyapunov exponents in the stability theory of differential and difference equations it is worth looking through Bylov, Vinograd, Grobman, Nemytskij [1] and Daletskij, Krein [1] (ch. III).

Section 26. The main idea of the proof of Theorem 26.6 (the use of a result of Bochner, Phillips [1]) is due to Semenjuta [1], where he used it in the case of $A \equiv C(S^1)$ and the regular representation of the circle group. Later it was also used by Hevelev [1, 2], Kurbatov [1, 2], Langhe, Rabinovich [1], and Lebedev [11, 16]. The general C^*–algebra and cross–product case was treated only by Lebedev. The exposition of **26.1–26.8** presented here follows Lebedev [16].

The results of **26.9–26.12** for regular group representations were obtained by Kurbatov [2]. We should mention here also the results of Shubin [2, 3]. The statements presented here are proved by Lebedev [16].

Section 27. It was Semenjuta [1] who first obtained results which showed that the conditions for invertibility of operators generated by irrational rotations of the circle can be expressed in terms of invertibility of operators generated by rational approximations of these rotations. (In fact he treated the problem posed by Antonevich: Can we find the conditions of invertibility in terms of small perturbations of shifts? In particular: Can we use the conditions for invertibility of operators generated by periodic shifts (see Section 15) for obtaining the conditions for invertibility of operators generated by shifts which can be approximated by periodic shifts? (see **6.2**).) Further investigations have been made by Hevelev [1, 2]. The present exposition is a generalization of their approach obtained by Lebedev [12, 16]. We would like to emphasize that in fact periodic approximations of dynamical systems and finite dimensional approximations of C^*–algebras are treated extensively in connection with various problems of analysis (see for example Vershik, Kerov [1], and Vershik, Kornfel'd [1]).

Chapter 4: Localization

'Если мысль не приходит в голову, она не приходит никуда'*

Though the spectrum $\hat{A}$ and the primitive ideal space Prim A are extremely natural objects that arise when dealing with C^*–algebras they are not the first objects one thinks of when solving functional differential equations. The reason for this (apart from historical motives) is, from our point of view, that when treating differential equations one thinks perhaps first of all of 'smooth' objects (while $\hat{A}$ and Prim A in natural situations are not only not smooth but even not Hausdorff) and it seems to be quite natural that non–smooth objects appeared in functional differential equations for the first time when a researcher (I.B. Simonenko) explored singular integral equations with discontinuous coefficients. But even in this situation he tried to solve the problem with the help of Hausdorff objects – commutative algebras – and arrived at the method which we now call the *localization procedure* and which is to some extent the most commutative procedure in the theory of non–commutative algebras.

On the other hand the purely C^*–algebraic approach (arising from different C^*–algebra problems), which in particular has led in the end to the same structure, was developed in the work by J. Dauns, K.H. Hofmann, R.G. Douglas, J. Varela, and N.L. Vasilevskij.

The ideas based on the work of I.B. Simonenko we call here analytical localization (Section 29) while the C^*–algebra ideas are named algebraic localization (Lemma 28.6). The coincidence of these approaches is established in Lemma 29.2.

We have to stress that while in C^*–algebra problems it may not be too natural to deal with a localization with respect to an algebra and an ideal (see Section 32) it is the main object to consider when treating functional differential equations and their normal solvability problem.

In the C^*–algebra theory $\hat{a}(x)$ play the natural role of the simplest things in terms of which elements $a \in A$ can be described (the 'non–commutative

---

*'If a thought does not come into a head it goes nowhere.'

Fourier images' of a). The same role in localization theory is played by the local representatives $b(m)$ (the term is due to I.B. Simonenko). Unfortunately the local representatives $b(m)$ cannot always be calculated effectively (and the same of course is true for $\hat{a}(x)$) but in a number of situations we can carry out the calculations. These are the situations when the local representatives are invariant under some 'local' group of automorphisms (such objects are often called equivariant) and thus we can apply the corresponding Fourier transform. This procedure is presented in Sections 31–33.

We finish this chapter with the consideration of algebras generated by local algebras and their automorphisms and obtain for these objects results corresponding to the similar objects considered in Chapter 2.

28. Localization – algebraic approach.

In this section we present the algebraic ideas of the so–called localization procedure in C^*–algebras. We want to emphasize at once that the main construction which is useful in the theory of functional differential equations (differing from the one presented here only by the addition of an ideal) is considered in Section 32.

28.1 Throughout this section we fix some commutative C^*–algebra $A \subset L(H)$ and our goal is to obtain information about the structure of C^*–algebras which belong to the commutant A' of A. Clearly A' is a weakly closed algebra and since $A = A^*$, A' is a C^*–algebra; in particular, A' is a full algebra. As A is commutative we have $A \subset A'$. We shall consider only those algebras $B \subset A'$ which contain A. In fact this situation is described by the following definition.

28.2 We shall say that a C^*–algebra B is of *local type* with respect to a commutative C^*–algebra A if $A \subset \mathbb{Z}(B)$ where $\mathbb{Z}(B)$ is the centre of B.

Thus we have

$$B \supset A$$

and $$\qquad (28.1)$$

$$ba = ab \qquad \text{for every} \quad a \in A \ , \quad b \in B \ .$$

The naturalness of the term 'local type' in this definition will be exposed by the forthcoming results and observations.

28.3 For any closed set $F \subset M_A$ (M_A is the maximal ideal space for A) with a (perhaps empty) interior $\overset{\circ}{F}$ we shall denote by P_F the projection onto the essential subspace $H_F \subset H$, that is, the subspace generated by the vectors

$$\{a\xi: \; a \in A, \; \hat{a}(m) = 0 \; m \bar{\in} \overset{\circ}{F}, \; \xi \in H\} \quad .$$

We know that $P_F \in A''$ (see Lemma 2.82) and the following lemma shows that local–type elements can be 'localized' by such projections.

Lemma 28.4 *If* $\mathrm{b} \in \mathrm{A}'$ *then for any two non–intersecting closed subsets* F_1, F_2 *of* M_A *we have*

$$\mathrm{P}_{\mathrm{F}_1}\mathrm{b}\mathrm{P}_{\mathrm{F}_2} = 0 \quad . \tag{28.2}$$

Proof Consider any elements $a_1, a_2 \in A$ such that

$$\hat{a}_1\big|_{F_1} \equiv 1 \quad , \qquad \hat{a}_2\big|_{F_2} \equiv 1 \quad , \qquad a_1 \cdot a_2 = 0 \quad .$$

Such elements exist in view of Urysohn's Lemma since M_A is Hausdorff. We have

$$a_1 b a_2 = a_1 \cdot a_2 b = 0$$

and thus

$$P_{F_1} b P_{F_2} = P_{F_1} a_1 b a_2 P_{F_2} = 0 \quad . \qquad \square$$

28.5 Now we shall show that the fixation of algebras B and A satisfying (28.1) implies the natural map φ: Prim $B \to M_A$ which is continuous and onto. For any ideal $J \in$ Prim B we denote by $\varphi(J)$ the ideal of A defined by

$$\varphi(J) = \{A \cap J\} \quad . \tag{28.3}$$

Lemma 28.6 (Localization) *The map* $J \to \varphi(J)$ *defines the continuous surjective mapping*

$$\varphi\colon \text{Prim } B \to M_A \quad .$$

Proof Let π be any irreducible representation of B such that $\text{Ker } \pi = J$. We shall show that $\text{Ker}\left(\pi|_A\right)$ is a maximal ideal in A. Clearly $\text{Ker}\left(\pi|_A\right)$ is an ideal in A, so there exists a closed set $N \subset M_A$ such that

$$\text{Ker}(\pi|_A) = \left\{a \in A\colon \hat{a}|_N \equiv 0\right\} \quad .$$

Suppose that $\text{Ker}(\pi|_A)$ is not a maximal ideal in A. Then N consists of more than one point. Since N is closed and M_A is Hausdorff and compact there exists a finite cover of N by open sets N_i, $i = \overline{1,k}$, such that $N \setminus \overline{N_i} \neq \emptyset$. Let $t_i \in N \setminus \overline{N_i}$, $i = \overline{1,k}$. By Urysohn's Lemma for every i there exists an element $a_i \in A$ such that

$$\hat{a}_i|_{\overline{N_i}} \equiv 0 \qquad \text{and} \qquad \hat{a}_i(t_i) = 1 \quad .$$

This implies:

$$\text{(1)} \quad a_i \,\bar{\in}\, \text{Ker}(\pi|_A), \text{ that is, } \pi(a_i) \neq 0 \quad ; \tag{28.4}$$

$$\text{(2)} \quad a = \prod_{i=1}^{k} a_i \in \text{Ker}(\pi|_A), \text{ that is, } \pi(a) = 0 \quad . \tag{28.5}$$

On the other hand since $a_i \in A$ in view of (28.1) $\pi(a_i)$ commutes with $\pi(b)$ for all $b \in B$, which means, as π is irreducible, that all $\pi(a_i)$ are scalars, that is (in view of (28.4)), $\pi(a_i) = \lambda_i I_{H_\pi}$, $\lambda_i \neq 0$ (I_{H_π} is the identity operator in H_π). Thus

$$\pi(a) = \left(\prod_{i=1}^{k} \lambda_i\right) I_{H_\pi} \neq 0$$

and we arrive at a contradiction of (28.5).

Consequently the set N consists of a single point and thus $\mathrm{Ker}(\pi|_A) = \varphi(J) \in M_A$.

On the other hand let $m \in M_A$ and π' be any irreducible representation of A such that $\pi'(a) = \hat{a}(m)$. Then by Theorem 2.81 there exists an irreducible representation π of B such that $\pi|_A = \pi'$. Thus $\varphi(\mathrm{Ker}\ \pi) = m$ and we have shown that the mapping φ: Prim $B \rightarrow M_A$ is onto.

Now let us observe that φ is continuous. Indeed let V be any closed set in M_A and

$$I_V = \left\{a \in A: \hat{a}|_V \equiv 0\right\} .$$

Then

$$\varphi^{-1}(V) = \left\{\mathrm{Ker}\ \pi,\ \pi \in \hat{B}: I_V \subset \mathrm{Ker}\ \pi\right\}$$

but this simply means (see 2.63) that $\varphi^{-1}(V)$ is a closed set in Prim B and the proof of lemma is complete. □

Corollary 28.7 *Let* Θ: $\hat{\mathrm{B}} \rightarrow$ Prim B *be the canonical mapping* ($\pi \rightarrow$ Ker π). *Then the mapping*

$$\Psi = \varphi \circ \Theta: \hat{\mathrm{B}} \rightarrow \mathrm{M_A} \tag{28.6}$$

is continuous and onto.

Proof Obvious by the previous lemma and the definition of the topology on $\hat{B}$ (see 2.64). □

28.8 For any $m \in M_A$ we denote by J_m the ideal in B defined by

$$J_m = \bigcap_{\varphi(J)=m} J . \tag{28.7}$$

Let ρ_m be the canonical mapping

$$B \xrightarrow{\rho_m} B/_{J_m} \overset{\text{def}}{=\!=} B(m) \quad . \tag{28.8}$$

The image $\rho_m(b)$ of an element $b \in B$ under the mapping ρ_m: $B \to B(m)$ we denote by $b(m)$ and call it the *local representative* of b at the point $m \in M_A$.

Remark Obviously for any $a \in A$, $b \in B$ we have

$$(ab)(m) = (ba)(m) = \hat{a}(m)b(m) \quad . \tag{28.9}$$

Theorem 28.9 *The set* $\{\rho_m\}_{m \in M}$ *is a sufficient set of representations of* B, *that is,* b *is invertible iff all* $\rho_m(b)$ *are invertible.*

Proof The necessity is evident so let us prove sufficiency.

Let all $\rho_m(b)$, $m \in M_A$, be invertible. Take any irreducible $\pi \in \hat{B}$. By Corollary 28.7 there exists a point $m \in M_A$ such that $\psi(\pi) = m$. By the definition of J_m (see 28.8) we have $J_m \subset \text{Ker } \pi$.

Thus there is defined a canonical homeomorphisms from $B(m)$ into $\pi(B)$ and the element $\pi(b)$ is invertible as the image of an invertible element $b(m)$. So we have proved that $\pi(b)$ is invertible for any $\pi \in \hat{B}$ which implies (see Naimark [1], §25.3) the invertibility of b. □

Remark 28.10

(1) Lemma 28.6 describes the correlation between an algebra $A \subset \mathbb{Z}(B)$ and some mapping φ: Prim $B \to X$ (where $X = M_A$ is a compact Hausdorff space). Suppose now that we have some mapping

$$\varphi_1\colon \text{Prim } B \to Y$$

where Y is compact Hausdorff and φ_1 is continuous and onto. The natural question arises: do these objects φ_1 and Y define any algebra $A_1 \subset \mathbb{Z}(B)$ such that $Y = M_{A_1}$ and φ_1 is defined by (28.3)? The answer is: yes.

In fact any continuous function $f \in C(Y)$ defines the function $\tilde{f} = f \circ \varphi_1 \in C(\text{Prim } B)$ and this function $\tilde{f}$ by the Dauns–Hofmann Theorem (Theorem 2.90) defines an element $a_{\tilde{f}} \in \mathbb{Z}(B)$ and consequently we obtain the algebra

$$A_Y = \{a_{\tilde{f}}: f \in C(Y)\} \subset \mathbb{Z}(B) \quad .$$

A routine argument now shows that φ_1 can be defined by (28.3).

In view of these considerations one can say that any triple

$$\{\text{Prim } B, \ \varphi, \ X\}$$

where X is compact Hausdorff and φ: Prim $B \to X$ is continuous and onto defines some localization of B.

Now for any $x \in X$ one has to set

$$J_x = \bigcap_{\varphi(J)=x} J \tag{28.10}$$

(cf. (28.7))

$$B \xrightarrow{\rho_x} B/J_x \stackrel{\text{def}}{=\!=} B(x) \tag{28.11}$$

(cf. (28.8))

and

$$b(x) = \rho_x(b) \quad .$$

(2) If $\mathbb{Z}(B) = \{\lambda I\}$ then clearly there is the unique trivial localization

$$\{\text{Prim } B, \ \varphi, \ \cdot\}$$

which is of no use.

29. Localization – analytical approach.

Here we present another approach to the definition of local representatives (see **28.8**) which naturally originates in the work of I.B. Simonenko.

29.1 We shall say that local–type elements $b_1, b_2 \in B$ are equivalent at a point $m_0 \in M_A$ if for every $\pi \in \hat{B}$ such that $\psi(\pi) = m_0$ (ψ is defined by (28.6)) we have

$$\pi(b_1) = \pi(b_2) \quad . \tag{29.1}$$

This equivalence relation will be denoted by

$$b_1 \overset{m_0}{\sim} b_2 \quad . \tag{29.2}$$

Obviously

$$b_1 \overset{m_0}{\sim} b_2 \Leftrightarrow b_1(m_0) = b_2(m_0) \quad . \tag{29.3}$$

Indeed

$$b_1 \overset{m_0}{\sim} b_2 \Leftrightarrow \pi(b_1 - b_2) = 0 \quad , \qquad \pi \in \psi^{-1}(m_0)$$
$$\Leftrightarrow (b_1 - b_2) \in J_{m_0} \Leftrightarrow b_1(m_0) = b_2(m_0) \quad .$$

Lemma 29.2 $\mathrm{b}_1 \overset{\mathrm{m}_0}{\sim} b_2$ *iff for any* $\varepsilon > 0$ *there exists a neighbourhood* $\mathrm{O(m_0)}$ *such that for any* $\mathrm{a} \in \mathrm{A}$ *satisfying the conditions*

$$\|\mathrm{a}\| \leq 1 \qquad \textit{and} \qquad \hat{\mathrm{a}}(\mathrm{m}) = 0 \quad , \qquad \mathrm{m} \,\bar{\in}\, \mathrm{O(m_0)} \tag{29.4}$$

the following inequalities are valid:

$$\|(\mathrm{b}_1 - \mathrm{b}_2)\mathrm{a}\| < \varepsilon \quad , \qquad \|\mathrm{a}(\mathrm{b}_1 - \mathrm{b}_2)\| < \varepsilon \quad . \tag{29.5}$$

Proof Since $b_i \in A'$, $i = 1,2$, the inequalities in (29.5) are equivalent so we shall use the left–hand one.

First we prove the sufficiency of (29.5).

Let $\pi \in \psi^{-1}(m_0)$. Fix $\varepsilon > 0$ and consider any $a \in A$ satisfying (29.4) and such that $\hat{a}(m_0) = 1$. Then $\pi(a) = I$ and

$$\begin{aligned}\|\pi(b_1 - b_2)\| &= \|\pi(b_1 - b_2)\pi(a)\| \\ &= \|\pi[(b_1 - b_2)a]\| \\ &\leq \|(b_1 - b_2)a\| < \varepsilon \quad .\end{aligned}$$

Since ε is arbitrary we conclude that $\|\pi(b_1 - b_2)\| = 0$, that is, $b_1 \overset{m_0}{\sim} b_2$.

Now we prove the necessity of (29.5).

Let $b_1 \overset{m_0}{\sim} b_2$, that is,

$$\pi(b_1 - b_2) = 0 \qquad \text{for every} \quad \pi \in \psi^{-1}(m_0) \quad . \tag{29.6}$$

Suppose that (29.5) is not true. This means that there exists an $\varepsilon > 0$ such that for any neighbourhood $O(m_0)$ there exists an $a \in A$ satisfying (29.4) such that

$$\|(b_1 - b_2)a\| \geq \varepsilon \quad .$$

It follows that there exists a $m \in O(m_0)$ and $\pi \in \psi^{-1}(m)$ such that

$$\|\pi[(b_1 - b_2)a]\| \geq \varepsilon$$

which implies

$$\|\pi(b_1 - b_2)\| \geq \varepsilon \quad . \tag{29.7}$$

Consider the set $V \subset M_A$ of all the points m such that (29.7) is valid for some $\pi \in \psi^{-1}(m)$. By Theorem 2.87 the set

$$W = \{\pi \in \hat{B}: \|\pi(b_1 - b_2)\| \geq \varepsilon\}$$

is quasicompact and since ψ is continuous, by Corollary 28.7 the set $V = \psi(W)$ is also quasicompact. Thus (as M_A is Hausdorff) V is compact and consequently closed. But m_0 is a limit point for V and it follows that $m_0 \in V$ which contradicts (29.6). The proof is finished. □

29.3 In a number of situations it is convenient to take the conditions of Lemma 29.2 as the definition of equivalence of two elements (not necessarily of local type) at a point m. Namely, let A be some commutative subalgebra of B (not necessarily $A \subset \mathbb{Z}(B)$). We shall say that $b_1, b_2 \in B$ are equivalent at a point $m_0 \in M_A$ if for any $\varepsilon > 0$ there exists $O(m_0)$ such that for any $a \in A$ satisfying (29.4) we have (29.5).

This equivalence relation will be also denoted by

$$b_1 \overset{m_0}{\sim} b_2 \quad . \tag{29.8}$$

In view of Lemma 29.2 in the case when $A \subset \mathbb{Z}(B)$ this definition is equivalent to **29.1**.

29.4 We say that an element b is *invertible at a point* $m_0 \in M_A$ if there exist element b_r^{-1}, b_l^{-1} such that

$$b_l^{-1} b \overset{m_0}{\sim} I \quad , \qquad b b_r^{-1} \overset{m_0}{\sim} I \tag{29.9}$$

and we call the elements $b_l^{-1} \left(b_r^{-1} \right)$ the *left* (*right*) *inverses of* b *at* m_0.

Lemma 29.5 *If* b *is invertible at* $m_0 \in M_A$ *then there exists a neighbourhood* $O(m_0)$ *and elements* b_l', b_r' *such that for any* $a \in A$ *satisfying*

$$\hat{a}(m) = 0 \qquad m \,\bar{\in}\, O(m_0)$$

we have

$$(b_l' b - I)a = 0 \tag{29.10}$$

$$a(bb'_r - I) = 0 \quad . \tag{29.11}$$

Proof Let b be invertible at m_0. Then (29.9) holds. Take $\varepsilon = 1$ and let U be the corresponding neighbourhood of m_0 considered in **29.3**. Since M_A is Hausdorff and compact there exist neighbourhoods W and V of m_0 such that

$$W \subset \overline{W} \subset V \subset U \quad . \tag{29.12}$$

Take any $a_0 \in A$ satisfying the conditions

$$\|a_0\| \le 1 \ ; \qquad \hat{a}_0\big|_W \equiv 1 \ ; \qquad \hat{a}_0(m) = 0 \ , \qquad m \,\bar{\in}\, U \ . \tag{29.13}$$

For this a_0 we have

$$\left\|b_l^{-1} b a_0 - a_0\right\| < 1 \quad . \tag{29.14}$$

For any $a \in A$ satisfying

$$\hat{a}(m) = 0 \quad , \qquad m \,\bar{\in}\, W$$

obviously

$$aa_0 = a_0 a = a \quad .$$

Thus for these a:

$$b_l^{-1} ba = (b_l^{-1} ba_0 - a_0)a + a = (C + I)a$$

where $C = b_l^{-1} ba_0 - a_0$ and in view of (29.14) $(C + I)$ is invertible.

It follows now that

$$b'_l ba \overset{\text{def}}{=\!=} \left[(C + I)^{-1} \ b_l^{-1}\right] ba = a$$

or

$$(b'_l b - I)a = 0 \quad .$$

So (29.10) is true. (29.11) can be proved by an analogous argument. □

29.6 This lemma leads to the following natural notion: we shall say that $b'_l(b'_r)$ is a *left (right) quasi–inverse for* b *at* m_0 if there exists a neighbourhood $O(m_0)$ such that for any $a \in A$ such that

$$\hat{a}(m) = 0 \quad , \qquad m \bar{\in} O(m_0)$$

the following equality is true

$$b'_l ba = a \qquad (abb'_r = a) \quad .$$

Recall again that here we do not mean that b is of local type with respect to A.

Lemma 29.7 *If* $\mathrm{b}_1 \overset{\mathrm{m}_0}{\sim} \mathrm{b}_2$ *and* b_1 *has left (right) quasi–inverse at* m_0 *then* b_2 *also has left (right) quasi–inverse at* m_0.

Proof Let b'_l be some left quasi–inverse of b_1. Since $b_1 \overset{m_0}{\sim} b_2$ there exists a neighbourhood $U(m_0)$ such that for every $a \in A$ satisfying

$$\|a\| \le 1 \quad , \qquad \hat{a}(m) = 0 \quad , \qquad m \bar{\in} U(m_0)$$

we have

$$\|(b_1-b_2)a\| < \frac{1}{\|b'_l\|} \quad , \qquad \|a(b_1-b_2)\| < \frac{1}{\|b'_l\|} \tag{29.15}$$

Let $O'(m_0) \subset U(m_0)$ be the corresponding neighbourhood for b'_l mentioned in **29.6.** Take any neighbourhood $O(m_0)$ such that

$\overline{O(m_0)} \subset O'(m_0)$ and any element $a_0 \in A$ with the property

$$\|a_0\| \leq 1 \; , \qquad \hat{a}_0\big|_{O(m_0)} \equiv 1 \; ; \qquad \hat{a}_0(m) = 0 \; , \qquad m \bar{\in} O'(m_0) \; .$$

Now for every $a \in A$ for which

$$\hat{a}(m) = 0 \quad , \qquad m \bar{\in} O(m_0)$$

we have

$$b_l'b_2a = b_l'b_2a_0a = b_l'(b_2 - b_1)a_0a + b_l'b_1a_0a$$
$$= b_l'(b_2 - b_1)a_0a + a_0a = (b_l'(b_2 - b_1)a_0 + I)a \quad .$$

In view of (29.15)

$$\|C\| \stackrel{\text{def}}{=} \|b_l'(b_2 - b_1)a_0\| < 1 \quad .$$

Thus $C + I$ is invertible and consequently

$$(C + I)^{-1}b_l'b_2a = a$$

which means that $(C + I)^{-1}b_l'$ is a left quasi–inverse of b_2 at m_0.

By similar reasoning one can find a right quasi–inverse of b_2. □

Lemma 29.8 *Let* b *be a local–type element and* $b_l'(b_r')$ *be a left (right) quasi–inverse of* b *at* m_0. *If* $a_0 \in A$ *satisfies the condition*

$$\hat{a}_0(m) \equiv 1 \qquad m \in U(m_0)$$

for some neighbourhood $U(m_0)$ *then* a_0b_l', $b_l'a_0$ $(a_0b_r', b_r'a_0)$ *are also left (right) quasi–inverses of* b.

Proof For any $a \in A$ such that

$$\hat{a}(m) = 0 \quad , \qquad m \bar{\in} O(m_0) \subset U(m_0)$$

where $O(m_0)$ is some neighbourhood mentioned in 29.6 we have

$$(b'_l a_0)ba = b'_l b a_0 a = a_0 a = a$$

which means that $b'_l a_0$ is a left quasi–inverse and

$$(a_0 b'_l)ba = a_0(b'_l ba) = a_0 a = a$$

which means that $a_0 b'_l$ is also a left quasi–inverse. The argument for right quasi–inverses is the same. □

Lemma 29.9 *Let* b *be a local–type element having left and right–quasi–inverses at* m_0. *Then there exist left and right–quasi–inverses of local type.*

Proof Let $b'_l(b'_r)$ be some left (right) quasi–inverse of b at m_0 and $O(m_0)$ be the corresponding neighbourhood of m_0 mentioned in **29.6**. Take any a_0 such that $\hat{a}_0(m) \equiv 1$ in some neighbourhood of m_0 and $\hat{a}(m) = 0$, $m \bar{\in} O(m_0)$. By the previous lemma $a_0 b'_l a_0$ is a left quasi–inverse of b. Let us show that it is of local type. Indeed for any $a \in A$

$$\begin{aligned} a(a_0 b'_l a_0) &= a a_0 b'_l (a_0 b b'_r) = a a_0 (b'_l b a_0) b'_r = a \cdot a_0 \cdot a_0 \cdot b'_r \\ &= a_0 \cdot a_0 \cdot a b'_r = a_0 (b'_l b a_0) a b'_r = a_0 b'_l a (a_0 b b'_r) = a_0 b'_l a a_0 \\ &= (a_0 b'_l a_0) a \quad . \end{aligned}$$

Similarly it can be shown that $a_0 b'_r a_0$ is of local type. □

29.10 By the lemma considered above if a local–type element b has left and right quasi–inverses at m_0 then there exists a local–type element c_l such that for some neighbourhood $O(m_0)$

$$c_l ba = a \tag{29.16}$$

for every $a \in A$ satisfying

$$\hat{a}(m) = 0 \quad , \qquad m \bar{\in} O(m_0) \quad .$$

Since c_l is of local type (29.16) implies

$$ac_l b = a$$

which in view of Lemma 29.2 shows that

$$c_l(m) = [b(m)]_l^{-1} \quad , \qquad m \in O(m_0) \quad .$$

By similar reasoning we can find a local–type element C_r such that

$$C_r(m) = [b(m)]_r^{-1} \qquad m \in O(m_0)$$

Thus we conclude (using **29.4** and Lemma 29.5) that if a local–type element is invertible at a point m_0 it is invertible in some neighbourhood of m_0.

30. Localization of a C^*–algebra acting in a subspace.

In some cases (for example when dealing with Toeplitz operators in a Hardy space H^2) we find ourselves in the situation of investigating local–type elements (Toeplitz operators) in a subspace ($H^2 \subset L^2$) and it is reasonable to discover to what extent this investigation can be reduced to the general case considered in the two previous sections. As we shall see, under some natural assumptions this can be done almost automatically.

Recall again that here we are looking through the conditions for invertibility and the main problem of finding the Fredholmity conditions will be considered later (in Section 32). The answers will follow easily from the results obtained in this section.

30.1 So we are considering the following situation: given some C^*–algebra $B \subset L(H)$ of operators acting on a Hilbert space H and a commutative C^*–algebra $A \subset \mathbb{Z}(B)$ (so B is a local–type algebra with respect to A), let $\tilde{H} \subset H$ be some subspace of H and P be the orthogonal projection onto $\tilde{H}$. What can be said about the structure of the set PBP considered as a subset of $L(\tilde{H})$?

The first main assumption here under which the theory goes smoothly is

$$P \in B \tag{30.1}$$

and naturally the corresponding subspace $\widetilde{H} = \operatorname{Im} P$ will be called a local subspace.

So one can say that a localization of a C^*–algebra with respect to a subspace can be done effectively if the subspace is local.

Under the assumption (30.1) let us introduce the following objects:

$$\widetilde{A} = PAP \qquad \text{and} \qquad \widetilde{B} = PBP$$

as C^*–algebras in $L(\widetilde{H})$.

Since $P \in B$, $\widetilde{A}$ is a commutative subalgebra of $\widetilde{B}$ and evidently $\widetilde{A} \subset \mathbb{Z}(\widetilde{B})$. Indeed for any $a \in A$, $b \in B$

$$PaPPbP = PabP = PbaP = PbPPaP \quad .$$

Thus $\widetilde{B}$ is a local–type algebra with respect to $\widetilde{A}$.

If $\widetilde{A}$ differs greatly from A the localization process for $\widetilde{B}$ will go with respect to a new object $M_{\widetilde{A}}$, the maximal ideal space of $\widetilde{A}$. In order not to introduce this new object we make the following second assumption:

$$\text{if } a \in A \text{ is a non–zero element then } aP \neq 0 \quad . \tag{30.2}$$

Just note that since P is of local type, $aP = PaP$.

We use this assumption to obtain the next statement.

Lemma 30.2 *Under the assumption* (30.2) $A \cong \widetilde{A}$.

Proof We have

$$a \neq 0 \Rightarrow PaP \neq 0 \quad , \qquad a \in A \quad .$$

Thus the homomorphism $A \to \widetilde{A}$ given by the formula $a \to PaP$ has zero kernel. By the definition of $\widetilde{A}$ it is onto. Thus this homomorphism is an isomorphism. □

The main technical result of this section is the following.

Lemma 30.3 *Let* π *be an irreducible representation of* B *on* H_π. *Then* $\pi_P = \pi|_{\tilde{B}}$ *is an irreducible representation of* $\tilde{B}$ *on* $H_{\pi_P} = \pi(P)H_\pi$.

On the other hand if π_P *is some irreducible representation of* $\tilde{B}$ *on* H_{π_P} *then there exists a Hilbert space* $H_\pi \supset H_{\pi_P}$ *and an irreducible representation* π *of* B *on* H_π *such that* $H_{\pi_P} = \pi(P)H_\pi$ *and* $\pi|_{\tilde{B}} = \pi_P$.

Proof The second statement of the lemma follows from the theorem on the extension of representations (Theorem 2.81). So we have to prove only the first part.

Let π be any irreducible representation of B on H_π. Its restriction π_P onto $\tilde{B}$ is a representation of $\tilde{B}$ on H_π and since

$$\pi(PbP) = \pi(P)\pi(b)\pi(P)$$

the right–hand part of this equality belongs to $L(H_{\pi_P})$ where $H_{\pi_P} = \pi(P)H_\pi$.

Fix any non–zero vector $\xi \in H_{\pi_P}$. Then

$$\overline{\left\{\pi_P(\tilde{B})\xi\right\}} = \overline{\{\pi(P)\pi(B)\pi(P)\xi\}} \quad .$$

Further since $\pi(P)$ is an orthogonal projection the subspace $\pi(P)\overline{(\pi(B)\pi(P)\xi)}$ is closed in H_π. This implies

$$\pi(P)\ \overline{\{\pi(B)\pi(P)\xi\}} \subset \overline{\{\pi(P)\pi(B)\pi(P)\xi\}} \subset \overline{\pi(P)\overline{\{\pi(B)\pi(P)\xi\}}}$$

$$= \pi(P)\overline{\{\pi(B)\pi(P)\xi\}}$$

which means that

$$\overline{\left\{\pi_P(\tilde{B})\xi\right\}} = \pi(P)\overline{\{\pi(B)\pi(P)\xi\}} = \pi(P)H_\pi$$

where the right–hand equality follows from the irreducibility of π. Thus every non–zero vector $\xi \in H_{\pi_P}$ is cyclic which means that π_P is an irreducible representation of $\tilde{B}$ on $H_{\pi_P} = \pi(P)H_\pi$. □

Now the local theory of local C^*–algebras acting on local subspaces reduces to the next statement.

Theorem 30.4 *Let* b(m) *be a local representative of* b ∈ B *at* m ∈ $M_A = M_{\tilde{A}}$ *(see* **28.8**). *Then* $\tilde{b}(m)$ = P(m)b(m)P(m) *is a local representative of* $\tilde{b}$ = PbP ∈ $\tilde{B}$ *at* m.

Proof For the algebra $\tilde{A}$ let us introduce the notions $\tilde{\psi}$, $\tilde{J}_m$, $\tilde{B}(m)$, $\tilde{\rho}(m)$ which have the same sense as the corresponding notions ψ, J_m, $B(m)$, ρ_m for A (see (28.6)–(28.8)). Now Lemma 30.3 implies

$$\tilde{J}_m = PJ_mP \quad . \tag{30.3}$$

Let us consider the sets Ker $\rho_m|_{\tilde{B}}$ and Ker $\tilde{\rho}_m$. We have

$$\text{Ker } \rho_m|_{\tilde{B}} = \{PbP\colon\ \pi(PbP) = 0 \quad , \quad \forall \pi \in \psi^{-1}(m)\} \tag{30.4}$$

$$\text{Ker } \tilde{\rho}_m = \left\{PbP\colon\ \pi_P(PbP) = 0 \quad , \quad \forall \pi_P \in \tilde{\psi}^{-1}(m)\right\} \quad . \tag{30.5}$$

By Lemma 30.3 for every $\pi_P \in \tilde{\psi}^{-1}(m)$ we have $\pi_P = \pi|_{\tilde{B}}$ for some $\pi \in \psi^{-1}(m)$ and, vice versa, for every $\pi \in \psi^{-1}(m)$ we have $\pi|_{\tilde{B}} = \pi_P \in \tilde{\psi}^{-1}(m)$. Thus (30.4) and (30.5) means

$$\text{Ker } \rho_m|_{\tilde{B}} = \text{Ker } \tilde{\rho}_m$$

and so

$$\tilde{B}(m) = \rho_m(\tilde{B}) \quad ,$$

that is, $\tilde{\rho}_m$ is the restriction of ρ_m onto $\tilde{B}$. It follows now that for $\tilde{b} = PbP \in \tilde{B}$

$$\rho_m(PbP) = P(m)b(m)P(m) = \tilde{b}(m)$$

and the theorem is proved. □

31. Local representatives. I. Equivariant representatives.

In the previous sections of this chapter we have developed a number of methods to construct a sufficient set of representations $b \to b(m)$ generated by the fixation of a commutative algebra $A \subset \mathbb{Z}(B)$ ($m \in M_A$, the maximal ideal space of A). None of these methods give us a tool to obtain an explicit expression for $b(m)$. Unfortunately, as the examples provided below show, there is no hope of obtaining such explicit expressions for $b(m)$ in the general situation. On the other hand there exists a rather large class of local–type algebras where the form of $b(m)$ can be written explicitly and we describe this class here.

31.1 Recall that for any C^*–algebra B the function $\pi \to \|\pi(b)\|$, $b \in B$, is lower semicontinuous on $\hat{B}$ (Theorem 2.87). The next statement shows that when dealing with local representatives the situation is quite opposite.

Lemma 31.2 *Let* B *be a local–type algebra. Then the function* m $\to$ ||b(m)||, b $\in$ B *is upper semicontinuous on* M_A.

Proof Let $V = (m: \|b(m)\| \geq K\}$, for some $K \geq 0$. Obviously

$$V = \bigcap_{\varepsilon>0} \psi(W_\varepsilon)$$

where $W_\varepsilon = \{\pi \in \hat{B}: \|\pi(b)\| \geq K - \varepsilon\}$ and ψ is defined by (28.6).

By Theorem 2.87 all W_ε are quasicompact. Thus, since ψ is continuous and M_A is Hausdorff, all $\psi(W_\varepsilon)$ are closed, which means that V is closed and consequently the function $m \to \|b(m)\|$ is upper semicontinuous. □

31.3 The upper semicontinuity of the function $m \to \|b(m)\|$, and the lower semicontinuity of the function $\pi \to \|\pi(b)\|$ suggest that in fact $m \to \|b(m)\|$ is continuous. The following example shows that unfortunately this is not so.

Example Let B be the C^*–algebra of continuous two–dimensional matrix functions $\left[b_{ij}(x)\right]$ on $[0, 1]$ which are diagonal at 0 and 1 and such that $b_{11}(0) = b_{11}(1)$. Let $A = \mathbb{Z}(B)$. Obviously all the elements in A have the form $a = a(x)E$ where $a(x) \in C[0, 1]$, $a(0) = a(1)$ and $E = \begin{bmatrix} 1 & 0 \\ 0 & 1 \end{bmatrix}$.

Thus M_A is $[0, 1]$ with glued ends $\{0\} \longleftrightarrow \{1\}$. Let m_0 be the point arising from these two glued points. Clearly we have

$$b(m_0) = \{b_{11}(0),\ b_{22}(0),\ b_{22}(1)\}$$

so

$$\|b(m_0)\| = \max\{\,|b_{11}(0)|,\ |b_{22}(0)|,\ |b_{22}(1)|\,\} \quad .$$

For any other point $m \in (0, 1)$ we have

$$b(m) = \left[b_{ij}(m)\right]$$

and consequently

$$\|b(m)\| = \left\|\left[b_{ij}(m)\right]\right\| \quad .$$

If for example $b_{11}(0) = b_{22}(0) = 0$ and $b_{22}(1) = 1$ then since $\left[b_{ij}(x)\right]$ is a continuous matrix function there exists a $\delta > 0$ such that for all $0 < m < \delta$

$$\|b(m)\| = \left\|\left[b_{ij}(m)\right]\right\| < \tfrac{1}{2}$$

and on the other hand $\|b(m_0)\| = 1$. So $\|b(m)\|$ is not continuous.

Remark If B is a Von Neumann algebra and $A = \mathbb{Z}(B)$ then $m \to \|b(m)\|$ is continuous (see Glimm [1]).

31.4 The example which will be provided now shows that even when we know all the local representatives $b(m)$, $m \in \{O(m_0)\backslash m_0\}$, for some neighbourhood $O(m_0)$, we cannot say much (in fact nothing) of $b(m_0)$ in the general situation.

Example Let $H = L^2[0, 1] \oplus H_2$, where H_2 is some Hilbert space, $B = B_1 \oplus B_2$, where B_1 is the C^*–algebra of multiplications by continuous functions $b_1 \in C[0, 1]$ and $B_2 = L(H_2)$. Let A consist of the elements of the form $a = a_1 \oplus a_2$, where $a_1 \in B_1$ and $a_2 = a_1(1)E$. E is the identity in $L(H_2)$. Thus $A \cong C[0, 1]$ and for any $b = b_1 \oplus b_2 \in B$ we have

$$b(m) = b_1(m) \quad , \qquad m \in [0, 1)$$

$$b(1) = \{b_1(1), b_2\} \quad .$$

So what can we say about $b(1)$ even if we know all $b(m)$, $m \in [0, 1)$?

31.5 The examples presented in 31.4 and 31.3 show that in general there is almost no correlation between $b(m_1)$ and $b(m_2)$ even if m_1 and m_2 are very close.

Now we shall look at the situation when the structure of $b(m)$ can be described explicitly. Since by (28.9)

$$b(m) = (ab)(m)$$

for all $a \in A$ such that $\hat{a}(m) = 1$, the structure of $b(m_0)$ can be obtained by considering the action of b on $H_{O(m_0)}$, where

$$H_{O(m_0)} = \overline{\left\{a\xi\colon \hat{a}(m) = 0 \ , \quad m \bar{\in} O(m_0)\right\}} \tag{31.1}$$

and $O(m_0)$ is an arbitrary neighbourhood of m_0. The main idea is to consider $H_{O(m_0)}$ as a subspace of some other space H_1 and to consider elements $b \in B$ as some local–type elements in $L(H_1)$. In this case the structure of $b(m_0)$ can be described if, among the elements $b \in L(H_1)$ having the same local representative $b(m_0)$, one can find an element with 'good' properties, namely invariant under some special group of automorphisms of $L(H_1)$.

So let us fix some (as small as we want) neighbourhood $O(m_0)$ and consider the subspace $H_{O(m_0)}$ defined by (31.1). Let θ be some unitary isomorphism

$$H_{O(m_0)} \overset{\theta}{\longleftrightarrow} \overline{H}_{O(m_0)} \subset H_1 \quad . \tag{31.2}$$

This isomorphism defines an isomorphism between $A_{0(m_0)}$

$$A_{O(m_0)} = \left\{ a \in A\colon \hat{a}(m) = 0 \ , \quad m \,\bar{\in}\, O(m_0) \right\} \tag{31.3}$$

and the C^*–algebra $\overline{A}_{O(m_0)} = \theta A_{O(m_0)} \theta^{-1}$ which can clearly be regarded as a C^*–subalgebra of $L(H_1)$ by setting $\bar{a}(H_1 \ominus \overline{H}_{O(m_0)}) = 0$, $a \in A_{O(m_0)}$.

As $ba = ab$, $a \in A$, it follows that $H_{O(m_0)}$ is invariant under b and consequently we can consider the operator $\bar{b} = \theta b \theta^{-1}$ on $\overline{H}_{O(m_0)}$ which is a local–type element with respect to $\overline{A}_{O(m_0)}$.

Note that

$$\|b(m_0)\| = \inf_{\substack{a \in A_{O(m_0)} \\ \hat{a}(m_0) = 1}} \|ab\| \ . \tag{31.4}$$

Indeed since $b \to b(m_0)$ is a representation of B (28.9) implies

$$\|b(m_0)\| = \|(ab)(m_0)\| \leq \|ab\| \tag{31.5}$$

for every a such that $\hat{a}(m_0) = 1$.

On the other hand by Lemma 31.2 the function $m \to \|b(m)\|$ is upper semicontinuous. Thus for any $\varepsilon > 0$ there exists a neighbourhood $U(m_0) \subset O(m_0)$ such that

$$\|b(m)\| < \|b(m_0)\| + \varepsilon \quad , \qquad m \in U(m_0) \quad .$$

This means that for any $a \in A_{U(m_0)}$ such that $\|a\| \leq 1$

$$\|ab\| = \sup_{U(m_0)} |\hat{a}(m)| \; \|b(m)\| < \|b(m_0)\| + \varepsilon$$

which, together with (31.5), proves (31.4). The definitions of $\bar{A}_{O(m_0)}$ and $\bar{b}$ show that

$$\|b(m_0)\| = \|\bar{b}(m_0)\| = \inf_{\substack{a \in A_{O(m_0)} \\ \hat{a}(m_0)=1}} \|\bar{a}\; \bar{b}\| \quad .$$

31.6 Now consider some group G and its unitary representation $g \to V_g \in L(H_1)$. For any neighbourhood $U(m_0) \subset O(m_0)$ let $P_{U(m_0)}$ be the projection onto $\bar{H}_{U(m_0)}$ where $\bar{H}_{U(m_0)}$ is defined in a way analogous to $\bar{H}_{O(m_0)}$.

We shall frequently use the next assumption:

there exists a sequence $\{V_{g_n}\}$ such that for any fixed $U(m_0)$

$$\forall \xi \in H_1 \qquad \left\| V_{g_n} \xi - P_{U(m_0)} V_{g_n} \xi \right\| \to 0 \tag{31.6}$$

which means that $\{V_{g_n}\}$ 'drives' the whole of H_1 into an 'arbitrarily small' subspace (over $U(m_0)$).

Lemma 31.7 *The assumption* (31.6) *is equivalent to the following statement: there exists a sequence* $\{V_{g_n}\}$ *such that for any fixed* $a \in A_{O(m_0)}$ *with* $\hat{a}(m_0) = 1$

$$\forall \xi \in H_1 \qquad \left\| V_{g_n}\xi - \bar{a}V_{g_n}\xi \right\| \to 0 \quad . \tag{31.7}$$

Proof

(i) (31.6) ⇒ (31.7)

Observe that $\bar{a}P_{U(m_0)} = P_{U(m_0)}\bar{a}$ and for any $\varepsilon > 0$ one can choose $U(m_0)$ in such a way that

$$\left\|\bar{a}P_{U(m_0)} - P_{U(m_0)}\right\| = \left\|\bar{a}P_{U(m_0)} - \hat{a}(m_0)P_{U(m_0)}\right\|$$

$$< \sup_{U(m_0)} |\hat{a}(m) - \hat{a}(m_0)| < \varepsilon \ . \tag{31.8}$$

Now routine computations show that (31.6) and (31.8) imply (31.7).

(ii) (31.7) ⇒ (31.6)

If $a \in A_{U(m_0)}$ then

$$\bar{a}V_{g_n}\xi = (P_{U(m_0)}\bar{a}V_{g_n})\xi \in \bar{H}_{U(m_0)} \quad .$$

Now (31.7) means that

$$\left\|\bar{a}V_{g_n}\xi - P_{U(m_0)}V_{g_n}\xi\right\|^2 + \left\|(I - P_{U(m_0)})V_{g_n}\xi\right\|^2$$

$$= \left\|\bar{a}V_{g_n}\xi - V_{g_n}\xi\right\|^2 \to 0 \tag{31.9}$$

and it follows that

$$\left\|(I - P_{U(m_0)})V_{g_n}\xi\right\| \to 0 \quad ,$$

thus (31.6) is true. □

The next statement in fact shows that local–type elements commuting with $\{V_g\}_{g\in G}$ (such elements are called equivariant (with respect of $\{V_g\}_{g\in G}$)) 'coincide' with their local representatives in m_0.

Theorem 31.8 *Let* $b' \in L(H_1)$ *be some local–type (with respect to* $\overline{A}_{O(m_0)}$ *element commuting with* $\left\{V_{g_n}\right\}$ *where* $\left\{V_{g_n}\right\} \subset \{V_g\}_{g\in G}$ *is a sequence satisfying the assumption* (31.6) *(or* (31.7)). *Then*

$$\|b'(m_0)\| = \|b'\| \quad . \tag{31.10}$$

Proof Since obviously $\|b'(m_0)\| \le \|b'\|$ we have to prove only

$$\|b'(m_0)\| \ge \|b'\| \quad . \tag{31.11}$$

Fix $\varepsilon > 0$ and choose $\xi \in H_1$, $\|\xi\| = 1$ such that

$$\|b'\xi\| \ge \|b'\| - \varepsilon \quad .$$

Take any $a \in A_{O(m_0)}$ with $\hat{a}(m_0) = 1$. Then for sufficiently large numbers $n \in N$ in view of (31.7) we have

$$\begin{aligned}\|b'\| - \varepsilon \le \|b'\xi\| &= \|V_{g_n} b'\xi\| \le \|\bar{a}V_{g_n}(b'\xi)\| + \varepsilon \\ &= \|\bar{a}b\,(V_{g_n}\xi)\| + \varepsilon \le \|\bar{a}b'\| + \varepsilon \quad .\end{aligned}$$

Thus

$$\|b'\| \le \|\bar{a}b'\| + 2\varepsilon$$

and as ε is arbitrary we conclude

$$\|b'\| \leq \|\bar{a}b'\|$$

which together with (31.4) finishes the proof. □

Corollary 31.9 *If* b'_1 *and* b'_2 *are two local–type elements in* $L(H_1)$ *commuting with* $\{V_{g_n}\}$ *and* $b'_1(m_0) = b'_2(m_0)$ *then* $b'_1 = b'_2$.

Proof By the theorem

$$\|b'_1 - b'_2\| = \|(b'_1 - b'_2)(m_0)\| = \|b'_1(m_0) - b'_2(m_0)\| = 0 \quad . \quad \square$$

Corollary 31.10 *Let there exist, for every* $b \in B$, $\bar{b}' \in L(H_1)$ *such that* $\bar{b}'(m_0) = \bar{b}(m_0)$ *and* $\bar{b}'$ *commutes with* $\{V_{g_n}\}$. *Then the mapping*

$$\bar{b}' \to b(m_0) \tag{31.12}$$

defines the isomorphism

$$\{\bar{b}'\}_{b \in B} \cong B(m_0) \quad . \tag{31.13}$$

Proof First of all by the previous corollary for every $b \in B$ there could exist only one $\bar{b}'$ with the desired properties (recall that $\bar{b}(m_0) = b(m_0)$ in view of the definition of $\bar{b}$). Evidently the set of elements $\bar{b}'$ which commute with $\{V_{g_n}\}$ forms a C^*–algebra (it is clearly a Banach algebra and

$$V_{g_n} a = aV_{g_n} \Rightarrow V^*_{g_n} aV_{g_n} = a \Rightarrow V^*_{g_n} a^* V_{g_n} = a^*$$
$$\Rightarrow a^* V_{g_n} = V_{g_n} a^*$$

thus it is a *–algebra).

So $\{\bar{b}'\}_{b \in B}$ is a C^*–algebra and in view of Theorem 31.8 and the fact that $\bar{b}(m_0) = b(m_0)$ the mapping (31.12) defines the isomorphism (31.13). □

Remark It is often convenient to consider a set of elements $\overline{b}' \in L(H_1)$ which are invariant under some unitary group $\{V_g\}_{g \in G} \supset \left\{V_{g_n}\right\}$, where $\left\{V_{g_n}\right\}$ satisfy the assumption (31.6) (or (31.7)). In fact it is clear that the larger the G we could consider, the more we shall know about the structure of $\overline{b}'$ and consequently of $b(m_0)$.

Examples

Here we present some examples of sets of operators satisfying the assumption (31.6).

31.11 Let $H_1 = L^2_\mu(X)$ where X is a topological space with a σ–finite Borel measure μ and $\{\alpha_g\}_{g \in G}$ is a group of homeomorphisms of X which preserves equivalence classes of μ (that is $\mu(\omega) = 0 \Leftrightarrow \mu(\alpha_g(\omega)) = 0$ $\forall_g \in G$).

By V_g we denote the operator defined by the formula

$$(V_g\xi)(x) = [\alpha_g'(x)]^{\frac{1}{2}}\xi(\alpha_g(x)) \quad , \qquad \xi \in H_1 \tag{31.14}$$

where α_g' is the Radon–Nikodim derivative of the measure $\alpha_g(\mu)$ with respect to the measure μ. By construction V_g is obviously unitary. For any $x_0 \in X$ and any neighbourhood $O(x_0)$ we denote by $\overline{A}_{O(x_0)}$ the set of operators of multiplication by continuous bounded functions vanishing on $X \backslash O(x_0)$.

We shall say that x_0 is a *globally attracting point* for a sequence of homeomorphisms $\left\{\alpha^{-1}_{g_n}\right\}$ if for any neighbourhood $U(x_0)$ and for any open set W having finite measure

$$\mu\left(W \backslash \alpha_{g_n}(U(x_0))\right) \to 0 \quad . \tag{31.15}$$

One can easily observe that (31.15) is equivalent to the following:

$$\forall_{\xi} \in H_1 \qquad \left\| \xi - P_{\alpha_{g_n}(U(x_0))}\xi \right\| \to 0 \tag{31.16}$$

where $P_{\alpha_{g_n}(U(x_0))}$ is a projection defined by the operator of multiplication by the characteristic function $\chi_{\alpha_{g_n}(U(x_0))}$.

Indeed if (31.15) is true then for any fixed ξ and any $\varepsilon > 0$ let us choose some open $W = W_\xi$ such that $\mu(W) < \infty$ and

$$\|\xi - P_W\xi\| < \varepsilon \quad . \tag{31.17}$$

Having this in mind we conclude

$$\begin{aligned} \left\| \xi - P_{\alpha_{g_n}(U(x_0))}\xi \right\| &\le \left\| \xi - P_{W \cup \alpha_{g_n}(U(x_0))}\xi \right\| \\ &+ \left\| P_{W \cup \alpha_{g_n}(U(x_0))}\xi - P_{\alpha_{g_n}(U(x_0))}\xi \right\| \\ &\le \varepsilon + \left\| P_{W \setminus \alpha_{g_n}(U(x_0))}\xi \right\| \quad . \end{aligned} \tag{31.18}$$

Since ε is arbitrary and the last term in (31.18) tends to zero (by (31.15)) it follows that (31.16) is true.

On the other hand if (31.16) is true then take $\xi = \chi_W$ and with this ξ (31.16) is equal to (31.15).

All this leads to the next statement:

if x_0 *is a globally attracting point for* $\left\{\alpha_{g_n}^{-1}\right\}$ *then* $\left\{V_{g_n}\right\}$ *satisfies the assumption* (31.6) *at* x_0.

In fact

$$\left\| V_{g_n}\xi - P_{U(x_0)} V_{g_n}\xi \right\| = \left\| V_{g_n}\xi - V_{g_n} P_{\alpha_{g_n}(U(x_0))}\xi \right\|$$

$$= \left\| \xi - P_{\alpha_{g_n}(U(x_0))}\xi \right\| \to 0 \quad .$$

31.12 Now we shall consider a special case of the previous situation which in fact is on the one hand rather common and on the other hand gives us a very simple instrument to describe all the elements commuting with the considered group of operators $\{V_g\}$ and thus provides us with a method of finding out the structure of local representatives (with the help of Corollary 31.10).

Suppose that the set X considered in the previous example has the following form:

$$X = \{x_0\} \cup [Y \times G]$$

where Y is a Hausdorff space and G is a locally compact topological group. The set $Y \times G$ has the product topology and the point x_0 has a base of neighbourhoods which is described as follows:

(1) we fix some open set $\omega \subset G$ and some element $g_0 \in G$ having the property

$$g_0\omega \subset \omega \tag{31.19}$$

where the inclusion is proper (evidently we cannot do this if g_0 has a finite order);

(2) the base of neighbourhoods of x_0 are the sets having the form

$$O(x_0) = \{x_0\} \cup \left[Y \times g_0^n\omega\right] \qquad n \in \mathbb{N} \quad . \tag{31.20}$$

We note in particular that under this definition a sequence $x_n = \left(y, g_0^n x\right)$, $y \in Y$, $x \in \omega$, tends to x_0. So one can think of x_0 as of some 'infinite' point glued to $Y \times \omega$.

Let $H_1 = L^2_\mu(X)$, where $\mu = \mu_1 \otimes \mu_2$ and μ_1 is some Borelean σ–finite measure on Y, while μ_2 is a left–invariant Haar measure on G; $\mu(x_0) = 0$.

By V_g, $g \in G$, we denote the operator defined by

$$(V_g\xi)(y, h) = \xi(y, g^{-1}h) , \qquad \xi \in H_1, (y, h) \in Y \times G . \tag{31.21}$$

Clearly V_g is unitary.

Suppose that in addition to (31.19) ω and g_0 satisfy the following condition

$$\bigcup_{n>0} g_0^{-n}\omega = G \quad . \tag{31.22}$$

In this case

> x_0 *is a globally attracting point for* $\{\alpha_{g_0^{-n}}\}$ *where* $\alpha_g(y,h) = (y,g^{-1}h)$, $\alpha_g(x_0) = x_0$ *and thus* $\{V_{g_0^n}\}$ *satisfies the assumption* (31.6) *at* x_0.

Indeed, for any $U(x_0)$ take some $O(x_0) \subset U(x_0)$ having the form

$$O(x_0) = \{x_0\} \cup \left[Y \times g_0^k\omega\right]$$

for some $k > 0$ (see (31.20)).

We have

$$\alpha_{g_0^n}(U(x_0)) \supset \alpha_{g_0^n}(O(x_0)) = \{x_0\} \cup \left[Y \times g_0^{k-n}\omega\right] . \tag{31.23}$$

In view of (31.19)

$$\bigcup_{n_0 \geq n > 0} g_0^{-n}\omega = g_0^{-n_0}\omega \quad .$$

Now (31.22) and (31.23) imply (31.15). Thus x_0 is a globally attracting point for $\{\alpha_{g_0^{-n}}\}$ which, having in mind the statement obtained in the previous example, means that $\{V_{g_0^n}\}$ satisfies the assumption (31.6) in x_0.

If G is a commutative group one can use the Fourier transformation to obtain a rather effective description of the set of operators commuting with all V_g, $g \in G$.

Indeed we clearly have

$$L^2_\mu(X) = L^2_{\mu_1 \otimes \mu_2}(Y \times G) \cong L^2_{\mu_2}(G, L^2_{\mu_1}(Y)) \qquad (31.24)$$

where the last term is the set of all μ_2 square integrable vector functions taking their values in $L^2_{\mu_1}(Y)$ and the isomorphism is established by the mapping (obviously defined on continuous functions and thus everywhere)

$$\left[C(Y \times G) \cap L^2_{\mu_1 \otimes \mu_2}(Y \times G)\right] \ni f \to$$

$$\tilde{f} \in \left[C(G, L^2_{\mu_1}(Y)) \cap L^2_{\mu_2}(G, L^2_{\mu_1}(Y))\right]$$

where

$$\tilde{f}(h) = f(y, h) \qquad h \in G \ , \quad y \in Y \ .$$

To reduce the notation we denote the last term in (31.24) by $L^2_\lambda(G, B)$ with $B = L^2_{\mu_1}(Y)$, $\lambda = \mu_2$.

The Fourier transformation (see for example Hewitt, Ross [2], Chapter 8) in view of the Plancherel theorem (Hewitt, Ross [2], 31.18) establishes the isomorphism

$$L^2_\lambda(G, B) \xrightarrow{\wedge} L^2_\theta(\hat{G}, B)$$

where $\hat{G}$ is a character group for G, and λ and θ are defined in Hewitt, Ross [2], 31.1 (in fact λ here may differ from μ_2 by a constant multiplier).

We are taking here the Fourier transform of $\bar{f}(h) = f(y, h)$ by h. For convenience we denote $\hat{\bar{f}}(\chi)$ by $F(\bar{f})(\chi)$, $\chi \in \hat{G}$.

The condition

$$bV_g = V_g b \qquad \text{for all} \quad g \in G$$

means that

$$(FbF^{-1})a = a(FbF^{-1}) \tag{31.25}$$

for every operator a of multiplication by a continuous bounded function $a(\chi)$, which means in turn that FbF^{-1} can be regarded as a bounded measurable operator–valued function on $\hat{G}$ taking values in $L(B)$.

Since the mapping $\bar{b}' \to F\bar{b}'F^{-1}$ is isometric we can use elements $(F\bar{b}'F^{-1})$ instead of $\bar{b}'$ to be local representatives while using Corollary 31.10.

Remark 31.13 Sometimes we have to consider not the space $L^2_{\mu_1 \otimes \mu_2}(Y \times G)$ as in the previous example but a space $L^2_{\mu_1 \otimes \mu_2}(Y \times G, E)$ of all the square measurable vector functions taking their values in E(E is some Hilbert space). Clearly this case can also be reduced to the consideration of the space $L^2_{\mu_2}(G, B)$ where $B = L^2_{\mu_1}(Y, E)$ and thus all the arguments of the foregoing example follow naturally.

Now we shall consider some examples of the concrete computation of the structure of elements commuting with a given unitary group satisfying assumption (31.6).

31.14 Consider the space $L^2(\mathbb{R}^n)$ and a unitary group $\{V_\lambda\}$, $\lambda > 0$, defined by

$$(V_\lambda \xi)(x) = \lambda^{-\frac{n}{2}} \xi\left(\frac{x}{\lambda}\right) \qquad \xi \in L^2(\mathbb{R}^n) \quad . \tag{31.26}$$

Let $\alpha_\lambda(x) = \frac{x}{\lambda}$ be the corresponding group of homeomorphisms of $\mathbb{R}^n$.

If we fix any $0 < \lambda_0 < 1$ then obviously 0 is a globally attracting point for the sequence of homeomorphisms $\left\{\alpha_{\lambda_0^{-n}}\right\}$ and thus $\left\{V_{\lambda_0^n}\right\}$ satisfies assumption (31.6) at 0.

Let $\gamma\colon L^2(\mathbb{R}^n) \to L^2(S^{n-1} \times \mathbb{R})$ be the isometry defined by

$$(\gamma\xi)(\varphi, t) = e^{\frac{nt}{2}}\,\xi(\varphi, e^t) \tag{31.27}$$

where $(\varphi, e^t) = (\varphi, \rho)$ are polar coordinates in $\mathbb{R}^n$.

One can easily verify that

$$\gamma V_\lambda \gamma^{-1} = T_{\ln\lambda} \tag{31.28}$$

where $T_h\colon L^2(S^{n-1} \times \mathbb{R}) \to L^2(S^{n-1} \times \mathbb{R})$ is a unitary operator of the form

$$(T_h f)(\varphi, t) = f(\varphi, t-h) \quad , \qquad h \in \mathbb{R} \quad . \tag{31.29}$$

Consider the topological space

$$X = \{x_0\} \cup [S^{n-1} \times \mathbb{R}]$$

where the bases of neighbourhoods of x_0 are the sets of the form

$$O(x_0) = \{x_0\} \cup [S^{n-1} \times (-\infty, s)] \quad .$$

Obviously if $h_0 < 0$ then $\left\{T_{h_0^n}\right\}$ satisfies the assumption (31.6) at x_0.

In view of (31.28) the mapping

$$b \to \gamma b \gamma^{-1}$$

maps the set of all the elements of $L(L^2(\mathbb{R}^n))$ commuting with all V_λ, $\lambda > 0$, into the set of all the elements of $L(L^2(S^{n-1} \times \mathbb{R}))$ commuting with all T_h, $h \in \mathbb{R}$, and thus we find ourselves in the situation described in example **31.12.** Consequently the mapping

$$b \to F\gamma b\gamma^{-1}F^{-1} \tag{31.30}$$

maps all the elements of $L(L^2(\mathbb{R}^n))$ commuting with $\{V_\lambda\}$, $\lambda > 0$, into the set of all the measurable bounded operator–valued functions on $\mathbb{R}$ $(= \hat{\mathbb{R}})$ having values in $L(L^2(S^{n-1}))$ (here F is the standard Fourier transformation with respect to $\mathbb{R}$).

Remark If instead of $L^2(\mathbb{R}^n)$ we consider a space $L^2(\mathbb{R}^n, E)$ (of vector–valued functions taking their values in some Hilbert space E) then (cf. Remark 31.13) the same mapping (31.30) transforms b into an operator–valued function on $\mathbb{R}$ having values in $L(L^2(S^{n-1}, E))$.

31.15 Now we present a special class of elements $b \in L(L^2(\mathbb{R}))$ which commute with all $\{V_\lambda\}$, $\lambda > 0$, considered in the preceding example. These elements will in fact turn out to be local representatives for singular integral operators with piecewise continuous coefficients (see **33.9**).

By $S_{\mathbb{R}}$ we denote the operator in $L^2(\mathbb{R})$ having the form

$$(S_{\mathbb{R}}f)(t) = \frac{1}{\pi i}\int_{-\infty}^{+\infty} \frac{f(\tau)}{\tau - t}\, d\tau \quad . \tag{31.31}$$

We denote by B the C^*–subalgebra of $L(L^2(\mathbb{R}))$ generated by $S_{\mathbb{R}}$ and all the operators of multiplication by piecewise constant functions on rays $R_{\pm} = \{t\colon \pm t > 0\}$:

$$c(t) = \begin{cases} c_+ \ , & t > 0 \\ c_- \ , & t < 0 \end{cases} \quad . \tag{31.32}$$

Our goal now is to describe the set

$$\tilde{B} = F\gamma B\gamma^{-1}F^{-1} \subset L(L^2(S^0 \times \mathbb{R})) = L(L^2(\mathbb{R}, \mathbb{C}^2))$$

(see (31.30)). In order to do this it is evidently enough to calculate the corresponding transformation for the generators of B.

Routine computation shows that for every operator c of multiplication by

$c(t)$ defined by (31.32) $F\gamma c\gamma^{-1}F^{-1}$ is the operator of $L^2(\mathbb{R}, \mathbb{C}^2)$ of multiplication by the constant matrix function

$$\tilde{c} = \begin{pmatrix} c_+ & 0 \\ 0 & c_- \end{pmatrix} . \tag{31.33}$$

Now we shall show that $F\gamma S_{\mathbb{R}}\gamma^{-1}F^{-1}$ is the operator of multiplication by the matrix function

$$\tilde{S}_{\mathbb{R}} = \begin{pmatrix} \operatorname{th}\pi t & i[\operatorname{ch}\pi t]^{-1} \\ -i[\operatorname{ch}\pi t]^{-1} & -\operatorname{th}\pi t \end{pmatrix} . \tag{31.34}$$

Indeed, in view of (31.27) having in mind that $S^0 = \{1, -1\}$ we obtain

$$(\gamma\xi)(t) = e^{\frac{t}{2}}(\xi(e^t),\ \xi(-e^t)) \tag{31.35}$$

where $\xi \in L^2(\mathbb{R})$, $\gamma\xi \in L^2(S^0 \times \mathbb{R}) = L^2(\mathbb{R}, \mathbb{C}^2)$ and consequently

$$(\gamma^{-1}g)(s) = \begin{cases} (s)^{-\frac{1}{2}}\ g_1(\ln s)\ , & s > 0 \\ (-s)^{-\frac{1}{2}}\ g_2(\ln(-s))\ , & s < 0 \end{cases} \tag{31.36}$$

where $g = (g_1, g_2) \in L^2(\mathbb{R}, \mathbb{C}^2)$.

Evidently

$$\left[\left(\gamma S_{\mathbb{R}}\gamma^{-1}\right)g\right](t) = \frac{e^{\frac{t}{2}}}{\pi i}\left(\int_{-\infty}^{0} \frac{(-\tau)^{-\frac{1}{2}}g_2(\ln(-\tau))}{\tau - e^t}\,d\tau + \int_0^{+\infty} \frac{\tau^{-\frac{1}{2}}g_1(\ln\tau)}{\tau - e^t}\,d\tau\ ;\right.$$

$$\left.\int_{-\infty}^{0} \frac{(-\tau)^{-\frac{1}{2}}g_2(\ln(-\tau))}{\tau + e^t}\,d\tau + \int_0^{+\infty} \frac{\tau^{-\frac{1}{2}}g_1(\ln\tau)}{\tau + e^t}\,d\tau\right)$$

$$= \frac{e^{\frac{t}{2}}}{\pi i}\left(-\int_{-\infty}^{+\infty} \frac{e^{-\frac{s}{2}} g_2(s) e^s}{e^s + e^t}\, ds + \int_{-\infty}^{+\infty} \frac{e^{-\frac{s}{2}} g_1(s) e^s}{e^s - e^t}\, ds\ ;\right.$$

$$\left.\int_{-\infty}^{+\infty} \frac{e^{-\frac{s}{2}} g_2(s) e^s}{e^t - e^s}\, ds + \int_{-\infty}^{+\infty} \frac{e^{-\frac{s}{2}} g_1(s) e^s}{e^t + e^s}\, ds\right)$$

$$= (K_{11}g_1 + K_{12}g_2,\quad K_{21}g_1 + K_{22}g_2) \tag{31.37}$$

where

$$(K_{ij}\varphi)(t) = \int_{-\infty}^{+\infty} k_{ij}(t - s)\varphi(s)ds$$

and

$$k_{11}(y) = \frac{1}{\pi i} \frac{e^{\frac{y}{2}}}{1 - e^y}$$

$$k_{12}(y) = \frac{1}{\pi i} \frac{-e^{\frac{y}{2}}}{e^y + 1}$$

$$k_{21} = -k_{12}\ , \qquad k_{22} = -k_{11}$$

and the integral is taken in the principal value sense.

Note that

$$(FK_{ij}F^{-1})g_j = \Phi(k_{ij} * (\Phi^{-1} g_j)) = (\Phi k_{ij}) \cdot g_j$$

where * is the convolution operation, and

$$(\Phi\psi)(t) = \int_{-\infty}^{+\infty} e^{-iyt}\ \psi(y)dy \tag{31.38}$$

and the integral in (31.38) is also taken in the principal value sense.

In view of (31.37) to calculate $F\gamma S_{\mathbb{R}}\gamma^{-1}F^{-1}$ it is enough to calculate Φk_{ij}. Let us calculate, for example, Φk_{11}

$$(\Phi k_{11})(t) = \frac{1}{\pi i}\int_{-\infty}^{+\infty}\frac{e^{y(-it+\frac{1}{2})}}{1-e^{y}}\,dy \quad . \tag{31.39}$$

Consider the integral

$$I_1(z) = \int_{-\infty}^{+\infty}\frac{e^{yz}}{1-e^{y}}\,dy \qquad z = -it+\tfrac{1}{2} \quad . \tag{31.40}$$

Since $\dfrac{e^{yz}}{1-e^{y}}$ is an analytic function when $y \neq 2k\pi i$ to calculate this integral one can use residue theory (Fig. 6).

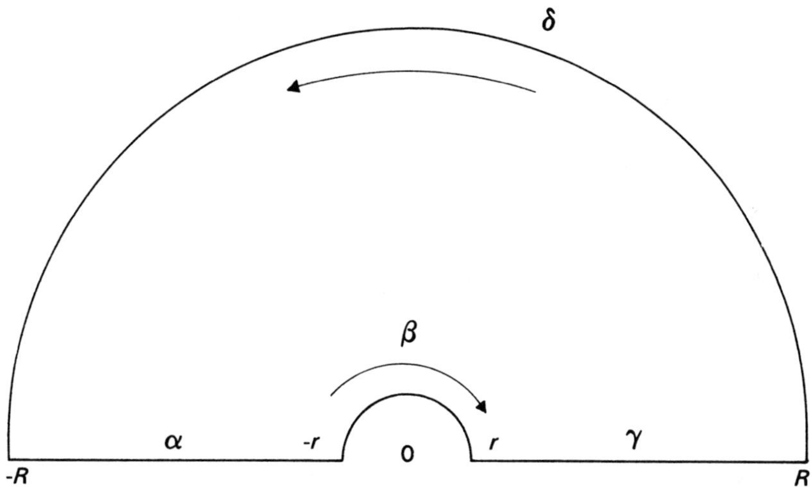

Fig. 6

We have

$$I_1(z) = \lim_{\substack{r\to 0\\ R\to\infty}}\left[\int_{\alpha+\beta+\gamma+\delta}\cdots - \int_{\delta}\cdots - \int_{\beta}\right]$$

where $\delta \;\bar{\ni}\; 2k\pi i$.

In view of Jordan's Lemma (see for example Shabat [1], §7.26)

$$\lim_{R\to\infty} \int_{\delta} \ldots = 0$$

and one can easily verify that

$$\lim_{r\to 0} \int_{\beta} \frac{e^{yz}}{1-e^{y}}\, dy = -i\pi \cdot \operatorname*{res}_{y=0} \frac{e^{yz}}{1-e^{y}} = i\pi \quad .$$

Thus

$$I_1(z) = 2\pi i \sum_{k=1}^{+\infty} \operatorname*{res}_{y=2k\pi i} \frac{e^{yz}}{1-e^{y}} - \pi i$$

$$= 2\pi i \sum_{k=1}^{+\infty} (-1)e^{2k\pi iz} - \pi i = -\pi i \left(2\sum_{k=1}^{+\infty} e^{2k\pi iz} + 1\right)$$

$$= -\pi i\left(\frac{2e^{2\pi iz}}{1-e^{2\pi iz}} + 1\right) = \pi i\left(\frac{e^{2\pi iz}+1}{e^{2\pi iz}-1}\right)$$

$$= \pi i \operatorname{cth}(\pi iz) \quad .$$

As $z = -it + \frac{1}{2}$ we obtain

$$I_1(-it + \tfrac{1}{2}) = \pi i \operatorname{cth}\left(\pi t + \frac{\pi i}{2}\right) = \pi i \operatorname{th} \pi t$$

and in view of (31.39) and (31.40) this implies

$$(\Phi k_{11})(t) = \operatorname{th} \pi t \quad .$$

Applying an analogous argument one can show that

$$(\Phi k_{12})(t) = i[\operatorname{ch} \pi t]^{-1}$$

and consequently (31.34) is proved.

Now let us describe the structure of the whole of the algebra $\tilde{B}$ generated by $\tilde{c}$, $\tilde{S}_{\mathbb{R}}$ ((31.33), (31.34)). Routine calculations show that the elements

$$\begin{pmatrix} 1 & 0 \\ 0 & 0 \end{pmatrix}, \quad \begin{pmatrix} 0 & 0 \\ 0 & 1 \end{pmatrix}, \quad \begin{pmatrix} \operatorname{th} \pi t & 0 \\ 0 & 0 \end{pmatrix}, \quad \begin{pmatrix} 0 & 0 \\ 0 & \operatorname{th} \pi t \end{pmatrix},$$

$$\begin{pmatrix} 0 & [\operatorname{ch} \pi t]^{-1} \\ 0 & 0 \end{pmatrix}, \quad \begin{pmatrix} 0 & 0 \\ [\operatorname{ch} \pi t]^{-1} & 0 \end{pmatrix}$$

belong to $\tilde{B}$.

In view of the Stone–Weierstrass theorem the functions 1 and $\operatorname{th}(\pi t)$ generate the whole of $c(\overline{\mathbb{R}})$, the algebra of continuous functions having limits at $+\infty$ and $-\infty$. Thus all the matrices of the form

$$\begin{pmatrix} \varphi(t) & 0 \\ 0 & \psi(t) \end{pmatrix}, \qquad \varphi(t)\ , \quad \psi(t) \in C(\overline{\mathbb{R}})$$

belong to $\tilde{B}$.

One can also verify that for any element of $\tilde{B}$ of the form

$$\begin{pmatrix} \varphi_{11}(t) & \varphi_{12}(t) \\ \varphi_{21}(t) & \varphi_{22}(t) \end{pmatrix}$$

we have

$$\lim_{t \to \pm\infty} \varphi_{ij}(t) = 0 \qquad \text{when} \quad i \neq j \ .$$

On the other hand the functions $[\operatorname{ch} \pi t]^{-1}$ and $\operatorname{th} \pi t \cdot [\operatorname{ch} \pi t]^{-1} = \dfrac{\operatorname{sh} \pi t}{[\operatorname{ch} \pi t]^2}$ separate points of $\mathbb{R}$ and thus generate the whole of $\overset{\circ}{C}(\overline{\mathbb{R}})$, the algebra of continuous functions tending to zero at $\pm\infty$.

So we conclude that $\tilde{B}$ consists of all the matrices of the form

$$\begin{pmatrix} \varphi_{11}(t) & \varphi_{12}(t) \\ \varphi_{21}(t) & \varphi_{22}(t) \end{pmatrix}$$

where $\varphi_{11},\varphi_{22} \in C(\overline{\mathbb{R}})$ and $\varphi_{12},\varphi_{21} \in \overset{\circ}{C}(\overline{\mathbb{R}})$.

31.16 Here we consider a generalization of the previous example – local representatives for the pseudodifferential operators with piecewise continuous coefficients (see **33.10**).

Let $\Phi(\varphi, \theta)$, $(\varphi, \theta) \in S^{n-1} \times S^{n-1}$, be some C^{∞} function and $\Phi(x, \xi)$, $(x, \xi) \in \mathbb{R}^n \times \mathbb{R}^n\backslash\{0\}$, be its zero–degree homogeneous extension.

Consider the operator $b = b_{\Phi} \in L(L^2(\mathbb{R}^n))$ of the form

$$(b_{\Phi}f)(x) = (2\pi)^{-\frac{n}{2}} \int e^{ix\xi}\, \Phi(x, \xi)\, \hat{f}(\xi)d\xi \tag{31.41}$$

where

$$\hat{f}(\xi) = (2\pi)^{-\frac{n}{2}} \int e^{-i\xi x}\, f(x)dx \quad .$$

Clearly we have

$$bV_{\lambda} = V_{\lambda}b$$

for every V_{λ} considered in example **31.14**. Let B be the C^{*}–algebra generated by operators defined by (31.41). In view of the argument used in **31.14** the mapping (31.30)

$$b \to F\gamma b\gamma^{-1}F^{-1} = \tilde{b}$$

maps B into the subset $\tilde{B}$ of the set of measurable operator–valued functions on $\mathbb{R}$ taking values in $L(L^2(S^{n-1}))$.

Let us look closer at this set $\tilde{B}$. Observe first that

$$(F\gamma\, f)(t, \varphi) = (Mf)\left(t + \frac{in}{2}, \varphi\right) \tag{31.42}$$

where $(t, \varphi) \in \mathbb{R} \times S^{n-1}$ and M is the Mellin transform (by ρ ($\rho = \|x\|$, $\varphi = \frac{x}{\|x\|}$ are polar coordinates in $\mathbb{R}^n$)):

$$(Mf)(\lambda,\ \varphi) = \frac{1}{\sqrt{2\pi}} \int_0^{\infty} \rho^{-i\lambda-1} f(\rho,\ \varphi)d\rho\ , \qquad \lambda \in \mathbb{C}\ . \tag{31.43}$$

Indeed

$$(F\gamma\ f)(t,\ \varphi) = F(e^{\frac{n\tau}{2}} f(e^{\tau},\ \varphi))$$

$$= \frac{1}{\sqrt{2\pi}} \int_{-\infty}^{+\infty} e^{-it\tau}\ e^{n\tau}\ f(e^{\tau},\ \varphi)d\tau$$

$$\frac{1}{\sqrt{2\pi}} \int_0^{+\infty} \rho^{-i\left(t\ +\ \frac{in}{2}\right)-1}\ f(\rho,\ \varphi)d\rho\ \ .$$

Recall that the inverse of the Mellin transform has the form

$$f(\rho,\ \varphi) = \frac{1}{\sqrt{2\pi}} \int_{\mathrm{Im}\lambda=\tau} \rho^{i\lambda}\ Mf(\lambda,\ \varphi)d\lambda\ \ . \tag{31.44}$$

By Plamenevskij [1], 2.3.1, the operator b_{Φ} defined by (31.41) can be written as

$$(b_{\Phi}f)(\rho,\ \varphi) = \frac{1}{\sqrt{2\pi}} \int_{-\infty}^{+\infty} \rho^{i\left(t\ +\ \frac{in}{2}\right)} \left\{U_{\psi\to\varphi}(t)(Mf)\left(t\ +\ \frac{in}{2},\ \psi\right)\right\}dt$$

which means, in view of (31.30), (31.42), (31.43), (31.44) that

$$F\gamma b_{\Phi}\gamma^{-1}F^{-1} = \tilde{b}_{\Phi} = \{U_{\Phi}(t)\} = \{U(t)\}$$

where $\{U(t)\}$ is an operator–valued function with values in $L(L^2(S^{n-1}))$ such that for every t $U(t)(= U_{\psi\to\varphi}(t)\}$ is a *canonical meromorphic pseudodifferential operator* (the term is due to B.A. Plamenevskij) defined by the formula (3.1.3) (Plamenevskij [1]):

$$U_{\psi\to\varphi}(t) = E_{\theta\to\varphi}(t)^{-1}\Phi(\varphi,\ \theta)E_{\psi\to\theta}(t)\ \ . \tag{31.45}$$

Here $E(t)$ is a unitary operator on $L^2(S^{n-1})$ defined via the formula 1.2.1 (Plamenevskij [1]) (see also (1.4.6) Plamenevskij [1]). (In this book we shall not use the explicit form of $E(t)$ and since the formula 1.2.1 (Plamenevskij [1]) and the corresponding notations are rather routine and take substantial room, we omit it).

$E(t)$ is a continuous operator–valued function. Thus $U(t)$ is a continuous operator–valued function as well.

We recall now some properties of the operators $U(t)$ and $\tilde{b}$ (all the results are borrowed from Plamenevskij [1]).

In all the forthcoming statements $\Phi(\varphi, \theta) \in C(S^{n-1} \times S^{n-1})$.

Proposition 31.17 *The algebra* $\mathcal{U}(t)$ *generated by the operators* $U(t)$ *of the form* (31.45) *is irreducible for all* $t \in \mathbb{R}$.

Proposition 31.18 $\mathcal{U}(t)$ *contains all the compact operators in* $L^2(S^{n-1})$.

Proposition 31.19 *Let* b_{Φ_i}, $i = 1,2$, *be two operators defined by* (31.41). *Then*

$$\left[\widetilde{b_{\Phi_1} \cdot b_{\Phi_2}}\right] - \tilde{b}_{\Phi_1 \cdot \Phi_2} = \{K(t)\}$$

where $\{K(t)\}$ *is a continuous operator function with compact values tending to zero when* $t \to \pm\infty$.

31.20 We define by $V(n, 2)$ the manifold of all the pairs $(\varphi, \theta) \in S^{n-1} \times S^{n-1}$ such that φ and θ are orthogonal. Notice that $V(1, 2) = \emptyset$.

Proposition 31.21

$$\inf_{k \in K} \|U_\Phi(t) - k\| = \max_{V(n,2)} |\Phi| \qquad n > 1$$

where K *is the set of all the compact operators in* $L^2(S^{n-1})$. *If* $n = 1$ *then* $\mathcal{U}(t) = K$.

This proposition together with Propositions 31.18 and 31.19 leads to the following two results.

Theorem 31.22

$$\mathcal{U}(t)/_{K} \cong C(V(n,2)) \qquad n > 1 \quad .$$

Theorem 31.23 *If* $n > 1$ *then every irreducible representation of* $\mathcal{U}(t)$ *is either one–dimensional or is equal to the identity representation. Every one–dimensional representation has the form*

$$\pi(\varphi,\ \theta)\colon\ U(t) \to \Phi(\varphi,\ \theta) \qquad (\varphi,\ \theta) \in V(n,\ 2)$$

where Φ *is the symbol of the class* $[U(t)]$ *in* $\mathcal{U}(t)/_{K}$.

If $n = 1$ *then every irreducible representation is equal to the identity representation.*

Proposition 31.24 *If* $t_1 \neq t_2$ *then the representations* $\pi(t_i)\colon\ \tilde{B} \to L(L^2(S^{n-1}))$ *defined by* $\pi(t_i)(\tilde{b}) = U(t_i)$, $i = 1,2$, (*see* (31.45)) *are not equivalent.*

31.25 Let K_0 be the C^*–algebra of all the continuous operator–valued functions $\{k(t)\}$ on $\mathbb{R}$ taking values in $L(L^2(S^{n-1}))$ such that $k(t)$ is a compact operator for every t and tending to zero at $\pm\infty$.

Proposition 31.26

$$K_0 \subset \tilde{B} \quad .$$

Proposition 31.27

$$\tilde{B}/_{K_0} \cong C(S^{n-1} \times S^{n-1}) \quad .$$

Corollary 31.28 *Let* π *be any irreducible representation of* $\tilde{B}$ *such that* $\pi(K_0) = 0$. *Then* π *is equivalent to a representation of the form*

$$\tilde{\mathfrak{b}} \to \Phi(\varphi_0, \theta_0)$$

where

$$\Phi = \left[\tilde{\mathfrak{b}}/K_0\right] \quad , \qquad (\varphi_0, \theta_0) \in S^{n-1} \times S^{n-1} \quad .$$

Theorem 31.29 *Let* $\tilde{\mathfrak{b}} \in \tilde{\mathrm{B}}$ *and* $\Phi = \left[\tilde{\mathfrak{b}}/K_0\right] \in C(S^{n-1} \times S^{n-1})$. *Then*

(1) *all the mappings*

$$\pi(\varphi, \theta)\colon \tilde{\mathfrak{b}} \to \Phi(\varphi, \theta) \qquad (\varphi, \theta) \in S^{n-1} \times S^{n-1}$$

and

$$\pi(t)\colon \tilde{\mathfrak{b}} \to U(t)$$

are mutually not equivalent;

(2) *every irreducible representation of* $\tilde{\mathrm{B}}$ *is equivalent to one of those mentioned in* (1).

31.30 Now we shall consider the algebra generated by operators of the form

$$\widetilde{P_{\mathbb{R}} c P_{\mathbb{R}}} = \tilde{P}_{\mathbb{R}} \tilde{c} \tilde{P}_{\mathbb{R}} \tag{31.46}$$

acting on Im $\tilde{P}_{\mathbb{R}}$. Here

$$P_{\mathbb{R}} = \frac{I + S_{\mathbb{R}}}{2} \tag{31.47}$$

where $S_{\mathbb{R}}$ is defined by (31.31) and c is defined by (31.32). In fact we shall see further that this algebra is the algebra of local representatives of Toeplitz operators with piecewise continuous coefficients.

Let us first calculate Im $\tilde{P}_{\mathbb{R}}$, that is, find vectors $f = \begin{pmatrix} f_1 \\ f_2 \end{pmatrix} \in L^2(\mathbb{R}, \mathbb{C}^2)$ such that $\tilde{P}_{\mathbb{R}} f = f$. In view of (31.33) and (31.34) this is equal to

$$\frac{1}{2}\begin{pmatrix} 1 + \text{th}\,\pi t & i[\text{ch}\,\pi t]^{-1} \\ -i[\text{ch}\,\pi t]^{-1} & 1 - \text{th}\,\pi t \end{pmatrix}\begin{pmatrix} f_1 \\ f_2 \end{pmatrix} = \begin{pmatrix} f_1 \\ f_2 \end{pmatrix} .$$

Solving this equation and denoting (for brevity) $\pi t = x$ we obtain

$$f_2 = (-\mathrm{i})(\text{th}x - 1)\text{ch}x f_1 = \mathrm{i}e^{-x} f_1 \tag{31.48}$$

which means that Im $\tilde{P}_{\mathbb{R}}$ consists of the vectors having the form

$$f = \lambda(x)\begin{pmatrix} 1 \\ \mathrm{i}e^{-x} \end{pmatrix} \tag{31.49}$$

where $\lambda(x)$ and $\lambda(x)e^{-x}$ are both square integrable.

To calculate Ker $\tilde{P}_{\mathbb{R}}$ we observe that if $\varphi = \begin{pmatrix} \varphi_1 \\ \varphi_2 \end{pmatrix} \in L^2(\mathbb{R}, \mathbb{C})$ is such that $\langle \varphi, f \rangle = 0$ for every f defined by (31.49) then

$$0 = \int_{\mathbb{R}} \varphi_1 \bar{\lambda}(x) - \mathrm{i}\varphi_2 \bar{\lambda}(x) e^{-x} dx = \int_{\mathbb{R}} \bar{\lambda}(x)\left[\varphi_1 - \mathrm{i}\varphi_2 e^{-x}\right] dx . \tag{31.50}$$

Taking $\lambda(x) = \chi_\Delta$ where χ_Δ is a characteristic function for some finite interval Δ we conclude that (31.50) implies

$$\varphi_1 = \mathrm{i}e^{-x}\varphi_2 \qquad \text{a.e.} \tag{31.51}$$

Thus Ker $\tilde{P}_{\mathbb{R}}$ consists of all the vectors of the form

$$\varphi = \delta(x)\begin{pmatrix} \mathrm{i}e^{-x} \\ 1 \end{pmatrix} \tag{31.52}$$

where $\delta(x)$ have the same properties as $\lambda(x)$.

Now let

$$\tilde{c} = \begin{bmatrix} c_+ & 0 \\ 0 & c_- \end{bmatrix} .$$

Having in mind that $\tilde{P}_{\mathbb{R}}\tilde{c}\tilde{P}_{\mathbb{R}}f = \tilde{P}_{\mathbb{R}}\tilde{c}f$ for any $f \in \operatorname{Im} \tilde{P}_{\mathbb{R}}$ and writing $\tilde{c}f$ in the form:

$$\tilde{c}f = \lambda(x)\begin{pmatrix} c_+ \\ c_-\mathrm{ie}^{-x} \end{pmatrix} = \gamma(x)\begin{pmatrix} 1 \\ \mathrm{ie}^{-x} \end{pmatrix} + \delta(x)\begin{pmatrix} \mathrm{ie}^{-x} \\ 1 \end{pmatrix}$$

a routine calculation gives

$$\gamma(x) = \lambda(x)\left[\frac{c_+ + c_-\mathrm{e}^{-2x}}{1 + \mathrm{e}^{-2x}}\right] = \lambda(x)\left[c_- + \frac{(c_+ - c_-)}{1 + \mathrm{e}^{-2x}}\right]$$

which means that on $\operatorname{Im} \tilde{P}_{\mathbb{R}}$ we have:

$$\tilde{P}_{\mathbb{R}}\tilde{c}\tilde{P}_{\mathbb{R}} = c_- + \frac{c_+ - c_-}{1 + \mathrm{e}^{-2\pi t}} \tag{31.53}$$

(recall that $x = \pi t$).

Remark 31.31 In the foregoing examples **31.15**, **31.16**, and **31.30** we have considered the spaces of scalar functions $L^2(\mathbb{R}^n)$. Having in mind Remark 31.13 one can easily consider corresponding examples for the spaces of vector functions $L^2(\mathbb{R}^n, E)$. A large class of local representatives satisfying assumption (31.6) can be constructed also as direct sums of elements considered in **31.11**, **31.12**, **31.15**, **31.16**, and **31.30** (see also Remark 31.33).

31.32 Now we consider a particular case of example **31.16** which has already been examined. Namely let

$$(b_\Phi f)(x) = (2\pi)^{-\frac{n}{2}} \int \mathrm{e}^{\mathrm{i}x\xi}\Phi(\xi)\hat{f}(\xi)\mathrm{d}\xi \tag{31.54}$$

where $\Phi(\xi) = \Phi\left(\frac{\xi}{\|\xi\|}\right) \in C(S^{n-1})$.

Thus here the function Φ does not depend on x (in fact this is a local representative for a pseudodifferential operator with continuous coefficients).

By **31.14** the sequence $\left\{V_{\lambda_0^n}\right\}$ considered there satisfies the assumption (31.6) at 0, but in this particular case it is reasonable not to use the group $\{V_\lambda\}$ (defined by (31.26)) to construct the 'symbol' for b_Φ but to use the group $\{W_h\}$, $h \in \mathbb{R}^n$, where

$$(W_h f)(x) = f(x - h) \qquad f \in L^2(\mathbb{R}^n) \quad . \tag{31.55}$$

Clearly

$$b_\Phi W_h = W_h b_\Phi \quad .$$

Thus regarding $\mathbb{R}^n$ as a group the formula (31.25) and the argument after it means that in this situation the mapping

$$b_\Phi \to \Phi$$

establishes an isomorphism between the algebra of local representatives of the form (31.54) and the algebra $C(S^{n-1})$.

Remark 31.33 This example shows that on the one hand if some set $\left\{V_{\lambda_n}\right\}$ satisfies the assumption (31.6) then the considered objects are really local representatives, but on the other hand the structure of these objects can be computed with the help of another group $\{W_h\}$.

32. Localization with respect to an algebra and an ideal.

This section is devoted to the main case (for the theory of functional differential equations) of localization with respect to an algebra and an ideal. The localization can be obtained easily because in fact the whole of the theory was already developed in Sections 28–30. But we want to stress that all the work can be done effectively only under a rather special assumption of interrelation between a localizing algebra A and an ideal I (**32.1**) (if this assumption is not satisfied we have to change (reduce) the algebra A). Fortunately for the Fredholm theory this assumption is not a burden (in view of Theorem 14.19).

32.1 Let A be some commutative C^*–algebra which is a subalgebra of the algebra $L(H)$ of linear bounded operators acting on a Hilbert space H. Consider also some closed *–ideal $I \subset L(H)$. We shall assume that A and I satisfy the condition

$$A \cap I = \{0\} \quad . \tag{32.1}$$

Remark Recall that this condition is satisfied automatically if I is the ideal of compact operators and the maximal ideal space of A does not contain an isolated point (see Theorem 14.19).

In view of Corollary 2.77 the condition (32.1) implies

$$[A] = {}^{A}/_{I} = {}^{A\,+\,I}/_{I} = {}^{A}/_{A\,\cap\,I} \cong A \quad . \tag{32.2}$$

We shall denote by $|||\ \cdot\ |||$ the factor norm

$$|||b||| = ||[b]|| = \inf_{i \in I} ||b + i|| \tag{32.3}$$

$b \in L(H)$.

We shall say that b_1 and b_2 are *I–equivalent* if $|||b_1 - b_2||| = 0$ or, which is the same, $b_1 - b_2 \in I$. We denote this fact by $b_1 \sim b_2 \pmod I$ or simply $b_1 \sim b_2$ if I is fixed.

Throughout this section we assume that A *and* I *satisfy the condition* (32.1).

32.2 An element $b \in L(H)$ is called an element of *local type* (with respect to A and I) if for every $a \in A$ we have

$$ab \sim ba \tag{32.4}$$

A C^*–algebra $B \subset L(H)$ will be called a *C^*–algebra of local type* if $B \supset A$ and every element $b \in B$ is of local type.

Clearly B is a local–type C^*–algebra iff $[A] \subset \mathbb{Z}([B])$ where $[B] = {}^{B}/_{I}$ (cf. **28.2**).

In view of (32.2) one can speak of the maximal ideal space M_A of A or of $[A]$ without ambiguity, and the results of Sections 28–30 lead immediately to the corresponding results for the situation we are considering now. So we begin to reproduce them.

We start from Section 28.

Algebraic approach

32.3 For any closed set $F \subset M_A$ with a (perhaps empty) interior $\overset{\circ}{F}$ we denote by P_F the projection onto the essential subspace $H_F \subset H$, that is,

$$\{a\xi :\ a \in A\ ,\ \ \hat{a}(m) = 0\ ,\ \ m \mathrel{\bar{\in}} \overset{\circ}{F}\ ,\ \ \xi \in H\}\quad .$$

Lemma 32.4 *If* b *is a local–type element (with respect to* A *and* I) *then for any two non–intersecting closed subsets* F_1, F_2 *of* M_A *we have*

$$|||P_{F_1} b P_{F_2}||| = 0\quad .$$

Proof Follows from the proof of Lemma 28.4. □

32.5 Let A, I, and B be any fixed objects described in **32.2**. For any ideal $J \in \mathrm{Prim}[B]$ we denote by $\varphi(J)$ the ideal of $[A]$ defined by

$$\varphi(J) = \{[A] \cap J\}\quad . \tag{32.5}$$

Lemma 32.6 (Localization) *The map* $J \to \varphi(J)$ *defines the continuous surjective mapping*

$$\varphi\colon \mathrm{Prim}[B] \to M_A\quad .$$

Proof cf. Lemma 28.6. □

Corollary 32.7 *Let* $\Theta\colon [\hat{B}] \to \mathrm{Prim}[B]$ *be the canonical mapping* $(\pi \to \mathrm{Ker}\ \pi)$. *Then the mapping*

$$\Psi = \varphi \circ \Theta : [\hat{B}] \to M_A \tag{32.6}$$

is continuous and onto.

32.8 For any $m \in M_A$ we denote by J_m the ideal in $[B]$ defined by

$$J_m = \bigcap_{\varphi(J)=m} J \quad . \tag{32.7}$$

Let ρ_m be the canonical mapping

$$\underbrace{B \to [B] \to [B]/_{J_m}}_{\rho_m} \stackrel{\text{def}}{=\!=\!=} B(m) \quad . \tag{32.8}$$

The image $\rho_m(b)$ of an element $b \in B$ under the mapping ρ_m: $B \to B(m)$ we denote by $b(m)$ and call it the *local representative* of b at the point $m \in M_A$.

32.9 An element $b \in L(H)$ will be called left (right) *I–Fredholm* if there exists an element $R_l(R_r) \in L(H)$ such that

$$R_l b \sim 1 \ (bR_r \sim 1)$$

(1 is the identity operator of $L(H)$).

An element b will be called *I–Fredholm* if it is left and right I–Fredholm simultaneously.

An element $R_l(R_r)$ (if it exists) will be called a left (right) *regularizer* for b.

Clearly if b is I–Fredholm and R_l, R_r are its regularizers then $R_l \sim R_r$. Note also that if I is the ideal of compact operators of $L(H)$ then the notion of I–Fredholmity coincides with the classical definition of Fredholmity.

Theorem 32.10 *An element* b *of a local–type* C^**–algebra* B *is* I*–Fredholm iff all* $\rho_m(b)$ *are invertible.*

Proof Cf. Theorem 28.9. □

Analytical approach.

32.11 We shall say that local–type elements (with respect to A and I satisfying (32.1)) $b_1, b_2 \in B$ are *I–equivalent at a point* $m_0 \in M_A$ if for every $\pi \in [\hat{B}]$ such that $\psi(\pi) = m_0$ (ψ was defined by (32.6)) we have

$$\pi([b_1]) = \pi([b_2]) \quad . \tag{32.9}$$

This equivalence relation will be denoted by

$$b_1 \overset{m_0}{\sim} b_2 \quad . \tag{32.10}$$

Obviously

$$b_1 \overset{m_0}{\sim} b_2 \Leftrightarrow b_1(m_0) = b_2(m_0)$$

(see (29.3) and (32.8)).

Lemma 32.12 $b_1 \overset{m_0}{\sim} b_2$ *iff for any* $\varepsilon > 0$ *there exists a neighbourhood* $O(m_0)$ *such that for any* $a \in A$ *satisfying the conditions*

$$\|a\| \le 1 \qquad \text{and} \qquad \hat{a}(m) = 0 \quad , \qquad m \,\bar{\in}\, O(m_0) \tag{32.11}$$

the following inequalities are valid

$$|||(b_1 - b_2)a||| < \varepsilon \quad , \qquad |||a(b_1 - b_2)||| < \varepsilon \quad . \tag{32.12}$$

Proof See Lemma 29.2 having in mind (32.2). □

32.13 It may happen that it is convenient to take the conditions of Lemma 32.12 as the definition of *I*–equivalence of two elements (not necessarily of local type) at a point m_0.

Namely let A and I satisfy the condition (32.1). We say that two elements b_1, b_2 (not necessarily of local type) are I–equivalent at a point $m_0 \in M_A$ if for any $\varepsilon > 0$ there exists $O(m_0)$ such that for any $a \in A$ satisfying (32.11) we have (32.12). This equivalence relation will be also denoted by

$$b_1 \overset{m_0}{\sim} b_2 \quad .$$

In view of Lemma 32.12 if b_1 and b_2 are of local type this definition is equivalent to (32.10).

32.14 We say that an element b is *I–Fredholm at a point* $m_0 \in M_A$ if there exist elements R_l, R_r such that

$$R_l b \overset{m_0}{\sim} 1 \quad , \qquad bR_r \overset{m_0}{\sim} 1$$

and we call the elements $R_l(R_r)$ the *left* (*right*) *regularizer of* b *at* m_0.

Lemma 32.15 *If* b *is I–Fredholm at* $m_0 \in M_A$ *then there exists a neighbourhood* $O(m_0)$ *and elements* R_1', R_r' *such that for any* $a \in A$ *satisfying*

$$\hat{a}(m) = 0 \qquad m \bar{\in} O(m_0)$$

we have

$$(R_1'b - 1)a \sim 0 \tag{32.13}$$

$$a(bR_r' - 1) \sim 0 \quad . \tag{32.14}$$

Proof See Lemma 29.5. □

32.16 As a result of this lemma, just as was done after Lemma 29.5, we arrive naturally at the following notion: we shall say that $R_l'(R_r')$ is a *left* (*right*) *quasiregularizer for* b *at* m_0 if there exists a neighbourhood $O(m_0)$ such that for any $a \in A$ such that

$$\hat{a}(m) = 0 \quad , \qquad m \bar{\in} O(m_0)$$

we have

$$R'_l ba \sim a \tag{32.15}$$

$$(abR'_r \sim a) \quad . \tag{32.16}$$

Recall again that we do not mean here that b is of local type.

Lemma 32.17 *If* $b_1 \overset{m_0}{\sim} b_2$ *and* b_1 *has a left (right) quasiregularizer at* m_0 *then* b_2 *also has a left (right) quasiregularizer at* m_0.

Proof This result can be extracted in a routine way from the proof of Lemma 29.7. □

Lemma 32.18 *Let* b *be a local–type element and* $R'_l(R'_r)$ *be a left (right) quasiregularizer of* b *at* m_0. *If* $a_0 \in A$ *satisfies the condition*

$$\hat{a}_0(m) \equiv 1 \qquad m \in U(m_0)$$

for some neighbourhood $U(m_0)$ *then* $a_0R'_l$, R'_la_0 $(a_0R'_r$, $R'_ra_0)$ *are also left (right) quasiregularizers of* b.

Proof See Lemma 29.8. □

Lemma 32.19 *Let* b *be a local–type element having left and right quasiregularizers at* m_0. *Then there exist left and right quasiregularizers of local type.*

Proof Routine from Lemma 29.9. □

32.20 Just as was done in 29.10 one can arrive at the following statement: if a local–type element is I–Fredholm at a point m_0 then it is I–Fredholm in some neighbourhood of m_0.

Localization of a C^*–algebra acting in a subspace.

32.21 Here we consider the following situation: given some commutative C^*–algebra A, an ideal $I \subset L(H)$ satisfying the condition (32.1), and a local–type (with respect to A and I) C^**–algebra* B, let $P \in B$ be some projection and $\widetilde{H} = \operatorname{Im} P$. Thus $\widetilde{H}$ is a 'local' subspace. We are investigating objects of the form

$$PbP \quad , \qquad b \in B$$

acting on $\widetilde{H}$.

Let us consider

$$\widetilde{A} = PAP \quad , \qquad \widetilde{B} = PBP \qquad \text{and} \qquad \widetilde{I} = PIP$$

as C^*–algebras and an ideal in $L(\widetilde{H})$.

It can easily be seen that $\widetilde{B}$ is of local type (with respect to $\widetilde{A}$ and $\widetilde{I}$). Indeed for any $b \in B$ and $a \in A$ we have

$$\begin{aligned}(PbP)(PaP) &= PbaP + Pi_1P = PabP + Pi_2P \\ &= (PaP)(PbP) + Pi_3P\end{aligned}$$

where $i_1, i_2, i_3 \in I$.

Let

$$[\widetilde{A}] = {}^{\widetilde{A}}/_{\widetilde{I}} \quad .$$

32.22 The following assumption will ensure that in fact $[\widetilde{A}]$ does not differ from A (which gives us an opportunity to use again the same maximal ideal space M_A). So we assume that

$$\text{if} \quad a \in A \quad \text{is a non–zero element then} \quad aP \,\bar{\in}\, I \tag{32.17}$$

(cf. (30.2)).

Lemma 32.23 *Under the assumption* (32.17) *we have*

$$[\widetilde{\mathrm{A}}] \cong \widetilde{\mathrm{A}} \cong \mathrm{A} \quad . \tag{32.18}$$

Proof If $a \in A$ then since P is of local type we have

$$PaP = aP + i \quad , \qquad i \in I \quad . \tag{32.19}$$

If $a \neq 0$ then in (32.19) we have

$$PaP = aP + i \bar{\in} I \tag{32.20}$$

and in particular

$$PaP \bar{\in} \tilde{I} \tag{32.21}$$

as $\tilde{I} = PIP \subset I$.

Thus $\tilde{A} \cap \tilde{I} = \{0\}$ and consequently

$$[\tilde{A}] \cong {}^{\tilde{A}}/_{\tilde{A} \cap \tilde{I}} = \tilde{A} \quad .$$

On the other hand (32.21) also means that the surjective homeomorphism

$$A \ni a \to PaP \in \tilde{A}$$

has zero kernel which shows that it is an isomorphism. □

32.24 Consider the C^*–algebra

$$[\tilde{B}] = {}^{\tilde{B}}/_{\tilde{I}} \quad .$$

Note that

$$\tilde{B} \cap I = \tilde{I} \quad ; \tag{32.22}$$

indeed if, for some $b \in B$, $PbP = i \in I$ then $i = P(PbP)P = PiP \in \tilde{I}$.

In turn (32.22) implies

$$\tilde{B}/_{I} \cong \tilde{B}/_{\tilde{B} \cap I} = \tilde{B}/_{\tilde{I}} = [\tilde{B}]$$

which means (as $\tilde{B} \subset B$) that we can consider $[\tilde{B}]$ as a C^*–subalgebra of $[B] = B/_I$.

Throughout the rest of this section we shall bear this in mind.

Lemma 32.25 *Let* π *be an irreducible representation of* $[\mathrm{B}]$ *on* H_π. *Then* $\pi_\mathrm{P} = \pi|_{[\tilde{\mathrm{B}}]}$ *is an irreducible representation of* $\tilde{\mathrm{B}}$ *on* $\mathrm{H}_{\pi_\mathrm{P}} = \pi([\mathrm{P}])\mathrm{H}_\pi$.

On the other hand if π_P *is some irreducible representation of* $[\tilde{\mathrm{B}}]$ *on* $\mathrm{H}_{\pi_\mathrm{P}}$ *then there exists a Hilbert space* $\mathrm{H}_\pi \supset \mathrm{H}_{\pi_\mathrm{P}}$ *and an irreducible representation* π *of* $[\mathrm{B}]$ *on* H_π *such that* $\mathrm{H}_{\pi_\mathrm{P}} = \pi([\mathrm{P}])\mathrm{H}_\pi$ *and* $\pi|_{[\tilde{\mathrm{B}}]} = \pi_\mathrm{P}$.

Proof See the proof of Lemma 30.3 assuming the preceding agreement on $[B]$ and $[\tilde{B}]$. □

Finally we obtain the main statement.

Theorem 32.26 *Let* $\mathrm{b(m)}$ *be a local representative of* $\mathrm{b} \in \mathrm{B}$ *at* $\mathrm{m} \in \mathrm{M}_\mathrm{A} = \mathrm{M}_{\tilde{\mathrm{A}}}$ *(see* (32.8)). *Then* $\tilde{\mathrm{b}}(\mathrm{m}) = \mathrm{P(m)b(m)P(m)}$ *is a local representative of* $\tilde{\mathrm{b}} = \mathrm{PbP} \in \tilde{\mathrm{B}}$ *at* m.

Proof See Theorem 30.4. □

33. Local representatives. II.

This section is a continuation of Section 31 and here we describe the structure of local representatives in the situation considered in Section 32, that is, when we carry out the localization with respect to an algebra A and an ideal I satisfying (32.1). In fact we shall find that the ideas of Section 31 can be extended up to this situation under some additional condition on A and I (see (33.2)) – namely the condition that no part of I is 'concentrated' at a single point $m_0 \in M_A$.

Lemma 33.1 *If* B *is a local–type algebra* (*see* 32.2) *then the function* $m \to \|b(m)\|$ (*see* 32.8) $b \in B$ *is upper semicontinuous on* M_A.

Proof Follows immediately from Lemma 31.2 and **32.8.** □

33.2 As the examples in Section 31 show the function $\|b(m)\|$ is not continuous in general and the structure of $b(m_0)$ can be calculated only under some additional assumptions on the behaviour of b in the neighbourhood $O(m_0)$ (Corollary 31.10).

When we consider localization with respect to an algebra and an ideal the situation is even worse and the structure of the ideal I at m_0 plays a significant role when investigating the structure of $b(m_0)$.

First of all we note that an argument analogous to that used in proving (31.4) shows that

$$\|b(m_0)\| = \inf_{\substack{a \in A_{O(m_0)} \\ \hat{a}(m_0)=1}} \||ab\|| = \inf_{\substack{a_i \in A_{O(m_0)} \\ \hat{a}_i(m_0)=1}} \||a_1ba_2\|| = \inf_{\substack{a_1 \in A_{O(m_0)} \\ \hat{a}_1(m_0)=1}} \||a_1ba_2\|| \tag{33.1}$$

where in the last term $a_2 \in A_{O(m_0)}$ is some fixed element with $\hat{a}_2(m_0) = 1$.

Remark One can also observe easily that in (33.1) we can use only elements a_1, a_2 such that $\hat{a}_i(m) \equiv 1$, $i = 1,2$, in some neighbourhoods $O'(m_0) \subset O(m_0)$ (depending on a_i).

The main assumption on A and I under which the theory of Section 31 can be used effectively is: for any fixed $i \in I$ and $a_2 \in A_{O(m_0)}$, for every $\varepsilon > 0$, there exists a neighbourhood $O'(m_0)$ such that

$$\|aia_2\| < \varepsilon \tag{33.2}$$

for every $a \in A_{O'(m_0)}$ with $\|a\| \leq 1$.

Remark The condition (33.2) means that no part of I is 'concentrated' at m_0. To obtain some idea of when this situation can occur it may be useful to look through the following.

Example Let I be the ideal of compact operators on $L^2(\mathbb{R})$ and A be the algebra of operators of multiplication by continuous bounded functions $a \in C(\mathbb{R})$. Let $m_0 = 0$ and $O(m_0)$ be some neighbourhood of 0. If we take a sequence of neighbourhoods $O'_n(0)$ tending to 0 then the condition

$$a \in A_{O'_n(0)} \quad , \qquad \|a\| \leq 1$$

means that the elements satisfying it are converging *strongly* to 0 which implies for every compact operator i

$$\|ai\| \to 0 \quad .$$

Thus (33.2) is fulfilled.

The next lemma shows that under condition (33.2) the ideal I does not play any essential role in the calculation of the structure of $b(m_0)$.

Lemma 33.3 *Under the assumption* (33.2)

$$\|b(m_0)\| = \inf_{\substack{a_1 \in A_{O(m_0)} \\ \hat{a}_1(m_0)=1}} \|a_1 b a_2\| \tag{33.3}$$

where $a_2 \in A_{O(m_0)}$ *is any fixed element with* $\hat{a}_2(m_0) = 1$.

Proof First since $|||a_1 b a_2||| \leq \|a_1 b a_2\|$ in view of (33.1) we have

$$\|b(m_0)\| \leq \inf_{\substack{a_1 \in A_{O(m_0)} \\ \hat{a}_1(m_0)=1}} \|a_1 b a_2\| \quad . \tag{33.4}$$

On the other hand for any fixed $a_1, a_2 \in A_{O(m_0)}$ and such that $\hat{a}_i(m) \equiv 1$, $i = 1,2$, in some neighbourhood $O'(m_0) \subset O(m_0)$ we have for every a_i' satisfying $a_i' a_i = a_i'$ and $\|a_i'\| \le 1$ (for example take $a_i' \in A_{O'(m_0)}$)

$$|||a_1 b a_2||| = \inf_{i \in I} \|a_1 b a_2 + i\| \ge \inf_{i \in I} \|a_1' b a_2' + a_1' i a_2'\|$$

$$\ge \inf_{\substack{a_1' \\ \hat{a}_1'(m_0)=1}} \inf_I \|a_1' b a_2' + a_1' i a_2'\| . \tag{33.5}$$

But clearly

$$\inf_{\substack{a_1' \\ \hat{a}_1'(m_0)=1}} \inf_I \|a_1' b a_2' + a_1' i a_2'\| \ge \inf_{\substack{a_1' \\ \hat{a}_1'(m_0)=1}} |||a_1' b a_2'||| \tag{33.6}$$

which together with (33.5) and (33.1) means

$$\|b(m_0)\| = \inf_{\substack{a_1 \in A_{O(m_0)} \\ \hat{a}_1(m_0)=1}} \inf_I \|a_1 b a_2 + a_1 i a_2\| \tag{33.7}$$

where $\|a_i\| \le 1$, $a_2 \in A_{O(m_0)}$ is any fixed element with $\hat{a}_2(m_0) = 1$.

Now for any $\varepsilon > 0$ take a_1, i, a_2 in (33.7) in such a way that

$$\|b(m_0)\| \le \|a_1 b a_2 + a_1 i a_2\| < \|b(m_0)\| + \frac{\varepsilon}{2} \tag{33.8}$$

where we can assume that $\hat{a}_1(m) \equiv 1$ in some $O'(m_0)$. Taking if necessary some neighbourhood $O''(m_0) \subset O'(m_0)$ and some $a_1' \subset A_{O'(m_0)}$, $\hat{a}'(m_0) = 1$ and $\|a_1'\| \le 1$. Applying the assumption (33.2) and (33.7), (33.8) we can obtain

$$\|b(m_0)\| \le \|a_1'ba_2 + a_1'ia_2\| < \|b(m_0)\| + \tfrac{\varepsilon}{2} \tag{33.9}$$

where $\|a_1'ia_2\| < \frac{\varepsilon}{2}$.

This means that

$$\big|\|b(m_0)\| - \|a_1'ba_2\|\big| < \varepsilon$$

and together with (33.4) this proves (33.3). □

Remark 33.4 If we don't assume the validity of the assumption (33.2) then Lemma 33.3 is not true.

Example Consider the space $L^2[0, 1] \oplus \mathbb{C} = H_1 \oplus H_2$. Let A be the algebra of operators a of the form

$$a = a_1 \oplus a_2$$

where $a_1 \in L(H_1)$ is an operator of multiplication by a continuous function $a_1(x) \in C[0, 1]$ and $a_2(z) = a_1(0)\cdot z$, $z \in \mathbb{C}$. Clearly $A \cong C[0, 1]$.

Let I be the ideal of compact operators in $L(H_1 \oplus H_2)$. Obviously

$$I \supset K_1 \oplus L(H_2)$$

where K_1 is the set of all the compact operators in H_1. Let

$$b = 0 \oplus 1 \quad .$$

Evidently b is local with respect to A. (Recall that since [0, 1] has no isolated points $I \cap A = \{0\}$ by Theorem 14.19 and thus (32.1) is fulfilled.)

As $b \in I$, applying (33.7) one has

$$\begin{aligned}\|b(0)\| &= \inf_{a_1} \inf_{I} \|a_1ba_2 + a_1ia_2\| \\ &\le \|a_1ba_2 + a_1(-b)a_2\| = 0 \quad .\end{aligned}$$

Thus $b(0) = 0$.

On the other hand for any a_1, a_2 such that $\hat{a}_i(0) = 1$

$$\|a_1ba_2\| \geq 1$$

which shows that (33.3) is not true.

33.5 If we assume the validity of (33.2) then we have everything for an almost automatic application of the ideas from Section 31.

Consider the space $H_{O(m_0)}$ defined by (31.1) and H_1, $\overline{H}_{O(m_0)}$, $A_{O(m_0)}$, $\overline{A}_{O(m_0)}$ defined in **31.5**. By Lemma 33.3 $b(m_0)$ is determined by the elements $a_1ba_2 \in L(H_{O(m_0)})$.

Let $J \subset L(H_1)$ be some ideal such that

$$\overline{P}_{O(m_0)}\, J\, \overline{P}_{O(m_0)} = \overline{P_{O(m_0)}\, I\, P_{O(m_0)}} \tag{33.10}$$

where $P_{O(m_0)}$ is the projection onto $H_{O(m_0)}$ and $\overline{(\quad\cdot\quad)} = \theta(\cdot)\theta^{-1}$ (see **31.5**).

Since $I \cap A_{O(m_0)} = \{0\}$ (33.10) implies $J \cap \overline{A}_{O(m_0)} = \{0\}$. Thus (32.1) is true for $\overline{A}_{O(m_0)}$ and J.

Moreover if the assumption (33.2) is true for $A_{O(m_0)}$ and I then by (33.10) the corresponding property is true for J and $\overline{A}_{O(m_0)}$ and we can easily deduce the analogue to Theorem 31.8.

Theorem 33.6 *Let* $\mathrm{b}' \in \mathrm{L(H_1)}$ *be some local–type (with respect to* $\overline{\mathrm{A}}_{\mathrm{O(m_0)}}$ *and* J) *element commuting with* $\left\{\mathrm{V}_{\mathrm{g_n}}\right\}$ *where* $\left\{\mathrm{V}_{\mathrm{g_n}}\right\} \subset \{\mathrm{V_g}\}_{\mathrm{g \in G}}$ *is a sequence satisfying the assumption* (31.6) (*or* (31.7)). *Then*

$$\|b'(m_0)\| = \|b'\| \quad . \tag{33.11}$$

Proof Clearly we have to prove only

$$\|b'(m_0)\| \geq \|b'\| \quad . \tag{33.12}$$

Fix $\varepsilon > 0$ and choose $\xi \in H_1$, $\|\xi\| = 1$, such that

$$\|b'\xi\| \geq \|b'\| - \varepsilon \quad .$$

Take any $a_1, a_2 \in A_{O(m_0)}$ with $\hat{a}(m_0) = 1$. Then for sufficiently large numbers $n \in \mathbb{N}$ in view of (31.7) one has

$$\|b'\| - \varepsilon \leq \|b'\xi\| = \|V_{g_n} b'\xi\| \leq \|\bar{a}_1 V_{g_n} b'\xi\| + \varepsilon$$
$$= \|\bar{a}_1 b' V_{g_n} \xi\| + \varepsilon \leq \|\bar{a}_1 b' \bar{a}_2 V_{g_n} \xi\| + 2\varepsilon$$

and since ε is arbitrary we conclude

$$\|b'\| \leq \|\bar{a}_1 b' \bar{a}_2\|$$

which by Lemma 33.3 proves (33.12) and thus (33.11). □

33.7 By Lemma 33.3 $b(m_0)$ is determined in terms of $a_1 b a_2 \in L(H_{O(m_0)})$. Thus, replacing if necessary b by $a_1 b a_2$ we can apply the mapping $b \to \bar{b} \in L(H_1)$ defined in **33.5**. In view of (33.10) one can identify $b(m_0)$ and $\bar{b}(m_0)$. So we obtain the following analogue of Corollary 31.10.

Corollary 33.8 *For every* $\mathrm{b} \in \mathrm{B}$ *let there exist* $\bar{\mathrm{b}}' \in \mathrm{L}(\mathrm{H}_1)$ *(which is of local type with respect to* $\bar{\mathrm{A}}_{\mathrm{O}(\mathrm{m}_0)}$ *and* J) *such that* $\bar{\mathrm{b}}'(\mathrm{m}_0) = \bar{\mathrm{b}}(\mathrm{m}_0)$ *and* $\bar{\mathrm{b}}'$ *commutes with* $\{\mathrm{V}_{\mathrm{g}_\mathrm{n}}\}$. *Then the mapping*

$$\bar{\mathfrak{b}}' \to \mathfrak{b}(\mathfrak{m}_0)$$

defines the isomorphism

$$\{\bar{\mathfrak{b}}'\}_{\mathfrak{b} \in \mathfrak{B}} \cong \mathfrak{B}(\mathfrak{m}_0) \quad .$$

Proof The same as in Corollary 31.10. □

Now applying the results of 31.11–31.32 we can present some examples of local–type operators and their local representatives.

33.9 Let Γ be some oriented, closed Lyapunov curve. We denote by B the C^*–algebra in $L(L^2(\Gamma))$ generated by the singular integral operator

$$(S_\Gamma f)(t) = \frac{1}{\pi i} \int_\Gamma \frac{f(\tau)}{\tau - t} \, d\tau \tag{33.13}$$

and operators of multiplication by piecewise continuous functions which are discontinuous on a finite set of points (depending on the function considered).

It is known that B contains all the compact operators and that for every operator a of multiplication by a continuous function $a(t)$ one has

$$aS_\Gamma - S_\Gamma a = k \tag{33.14}$$

where k is a compact operator (see 2.49).

Consider the subalgebra $A \subset B$ of all the operators of multiplication by continuous functions.

(33.14) means that B is of local type with respect to A and K (where K is the ideal of compact operators in $L(L^2(\Gamma))$). The assumption (32.1) here is satisfied automatically by Theorem 14.19.

Let us fix some point $t_0 \in \Gamma$ and some neighbourhood $O(t_0)$. We denote by φ_{t_0} the mapping $\varphi_{t_0}: O(t_0) \to \mathbb{R}$ which is isometric, maps to t_0 into O and preserves the orientation.

The mapping φ_{t_0} generates the isomorphism

$$L^2(\Gamma) \supset H_{O(t_0)} \xrightarrow{\theta} \overline{H}_{O(t_0)} = H_{\varphi_{t_0}(O(t_0))} \subset L^2(\mathbb{R})$$

where $H_{O(t_0)}$ is the subspace of $L^2(\Gamma)$ of all the functions vanishing on $\Gamma \setminus O(t_0)$ and $H_{\varphi_{t_0}(O(t_0))}$ is the corresponding subspace of $L^2(\mathbb{R})$.

Consider the ideal $\overline{K}$ of all the compact operators in $L(L^2(\mathbb{R}))$. Clearly we have

$$P_{\varphi_{t_0}(O(t_0))} \overline{K} P_{\varphi_{t_0}(O(t_0))} = \overline{P}_{O(t_0)} \overline{K} \overline{P}_{O(t_0)} = \overline{P_{O(t_0)} K P_{O(t_0)}}$$

(cf. (33.10)). Thus we can use all the previous results of this section.

Routine calculation shows that the mapping

$$S_\Gamma \to S_\mathbb{R} \tag{33.15}$$

($S_\mathbb{R}$ is defined in (31.31)) and

$$a \to \overline{a}_{t_0} \tag{33.16}$$

where a is an operator in $L^2(\Gamma)$ of multiplication by a piecewise continuous function $a(t)$ and $\overline{a}_{t_0}$ is the operator in $L^2(\mathbb{R})$ of multiplication by the function

$$\overline{a}_{t_0}(t) = \begin{cases} a_+ , & t > 0 \\ a_- , & t < 0 \end{cases} \tag{33.17}$$

where $a_+(a_-)$ are the corresponding one–side limits (with respect to the orientation on Γ) of $a(t)$ at t_0, is the one considered in Corollary 33.8 (the set $\{V_\lambda\}$ of operators commuting with these elements is defined in (31.26)).

Thus $S_\mathbb{R}$ and $\overline{a}_{t_0}$ generate the algebra $B(t_0)$ of local representatives of B at t_0. The structure of this algebra was explicitly described in **31.15**.

33.10 As a generalization of the foregoing item we consider now the pseudodifferential operators with piecewise continuous coefficients.

Let M be a compact C^∞ manifold without a boundary. We fix some atlas (V, χ), that is, a finite open cover $\{V\}$ and mutually C^∞ compatible coordinate mappings $\chi: V \to \mathbb{R}^n$. Let $\{\zeta\}$ be some partition of identity subordinate to this cover. These objects (V, χ) and $\{\zeta\}$ generate in a standard way the space $L^2(M)$.

Now we describe the objects which are in fact operators generated by the classical pseudodifferential operators of zero order in $L^2(M)$ and operators of multiplication by piecewise continuous functions on M having finite sets of singular points (depending on the function considered) and such that all the singularities are of the first genus.

Following Plamenevskij [1] we shall define these operators in a form convenient for our purpose.

(i) First we introduce the standard pseudodifferential operators of order 0 in $L^2(\mathbb{R}^n)$ having a first–genus singularity of 'symbols' at the origin.

Let $\Phi(x, \xi) \in C^\infty(\mathbb{R}^n \times \mathbb{R}^n\backslash\{0\})$ be a zero–degree homogeneous function of ξ having the form (in polar coordinates $x = (\rho, \varphi)$, $\theta \in S^{n-1}$)

$$\Phi(\rho, \varphi; \theta) = \Phi_0(\varphi, \theta) + \Phi_1(\rho, \varphi, \theta) \tag{33.18}$$

where

$$\Phi_0 \in C^\infty(S^{n-1} \times S^{n-1})$$

and Φ_1 satisfies the condition

$$\left\| |\ln \rho|^q (\rho D_\rho)^k \Phi_1(\rho, \cdot) \right\|_{C^p(S^{n-1} \times S^{n-1})} \le C_{p,q,k} \tag{33.19}$$

for all non–negative integers p, q, r.

Then we call the operator

$$(B_\Phi f)(x) = (2\pi)^{-\frac{n}{2}} \int e^{ix\xi}\, \Phi(x, \xi)\hat{f}(\xi) d\xi \tag{33.20}$$

a standard pseudodifferential operator which has a first–genus singularity of the 'symbol' $\Phi(x, \xi)$ at the origin.

Remark Here we assume the satisfaction of the condition (33.19) for $\Phi(x, \xi)$ which is a bit non–standard (see, for example, Treves [1], Section 2). In fact this is done simply because we have in mind the use of the transformation (31.30) (see also 31.16). Following Plamenevskij we define a standard pseudodifferential operator in such a way that all these transformations will go smoothly (see also Plamenevskij [1] Example 4.2.10 and Definition 4.2.11). In the end the only essential object for the construction of the local representative corresponding to this operator (at the origin) will be Φ_0.

(ii) We say that an operator b: $L^2(M) \to L^2(M)$ is a pseudodifferential operator of zero order which has the 'symbol' having first genus singularities in a finite set of points if the next conditions are satisfied.

(a) For any η, $\zeta \in C^\infty(M)$ and having supports in a small coordinate neighbourhood V

$$\eta b \zeta = \eta^\chi B_V \zeta_\chi + \eta k \zeta$$

where

$$\eta^\chi(f) = \eta(f \circ \chi)$$

$$\zeta_\chi(f) = (\zeta f) \circ \chi^{-1}$$

and B_V is a standard pseudodifferential operator in $L^2(\mathbb{R}^n)$ defined by (33.20). (Here we assume that V is a sufficiently small neighbourhood having not more than one singular point x_0 and $\chi(x_0) = 0$.) k is a compact operator.

(b) If $\operatorname{supp} \eta \cap \operatorname{supp} \zeta = \emptyset$ then

$$\eta b \zeta$$

is compact.

Let B be the C^*–algebra of operators generated by the operators defined in (ii).

Since B in fact is generated by the classical pseudodifferential operators in $L^2(M)$ and operators of multiplication by piecewise continuous functions for every $b \in B$ and for every operator a of multiplication by a continuous function, one has

$$aB - Ba = k \tag{33.21}$$

where k is a compact operator which shows that B is the local–type algebra with respect to A and K (here A is the algebra of operators of multiplication by all the continuous functions on M and K is the ideal of compact operators). The assumption (32.1) here is satisfied automatically by Theorem 14.19.

For every point $m_0 \in M$ one can choose a sufficiently small neighbourhood $V = O(m_0)$ such that the mapping $\chi\colon V \to \mathbb{R}^n$ generates the corresponding isomorphism

$$L^2(M) \supset H_V = H_{O(m_0)} \xrightarrow{\theta} \overline{H}_{O(m_0)} = H_{\chi(V)} \subset L^2(\mathbb{R}^n)$$

where $H_{O(m_0)}$ is the subspace of $L^2(M)$ of all the functions vanishing on $M \setminus O(m_0)$, $H_{\chi(V)}$ is the corresponding subspace of $L^2(\mathbb{R}^n)$ and we assume that $\chi(m_0) = 0$.

Let $\overline{K}$ be the ideal of compact operators in $L(L^2(\mathbb{R}^n))$.

Evidently

$$P_{\chi(V)}\ \overline{K}\ P_{\chi(V)} = \overline{P}_V\ K\ \overline{P}_V = \overline{P_V\ K\ P_V}$$

(cf. (33.10)) and consequently we can use Corollary 33.8.

Now routine calculation using (33.18) and (33.19) shows that the mapping

$$b \to \overline{b}_{m_0} = b_{\Phi_0} \tag{33.22}$$

where Φ_0 is defined by (33.18) and b_{Φ_0} is defined by (31.41) is the one considered in Corollary 33.8 (here the set $\{V_\lambda\}$ of operators commuting with these elements is defined in (31.26)).

The structure of the algebra $B(m_0)$ generated by the operators of the form b_{Φ_0} was described in 31.16.

Let $S(M)$ and $S^*(M)$ be the bundles of tangent and cotangent unit vectors. For every operator b defined by (ii) the function Φ_0 defines the function Φ on the Whitney sum $S(M) \oplus S^*(M)$. Since the set of singular points for b is finite (depending on b) Φ is a constant function on every $S(M)_m$ (except the set of singular points) and in every singular point it has a first–genus singularity.

In view of the foregoing argument we arrive at

Statement 33.11 *The mapping*

$$b \to \{\bar{b}_m\}_{m \in M} \tag{33.23}$$

defined by (33.22) *and this function* Φ *gives rise to the isomorphism*

$$B/K \cong \{B(m)\}_{m \in M} \tag{33.24}$$

such that an operator $b \in B$ *is Fredholm iff all* $\bar{b}_m$, $m \in M$, *are invertible.*

In fact we shall need a bit more information about B/K, namely, the description of its primitive ideal space. Since this description is obtained in full in Plamenevskii [1] we simply borrow the results we need from there.

33.12 Consider the set

$$\sigma = [S(M) \oplus S^*(M)] \cup \mathcal{J}$$

where $\mathcal{J}$ is the disjoint union of the set of straight lines $\{l_m\}_{m \in M}$. Let us define the topology on σ. The part of $S(M) \oplus S^*(M)$ based over a neighbourhood V of m_0 we shall consider as a set of triples (m, φ, θ), $m \in V$; φ, θ are unit tangent and cotangent vectors. Let $U(\varphi_0)$ and $W(\theta_0)$ be some neighbourhoods of φ_0 and θ_0 on the unit spheres. By $K(\varphi_0)$ we denote the subset of V such that its image under some coordinate map $\chi: V \to \mathbb{R}^n$ is the intersection of the open cone with the peak at $\chi(m_0)$ containing all the directions from $\chi'(m_0)$ $(U(\varphi_0))$ with some open ball with the origin at $\chi(m_0)$.

The base of the topology on σ will be defined in the following way:

(a) if $\varphi_0 \cdot \theta_0 = 0$ then a neighbourhood of $(m_0, \varphi_0, \theta_0)$ is the unity of sets

$$(x_0 \times U(\varphi_0) \times W(\theta_0)) \cup (K(\varphi_0) \times S^{n-1} \times W(\theta_0)) \cup \{l_m\}_{m \in K(\varphi_0)} \cup l_{m_0} \tag{33.25}$$

(b) if $\varphi_0 \cdot \theta_0 > 0$ (< 0) then a neighbourhood of $(m_0, \varphi_0, \theta_0)$ is given by the same expression (33.25) where the last term l_{m_0} is replaced by the set of the form

$$\left\{ t \in l_{m_0} : t > N \right\} \qquad (< N) \tag{33.26}$$

(c) finally a neighbourhood of a point $t \in l_m$ is an open interval on l_m containing m.

The set σ with the above–defined topology is in fact the spectrum of ${}^{B}/_{K}$ which is shown by the following result.

Theorem 33.13 (Plamenevskij [1]) *Let* b *be a zero–order pseudodifferential operator (defined in* 33.10), Φ *its 'symbol' (defined by* Φ_0 (33.18) *on* $S(M) \oplus S^*(M)$) *and* $b(m) \simeq \{U(t)\}$ *be its local representative at a point* $m \in M$ *(see* (33.22) *and* 31.16). *Then*

(1) *The mappings* $\pi(m, \varphi, \theta)$: $b \to \Phi(m, \varphi, \theta)$ $(\varphi \in S(M)_m, \theta \in S^*(M)_m)$ *and* $\pi(m, t)$: $b \to U(t)$ $(t \in 1_m)$ *generate the mutually inequivalent irreducible representations of* ${}^B/_K$;

$\pi(m, \varphi, \theta)$ *is a one–dimensional representation*
$\pi(m, t)$ *is a representation in* $L(L^2(S(M)_m))$.

(2) *Every irreducible representation of* ${}^B/_K$ *is equivalent to one of those mentioned in* (1).

(3) *The mapping*

$$\pi(m, \varphi, \theta) \to (m, \varphi, \theta)$$
$$\pi(m, t) \to t \in 1_m$$

is a bijection of the spectrum $\left\{\widehat{{}^B/_K}\right\}$ *and* σ. *The topology on* σ (*defined in* 33.12) *coincides with the Jacobson topology.*

Theorem 33.14 *The algebra* ${}^B/_K$ *is* GCR *and thus* $\left\{\widehat{{}^B/_K}\right\}$ = Prim ${}^B/_K$.

Proof Follows immediately from the previous theorem, Proposition 31.18 and Theorem 6.8.7 (Pedersen [1]). □

Remark 33.15 The description of the topology given in 33.12 shows that it does not contain a countable base.

33.16 From the definition of the topology we see that the part of Prim ${}^B/_K$ based 'over' a single point $m_0 \in M$ has a non–empty interior and for our future purposes (to use the results of Section 16) we need the description of the ideals in ${}^B/_K$ consisting of the elements b which are zero everywhere except one 'point' m_0, that is, a b for which $b(m) = 0$ when $m \neq m_0$.

Let us denote the set of such elements by $K_0(m_0)$.

Lemma 33.17 *If* b ∈ $K_0(m_0)$ *then* $\pi(m, \varphi, \theta) = \pi(m, t) = 0$ *for every* $m \neq m_0$ *and* $b(m_0) = \{U(t)\} \in K_0$ *where* K_0 *is the* C^**–algebra of all the continuous operator–valued functions* $\{k(t)\}$ *having compact values and tending to zero at* $\pm\infty$.

Proof Follows from Theorem 33.13 and Proposition 31.27. □

33.18 If we consider pseudodifferential operators with continuous coefficients, that is, when $\Phi_0(\varphi, \theta)$ defined in (33.18) does not depend on φ then by **31.32** we arrive at the standard symbolic calculus of pseudodifferential operators on compact manifolds

$$B/_K \cong C(S^*(M)) \quad .$$

33.19 Consider now the algebra of Toeplitz operators with piecewise continuous coefficients.

Recall that the Hardy space H^2 is the space of analytic functions f on the disk $D = \{z: |z| < 1\}$ such that

$$\|f\|_{H^2} = \sup_{0<r<1} (2\pi)^{-1} \left[\int_0^{2\pi} |f(re^{it})|^2 dt \right]^{\frac{1}{2}} < \infty \quad .$$

Another description of the Hardy space is that H^2 is the subspace of $L^2(S^1)$ consisting of all the functions f such that their negative Fourier coefficients are zero, that is,

$$f(t) = \sum_{k=0}^{+\infty} c_n e^{int} \quad .$$

If S is the singular integral operator in $L^2(S^1)$ given by (33.13) then it is well known that

$$P = \frac{I + S}{2} \tag{33.27}$$

is the orthogonal projection onto H^2. If a is an operator in $L^2(S^1)$ of multiplication by a continuous function then another well–known fact is

$$a \neq 0 \Rightarrow aP \text{ is not a compact operator} \quad . \tag{33.28}$$

Thus (32.17) is satisfied for the algebra A of operators of multiplication by continuous functions and the ideal K of compact operators in $L^2(S^1)$.

Let $\varphi \in L^\infty(S^1)$. The *Toeplitz operator* T_φ is the operator in H^2 defined by the formula

$$T_\varphi f = P(\varphi f) \qquad f \in H^2 \quad . \tag{33.29}$$

As $Pf = f$, T_φ can be written in the form

$$T_\varphi f = (P\varphi P)f \quad . \tag{33.30}$$

Since P is a local–type operator with respect to A and K (33.30) means that we find ourselves in the situation already considered in **32.21–32.26.** If φ belongs to the set of piecewise continuous functions on S^1 then we say that T_φ is a *Toeplitz operator with a piecewise continuous coefficient.* The C^*–algebra generated by such operators will be called the *algebra of Toeplitz operators with piecewise continuous coefficients*, and denoted by T.

In view of **33.9** and **31.30** the Fredholm theory and the form of local representatives for the algebra $^T/_K$ with respect to A and K can be derived immediately. Namely we have the following.

Proposition 33.20 *For every* $T_\varphi \in T$ *having the form* (33.29) *and every* $t_0 \in S^1$ *the mapping*

$$T_\varphi \to T_\varphi(t_0) = \varphi_- + \frac{\varphi_+ - \varphi_-}{1 + e^{-2\pi t}} \in C(\mathbb{R}) \tag{33.31}$$

where $\varphi_+(\varphi_-)$ *are one–sided limits (corresponding to the orientation of* S^1) *of* φ *at* t_0*, give rise to the mapping*

$$T \to \left[{}^{T}/_{K}\right] \to T(t_0) \qquad (33.32)$$

and describes the algebra $T(t_0)$ *of local representatives of* $\left[{}^{T}/_{K}\right]$ *at* t_0.

An element $\tau \in T$ *is Fredholm iff all* $\tau(t)$ $t \in S^1$ *are invertible.*

34. C^*–algebras generated by local–type algebras and their automorphisms. I.

Here we investigate a C^*–algebra generated by a local–type C^*–algebra and unitary operators inducing automorphisms of this algebra. The investigation is carried out in the spirit of the general approach presented in Chapter 2. The only difference is that we use M_A instead of Prim B and $b(m)$ instead of $\pi(x)$. This approach offers an opportunity in the case when we can calculate $b(m)$ because M_A is as a rule simpler than Prim B. We begin with the consideration of a local–type C^*–algebra described in Sections 28–31 (i.e. the localization with respect to a central subalgebra not having in mind any ideal). The general case examined in Sections 32–33 will follow later with minimal difficulties.

34.1 Let $B \subset L(H)$ be some local–type C^*–algebra with respect to A (that is, $A \subset \mathbb{Z}(B)$ see **28.2**).

Consider some unitary representation $\{T_g\}_{g\in G}$ of a group G in $L(H)$ such that

(i) $T_g B T_g^* = B$

(ii) $T_g A T_g^* = A \qquad g \in G$.

Remark If $A = \mathbb{Z}(B)$ then clearly (ii) follows from (i).

In view of (ii) $\hat{T}_g(a) = T_g a T_g^*$, $a \in A$, is an automorphism of A and thus it generates the homeomorphism t_g of the maximal ideal space M_A of A by the formula

$$\widehat{[\hat{T}_g(a)]}(m) = \hat{a}(t_{g^{-1}}(m)) , \qquad m \in M_A \quad . \qquad (34.1)$$

Consider the C^*–algebra $\mathbb{C} = C^*(B, T_g)$ generated by B and $\{T_g\}_{g\in G}$.

Our investigation of the algebra $\mathbb{C}$ will in fact follow the ideas of Section 16.

34.2 For any $g \in G$ let us introduce the set M_g as

$$M_g = \{m \in M_A: t_g(m) = m\}$$

and let (34.2)

$$I_g = \{b \in B: b(m) = 0 \ , \quad m \bar{\in} M_g\}$$

where $b(m)$ is a local representative of b at m (see **28.8**).

By M_∞ we denote the set

$$M_\infty = \{m \in M_A: \forall g \neq e, \quad \forall b \in I_g \ , \quad b(m) = 0\}$$

and let (34.3)

$$I^\infty = \{b \in B: b(m) = 0 \ , \quad m \in M_\infty\} \quad .$$

Remarks

(1) Note that in general $M_\infty \neq \left[M_A \setminus \left(\bigcup_{g\neq e} M_g\right)\right]$ (see the example in **16.11**).

(2) Note also that we do not assume that M_g has an interior (in comparison with X_g in **16.1**) because when determining I_g here we in fact consider the set $\varphi^{-1}(M_g)$ (φ is defined by (28.3)) which may have an interior even when $\overset{\circ}{M}_g = \emptyset$. Obviously if $\varphi^{-1}(M_g)$ has an empty interior then $I_g = 0$. We notice also that the induced action of $\hat{T}_g$ on $\varphi^{-1}(M_g)$ is not necessarily trivial and thus in general $\varphi^{-1}(M_g) \neq X_g$.

(3) Just as in **16.1** (see Remarks) we observe that $\left[\bigcup_{g\neq e} M_g\right]$ and M_∞ are invariant under all t_g, $g \in G$.

Analogous to the considerations of Section 16 let us introduce the ideals

$J^\infty(J_g)$ in $C^*(B, T_g)$ generated by $I^\infty(I_g)$ and define also the ideals $J^h \subset C^*(B, T_g)$, $h \in G$, by

$$J^h = \left\{\cap \text{ Ker } \pi \colon \pi \in \widehat{C^*(B, T_g)} \,, \quad \pi(I_h) \neq 0\right\} \,. \tag{34.4}$$

We denote by P_{g_0}, P_∞ the canonical homeomorphisms

$$C^*(B, T_g) \xrightarrow{P_{g_0}} C^*(B, T_g)/J^{g_0} = \mathbb{C}_{g_0} \tag{34.5}$$

$$C^*(B, T_g) \xrightarrow{P_\infty} C^*(B, T_g)/J^\infty = \mathbb{C}_\infty \,. \tag{34.6}$$

Theorem 34.3 $\{P_\infty, P_g\}_{\substack{g\in G\\ g\neq e}}$ *is a sufficient set of homeomorphisms for* $C^*(B, T_g)$.

Proof Analogous to that of Theorem 16.3. □

34.4 Just as in Section 16 we can reformulate this result.

Consider the Banach space J_{g_0} and the representation $\nu_{g_0} \colon C^*(B, T_g) \to L(J_{g_0})$ defined by

$$\nu_{g_0}(d)j = dj \quad , \qquad d \in C^*(B, T_g) \,, \quad j \in J_{g_0} \quad . \tag{34.7}$$

Theorem 34.5 $\mathbb{C}_{g_0} \cong \nu_{g_0}(C^*(B, T_g))$ *and the isomorphism is given by*

$$\nu_{g_0}(d) \longleftrightarrow P_{g_0}(d) \qquad d \in C^*(B, T_g)$$

($\mathbb{C}_{g_0}$ *is defined by* (34.5)).

Proof Just as the proof of Theorem 16.5. □

Corollary 34.6 $\{P_\infty, \nu_g\}_{\substack{g\in G\\ g\neq e}}$ *is a sufficient set of representations of* $C^*(B, T_g)$.

In addition to Theorem 34.5 we can describe, just as in Section 16, the structure of $\mathbb{C}_g$ in another way.

Proposition 34.7 *Let* K_{g_0} *be the essential subspace for* J_{g_0}. *Then* $\nu_{g_0}(C^*(B, T_g)) \cong C^*(B, T_g)|_{K_{g_0}}$.

Proof The same as the proof of Proposition 16.9. □

34.8 An argument analogous to that used in **16.11** shows that

$$B \cap J^\infty = I^\infty \quad . \tag{34.8}$$

Indeed for any fixed $b \in B$ and $\varepsilon > 0$ let $\pi \in \hat{B}$ be an irreducible representation such that $\operatorname{Ker}\pi \in \varphi^{-1}(M_\infty)$ and

$$\|\pi(b)\xi\| \geq \left[\sup_{\varphi^{-1}(M_\infty)} \check{b}(x)\right] - \varepsilon$$

where ξ, $\|\xi\| = 1$ is some vector in H_π.

Let $\nu \in \widehat{C^*(B, T_g)}$ be some extension of π up to an irreducible representation of $C^*(B, T_g)$. Just as in **16.11** denote

$$\operatorname{supp}\nu = \{x \in \operatorname{Prim} B\colon \check{b}(x) = 0 \text{ for } b \in \operatorname{Ker}\nu\} \quad .$$

Obviously $\operatorname{supp}\nu = \overline{\{t_g(\operatorname{Ker}\pi)\}}_{g\in G}$ and since $\varphi^{-1}(M_\infty)$ is invariant with respect to t_g we have

$$\{t_g(\text{Ker }\pi)\}_{g\in G} \subset \varphi^{-1}(M_\infty) \quad .$$

By the definition (see (34.3))

$$j(t_g(\text{Ker }\pi)) = 0 \qquad j \in I^\infty \ , \quad g \in G$$

we conclude (in view of the lower semicontinuity of ν) that

$$\nu(I^\infty) = 0$$

and this, just as in **16.11**, implies (34.8).

34.9 Let us introduce the algebras

$$B_\infty = {}^B/_{J^\infty} = {}^{B+J^\infty}/_{J^\infty} = {}^B/_{J^\infty\cap B} = {}^B/_{I^\infty} \tag{34.9}$$

and

$$A_\infty = {}^A/_{J^\infty} = {}^{A+J^\infty}/_{J^\infty} = {}^A/_{I^\infty\cap A} \cong C(\overline{M_\infty}) \tag{34.10}$$

where the last equality in (34.10) follows from the fact that

$$I^\infty \cap A = \{a \in A\colon \hat{a}(m) = 0 \ , \quad m \in M_\infty\} \quad .$$

Let $\hat{B}_{I^\infty}$ be the set of elements in $\hat{B}$ (the spectrum of B) which are zero on I^∞. By **2.85** and (34.9) the sets $\hat{B}_\infty$, $\left({}^B/_{J^\infty}\right)^\wedge$ and $\hat{B}_{I^\infty}$ are homeomorphic so we shall not distinguish them when it does not cause ambiguity.

Consider also the set $\text{Prim}_{I^\infty}(B)$ of all the primitive ideals of B which contain I^∞. Again by **2.85** and (34.9) we have that Prim B_∞, $\text{Prim}\left({}^B/_{J^\infty}\right)$ and $\text{Prim}_{I^\infty}B$ are homeomorphic.

Notice that clearly

$$\varphi^{-1}(M_\infty) \subset \text{Prim}_{I^\infty}B \tag{34.11}$$

and

$$\psi^{-1}(M_\infty) \subset \hat{B}_{I^\infty} \tag{34.12}$$

where ψ is defined by (28.6).

Lemma 34.10

$$\overline{\varphi^{-1}(\mathrm{M}_\infty)} = \mathrm{Prim}_{\mathrm{I}^\infty}\mathrm{B} \tag{34.13}$$

$$\overline{\psi^{-1}(\mathrm{M}_\infty)} = \hat{\mathrm{B}}_{\mathrm{I}^\infty} \quad . \tag{34.14}$$

Proof Since by the definition of the topology on the spectrum and the definition of ψ (34.14) follows from (34.13) we shall prove only the last one.

Suppose on the contrary that (34.13) is not true. Then there exists an open set $W \subset \mathrm{Prim}_{I^\infty}B$ such that $\varphi^{-1}(M_\infty) \cap W = \emptyset$. Take any $x_0 \in W$. By the definition of $\mathrm{Prim}_{I^\infty}B$ we have

$$x_0 \supset I^\infty \quad . \tag{34.15}$$

As x_0 is not contained in a closed set $(\mathrm{Prim}_{I^\infty}B)\backslash W$ then by the definition of the Jacobson topology

$$x_0 \not\supset I \tag{34.16}$$

where

$$I = \{\cap\ x\colon\ x \in [(\mathrm{Prim}_{I^\infty}B)\backslash W]\} \neq \emptyset \quad .$$

On the other hand since

$$\varphi^{-1}(M_\infty) \subset [(\mathrm{Prim}_{I^\infty}B)\backslash W]$$

one has

$$I \subset \widetilde{I^{\infty}} = \bigcap_{x \in \varphi^{-1}(M_\infty)} x = \bigcap_{\varphi(x) \in M_\infty} x = I^{\infty}$$

(by the definition of φ and I^{∞}).

But this means that (34.16) contradicts (34.15). □

34.11 We shall say that $\{\hat{T}_g\}_{g\in G}$ acts *freely* on M_∞ if for any $m \in M_\infty$ and any finite set $\{g_1,\dots,g_n\} = F \subset G$ with $g_i \neq e$, $i = 1,n$, we have

$$t_{g_i}(m) \neq m \qquad i = 1,n \quad .$$

One can easily verify that if $\{\hat{T}_g\}_{g\in G}$ acts freely on M_∞ then $\{\hat{T}_g\}_{g\in G}$ acts topologically freely on $\overline{\varphi^{-1}(M_\infty)}$. Thus reasoning as in **16.11** we conclude that *if* $\{\hat{T}_g\}_{g\in G}$ *acts freely on* M_∞ *then*

$$\mathfrak{C}_\infty \cong B_\infty \times_{[\hat{T}]} G \tag{34.17}$$

where $[\hat{T}]$ *is the group of automorphisms of* B_∞ *induced by the automorphisms* $\hat{T}_g$, $g \in G$.

Remarks 34.12

(1) The example in 16.11 shows that in general the group action on M_∞ is not necessarily free and (34.17) in general is not true.

(2) Note that B_∞ is a local–type algebra with respect to $A_\infty \cong C(\overline{M_\infty})$ (see (34.9), (34.10)). Observe also that

$$\begin{aligned} &\text{(i)} \quad [T_g]\, B_\infty [T_g^*] = B_\infty \\ & \qquad\qquad\qquad\qquad\qquad g \in G \\ &\text{(ii)} \quad [T_g]\, A_\infty [T_g^*] = A_\infty \end{aligned} \tag{34.18}$$

(cf. 34.1), and that $[T_g]$ induces the homeomorphism of $\overline{M_\infty}$ which coincides with the restriction of t_g on $\overline{M_\infty}$.

34.13 In view of this remark we can speak of the local representatives of $b_\infty \in B_\infty$ with respect to A_∞. We denote these local representatives by

$$b_\infty(m) \quad , \qquad m \in \overline{M_\infty} \quad .$$

The property (i) (34.18) guarantees that the algebras $B_\infty(m)$ and $B_\infty(t_g(m))$, $g \in G$, are isomorphic. Thus we can assume that there exists a Hilbert space H_m such that

$$b_\infty(t_g(m)) \in L(H_m) \qquad \text{for all} \quad g \in G \quad .$$

For every $m \in \overline{M_\infty}$ let us define a representation π_m of $\mathfrak{C}_\infty$ (in a way analogous to that done in 21.1) in the space $l^2(G, H_m)$ by the equations:

$$\pi_m(b_\infty)(g \otimes h) = g \otimes [\hat{T}_g(b)_\infty](m)h \quad , \qquad b_\infty \in B_\infty \tag{34.19}$$

$$\pi_m([T_{g_0}])(g \otimes h) = (g \cdot g_0^{-1}) \otimes h \quad , \qquad g_0 \in G \tag{34.20}$$

where $g \in G$, $h \in H_m$.

The next statement is the exact analogue of Theorem 21.2.

Theorem 34.14 *Let* G *be an amenable group and* $\overline{\mathrm{M}_\infty}$ *be separable compact. If* $\mathfrak{C}_\infty$ *possesses the property* (*) *then* d $\in \mathfrak{C}_\infty$ *is invertible iff all the elements* $\pi_{\mathrm{m}}(\mathrm{d})$, m $\in \overline{\mathrm{M}_\infty}$, *are invertible.*

Proof The same as the proof of Theorem 21.2. Just use $\overline{M_\infty}$ instead of Prim B_∞ and $b_\infty(m)$ instead of $\pi_x(b_\infty)$ $(x \in \hat{B}_\infty)$. □

Corollary 34.15 *If the group action on* M_∞ *is free and* $\overline{\mathrm{M}_\infty}$ *is separable then any* $\mathrm{d} \in \mathbb{C}_\infty$ *is invertible iff all the elements* $\pi_{\mathrm{m}}(\mathrm{d})$ $\mathrm{m} \in \mathrm{M}_\infty$ *are invertible.*

Proof Follows from the previous theorem and (34.17). □

Remarks 34.16

(1) The algebra B_∞ may not be separable even when $\overline{M_\infty}$ is separable so in a number of situations we can use this result even when the application of Theorem 21.2 is impossible.

(2) If $M_A = M_\infty$ and the group action is free then Corollary 34.15 describes a sufficient set of representations of $C^*(B, T_g)$.

34.17 Finally we obtain local–type algebra analogues of Theorems 16.13 and 16.14. By S_{g_0} we denote the ideal in B generated by all the ideals $I_{\tau g_0 \tau^{-1}}$, $\tau \in G$ (cf. the remarks in 16.1). If G is commutative then we have $S_{g_0} = I_{g_0}$.

Theorem 34.18 *Let* $\mathrm{C}^*(\mathrm{B}, \mathrm{T}_g)$ *and* $\mathrm{C}^*(\mathrm{B}_1, \mathrm{V}_g)$ *be two* C^**–algebras such that* $\mathrm{B}(\mathrm{B}_1)$ *is a local–type algebra with respect to* $\mathrm{A}(\mathrm{A}_1)$.

Suppose that an isomorphism $\Phi\colon \mathrm{B} \to \mathrm{B}_1$ *has the properties*

(i) $\Phi(\mathrm{A}) = \mathrm{A}_1$

(ii) $\Phi(\mathrm{T}_g \mathrm{b} \mathrm{T}_g^*) = \mathrm{V}_g\, \Phi(\mathrm{b}) \mathrm{V}_g^*\ , \qquad g \in \mathrm{G}\ ,\quad \mathrm{b} \in \mathrm{B}$

(G *is amenable*).

If $\mathrm{P}_\infty(\mathrm{C}^*(\mathrm{B}, \mathrm{T}_g)) \cong \mathrm{P}_\infty(\mathrm{C}^*(\mathrm{B}_1, \mathrm{V}_g))$ *where* P_∞ *is defined by* (34.6). *Then the mappings* $\Phi\colon \mathrm{B} \to \mathrm{B}_1$, $\mathrm{T}_g \to \mathrm{V}_g$ *give rise to the isomorphism* $\mathrm{C}^*(\mathrm{B}, \mathrm{T}_g) \cong \mathrm{C}^*(\mathrm{B}_1, \mathrm{V}_g)$ *iff for any* $g_0 \neq \mathrm{e}$ *and any finite set* $\mathrm{F} \subset \mathrm{G}$

$$\left\| \sum_{g \in F} \Phi(b_g)V_g \right\| = \left\| \sum_{g \in F} b_g T_g \right\| , \qquad b_g \in S_{g_0} .$$

Proof The same as for Theorem 16.13 using the results of this paragraph instead of the corresponding results of Section 16. □

Theorem 34.19 *Let* $C^*(B, T_g)$ *and* $C^*(B_1, V_g)$ *be two* C^**–algebras such that* $B(B_1)$ *is a local–type algebra with respect to* $A(A_1)$.

Suppose that an isomorphism $\Phi: B \to B_1$ *has the properties*

(i) $\Phi(A) = A_1$,

(ii) $\Phi(T_g b T_g^*) = V_g \Phi(b) V_g^*$, $g \in G$, $b \in B$,

(*G is amenable*).

If $\{\hat{T}_g\}_{g \in G}$ *acts freely on* M_∞ *then the mappings* $\Phi: B \to B_1$, $T_g \to V_g$ *give rise to the isomorphism* $C^*(B, T_g) \cong C^*(B_1, T_g)$ *iff for any* $g_0 \neq e$ *and any finite set* $F \subset G$

$$\left\| \sum_{g \in F} \Phi(b_g)V_g \right\| = \left\| \sum_{g \in F} b_g T_g \right\| , \qquad b_g \in S_{g_0} .$$

Proof Follows from the previous theorem and (34.17). □

35. C^*–algebras generated by local–type algebras and their automorphisms. II.

Here we consider the general case of a C^*–algebra $C^*(B, T_g)$ generated by a local–type algebra B (with respect to A and I (see Section 32)) and a unitary group inducing automorphisms of B. Since the results of this section follow in a straightforward way from the results of Sections 34 and 32 we shall simply sketch them.

35.1 We will examine the following situation. Let $B \subset L(H)$ be some C^*–algebra which is of local–type with respect to a subalgebra A and an ideal $I \subset B$ (see Section 32). Recall that in particular this means that $A \cap I = \{0\}$ (see (32.1)) and so $A \cong {}^{A}/_{I} = [A]$.

Consider some group G and a set of operators $\{T'_g\}_{g \in G}$ such that:

(i) for every $g \in G$ there exists $i_g \in I$ satisfying

$$T'_g + i_g = T_g$$

where $\{T_g\}_{g \in G}$ is a unitary representation of G having the properties:

(ii) $T_g B T_g^* = B$;

(iii) $T_g I \subset I$;

(iv) $[T_g A T_g^*] = [A]$.

Remarks

(1) Since (iii) is valid for every $g \in G$ we have

$$I = T_{g^{-1}}(T_g I) \subset T_{g^{-1}} I \subset I$$

and consequently $T_g I = I$, $g \in G$.

(2) (ii) and the previous remark means that $[T_g B T_g^*] = [B] = {}^{B}/_{I}$.

(3) If $[A] = \mathbb{Z}([B])$ then (iv) follows from (ii) and the first remark.

Consider the C^*–algebra $\mathbb{C} = C^*(B, T_g) = C^*(B, T'_g)$. We denote by $[\mathbb{C}]$ the factor algebra ${}^{\mathbb{C}}/_{I}$. Obviously $[\mathbb{C}] = C^*([B], [T_g])$.

In view of (iv) the automorphism $[\hat{T}_g] = [T_g(\cdot)T_g^*]$ generates (just as in (34.1)) the homeomorphism t_g of M_A (recall that since $[A] \cong A$ we do not distinguish $[a]$ and $a \in A$).

35.2 We introduce the sets M_g, M_∞ and the ideals I_g, I^∞ by the same formulae (34.2) and (34.3) where $b(m)$ is defined by (32.8), and in an obvious way we introduce J_g and J^∞.

The ideals $J^h \subset [\mathbb{C}]$, $h \in G$, are defined by

$$J^h = \{\cap \text{ Ker } \pi\colon \pi \in [\hat{\mathbb{C}}] \;,\quad \pi[I_h]) \neq 0\}$$

cf. (34.4).

We denote by P_{g_0}, P_∞ here the canonical homeomorphisms

$$\underbrace{\mathbb{C} \to [\mathbb{C}] \to {}^{[\mathbb{C}]}/_{J^{g_0}}}_{P_{g_0}} = \mathbb{C}_{g_0}$$

$$\underbrace{\mathbb{C} \to [\mathbb{C}] \to {}^{[\mathbb{C}]}/_{[J^\infty]}}_{P_\infty} = \mathbb{C}_\infty \quad .$$

Now the analogue of Theorem 34.3 is

Theorem 35.3 *An element* d $\in$ $\mathbb{C}$ *is* I–*Fredholm iff all* $\mathrm{P_g(d)}$, $\mathrm{g} \neq \mathrm{e}$, *and* $\mathrm{P_\infty(d)}$ *are invertible.*

The analogues of the other results of Section 34 can be formulated in an obvious way so we dwell only on the last theorem of Section 34.

Theorem 35.4 *For the algebra* $\mathrm{C^*(B,\ T_g)}$ *let there exist* $\mathrm{B_1} \supset \mathcal{Z}(\mathrm{B_1}) \supset \mathrm{A_1}$, *a unitary group* $\{\mathrm{V_g}\}_{\mathrm{g}\in\mathrm{G}}$, *and an isomorphism* Φ: $[\mathrm{B}] \to \mathrm{B_1}$ *having the properties*

(i) $\Phi([\mathrm{A}]) = \mathrm{A_1}$

(ii) $\Phi[\mathrm{T_g b T_g^*}] = \mathrm{V_g}\Phi([\mathrm{b}])\mathrm{V_g^*} \quad , \qquad \mathrm{g} \in \mathrm{G}\ , \quad \mathrm{b} \in \mathrm{B} \quad .$

If $\{[\hat{T}_g]\}_{g\in G}$ *acts freely on* M_∞ *then the mappings* Φ: $[B] \to B_1$, $T_g \to V_g$, *give rise to the isomorphism*

$$[\mathbb{C}] = {C^*(B, T_g)}/{I} \cong C^*(B_1, V_g)$$

iff for any $g_0 \neq e$ *and any finite set* $F \subset G$

$$\left\| \sum_{g\in F} \Phi([b_g])V_g \right\| = \left\| \left[\sum_{g\in F} b_g T_g \right] \right\| , \qquad b_g \in S_{g_0} . \tag{35.1}$$

Proof Obvious by Theorem 34.19. □

Corollary 35.5 *An element* $d \in C^*(B, T_g)$ *is* I–*Fredholm iff* $\Phi([d])$ *is invertible.*

In this situation the element $\Phi([d])$ is called the *symbol* of d (with respect to I).

Remark 35.6 Note finally that we don't have to check the equality (35.1) when

(1) $M_\infty = M_A$ and the group action is free on M_A (in view of (34.17));

(2) the mapping φ (28.3) has the property

$$\overset{\circ}{\omega} = \emptyset \Rightarrow \left[\varphi^{-1}(\overset{\circ}{\omega)}\right] = \emptyset$$

and $\{t_g\}_{g\in G}$ acts topologically freely on M_A (in this case the induced action of $\hat{T}_g$ on Prim $[B]$ is evidently also topologically free);

(3) the norm $\|b(m)\|$ is continuous (lower semicontinuous) on M_A and $\{t_g\}_{g\in G}$ acts topologically freely on M_A (in this case one can prove Theorem 35.4 using an argument analogous to that used in the proof of Corollary 12.17).

Concerning this remark it is worth recalling the Remark in 31.3.

We finish this chapter with an observation that the ideas of Chapter 3 can be naturally treated with respect to local–type algebras and their automorphisms. We do not intend here to go through this and leave the job to anyone who is interested.

Notes and remarks

Section 28. It was already mentioned that the main C^*–algebraic ideas of localization appeared in the work by Dauns, Hofmann [1], Hofmann [1], Douglas [1, 2], Varela [1], and Vasilevskij [1] and were mainly used for the purposes of internal C^*–algebra problems interrelated with representations of algebras by continuous sections. We should mention here that the first application of C^*–localization to certain problems of the theory of Toeplitz operators with discontinuous symbols was obtained by Douglas [1] (Chapter 7). Our presentation in this section is close to that given by Varela [1] (see also Lebedev, Dangh Suan Tkhan′ [1]).

Section 29. The ideas of analytical localization presented here originate in fact in the work by Simonenko [1] (a detailed description is offered in Simonenko, Chin′ Ngok Min′ [1]) where the main framework of it (attached tightly to the geometry of L^P spaces and not involving C^*–algebra methods) was used for the description of symbols of singular integral operators with piecewise continuous coefficients. Some ideas close to those presented in the section can also be found in the book by Gohberg, Krupnik [1]. Later, a number of modifications and generalizations of the local method appeared in investigations of different classes of local–type operators (like those investigated by Simonenko). Here it is worth mentioning for example the work by Duduchava [1], Kozak [1], Soldatov [1, 2], etc. The coincidence of analytical and algebraic localization for local–type operators considered by Simonenko was established by Semenjuta and Kozak (see Simonenko, Chin′ Ngok Min′ [1], Section 4). The general C^*–scheme for analytical localization and the coincidence of the algebraic and analytical approaches in this case is due to Lebedev, Dangh Suan Tkhan′ [1]. The exposition of this section follows this paper. In connection with the results considered here we should also mention the paper by Semenjuta [2].

Section 30. The material presented here shows that in fact there is no difference (from the local point of view) between the application of local methods on the initial space and on any of its local subspaces. Thus for example there is no (local) difference between the investigation of singular integral and Toeplitz operators.

The results here are borrowed from Lebedev, Dangh Suan Tkhan′ [1].

Sections 31–33. The ideas explored in these sections have their roots in the work by Simonenko [1] and Simonenko, Chin′ Ngok Min′ [1] (where the algebra of singular integral operators with piecewise continuous coefficients and its localization with respect to the ideal of compact operators was described). The C^*–scheme presented in Sections 31–33 in essence is due to Lebedev, Dangh Suan Tkhan′ [1]. The calculations in **31.15** are borrowed from Simonenko, Chin′ Ngok Min′ [1]. Anyone who is interested in the theory of pseudodifferential operators with discontinuous symbols should search in the first place the work of Plamenevskij, Senichkin, Vasilevskij, Rabinovich, Kokotov and Batkhin. A thorough exposition of the algebra of pseudodifferential operators having first–genus singularities in symbols is given by Plamenevskij [1]. The theory of pseudodifferential operators on manifolds with singularities is presented in Schulze [1].

The amount of work on Toeplitz operator theory is immense. One can refer for example to the monograph by Böttcher, Sillberman [1].

Thus the examples of local–type operators and calculations of their local representatives considered in Sections 31–33 in fact form a rather small part of the whole range of already investigated local–type operators related to singular integral and pseudodifferential operators.

Nevertheless we hope that the main C^*–frame of the approach to the investigation of such operators can be extracted from them.

We shall continue the investigation of an algebra generated by these operators and transformation operators of an argument in the second part of the book.

Sections 34–35. Here we follow in essence Lebedev, Dang Suan Tkhan′ [2]. In connection with the results of these sections we should also mention the paper by Karlovich [1] where some local–type algebras and topologically free acting groups of automorphisms were investigated and the corresponding version of the isomorphism theorem was obtained.

Bibliography

M.S. Agranovich

[1] Elliptic singular integro–differential operators. *Uspekhi mat. nauk*, 1965, v.20, N5, p.3–120 (Russian).

A.B. Antonevich

[1] On the index of a pseudodifferential operator with a finite group of shifts. *DAN SSSR*, 1970, v.190, N4, p.751–752 (Russian).

[2] The Gelfand representation of a certain ring of operators. *Vestn. Belorus, univ. Ser. I*, 1973, N2, p.3–7 (Russian).

[3] Elliptic pseudodifferential operators with a finite group of shifts. *Izv. AN SSSR, Ser. Mat.*, 1973, v.37, N3, p.663–675 (Russian).

[4] On a class of pseudodifferential operators with deviating argument on the torus. *Diff. Uravnenija*, 1975, v.11, N9, p.1550–1557 (Russian).

[5] On variations of spectrum under small perturbations of an operator. *Vestn. Belorus univ. Ser. I*, 1976, N3, p.60–62 (Russian).

[6] On operators generated by linear extensions of diffeomorphisms. *DAN SSSR*, 1978, v.243, N4, p.825–828 (Russian).

[7] On a certain factorization of a difference operator. *Vestsi Akad Navuk BSSR, Ser. Fiz.–Mat. navuk*, 1978, N3, p.10–15 (Russian).

[8] Operators with a shift generated by the action of a compact Lie group. *Sibirsk. Mat. Zh.*, 1979, v.20, N3, p.467–478 (Russian).

[9] The conditions of invertibility of operators with a convex rationally independent system of shifts. *DAN SSSR*, 1981, v.256, N1, p.11–14 (Russian).

[10] On the spectrum of a weighted shift operator in the space $W^1{}_p(X)$. *DAN SSSR*, 1982, v.264, N5, p.1033–1035 (Russian).

[11] On two methods of studying the invertibility of operators in C^*–algebras induced by dynamical systems. *Mat. Sborn.*, 1984, v.124 (166), N1, p.3–23 (Russian).

[12] Operators with a convex rationally independent system of shifts in vector–function spaces. *IX Shkola poteorii operat. v funkts. prostr., Ternopol'*, 1984, p.7 (Russian).

[13] Two–point problem for the string oscillations equation and related to it functional equations. *Diff. uravnenija*, 1985, v.21, N3, p.426–436 (Russian).

[14] *Linear functional equations. Operator approach.* Minsk, Universitetskoe, 1988 (Russian).

[15] Boundary value problems with strong nonlocalness for elliptic equations. *Izv. AN SSSR*, 1989, v.53, N1, p.3–24 (Russian).

A.B. Antonevich, V.V. Brenner

[1] On algebras of operators with a finite group of shifts. *Mat. issled., Karaghanda*, 1977, v.4, p.3–10 (Russian).

[2] On symbol of pseudodifferential operator with locally independent shifts. *Dokl. AN BSSR*, 1980, v.24, N10, p.884–887 (Russian).

A.B. Antonevich, Yu. A. Bykadorov
[1] On operators $(Sy)(t) = a(t)y(\alpha(t))$ related to functional–differential equations. *Dep. VINITI*, 1985, N2043–85, 53p. (Russian).

A.B. Antonevich, A.V. Lebedev
[1] Extension of operator algebras by unitary operators generating automorphisms. *Dokl. AN BSSR*, 1980, v.24, N5, p.404–407 (Russian).
[2] On Noetherity of functional–differential operator with partial derivatives containing a linear transformation of an argument. *Diff. uravnenija*, 1982, v.18, N6, p.987–996 (Russian).
[3] On spectral properties of operators with a shift. *Izv. AN SSSR. Ser. Mat.*, 1983, v.47, N5, p.915–941 (Russian).

A.B. Antonevich, V.B. Ryvkin
[1] On conditions for a certain singular integral operator with a shift to be Fredholm. *Dokl. AN BSSR*, 1975, v.19, N1, p.5–7 (Russian).
[2] On the normal solvability of a problem on periodic solutions of a linear differential equation with a deviating argument. *Diff. uravnenija*, 1974, v.10, N3, p.1347–1353 (Russian).
[3] Operators generated by homeomorphisms of the circle conjugate to a rotation. *Mat. zametki*, 1982, v.31, N5, p.773–783 (Russian).

A.B. Antonevich, Yu. V. Vol'pert
[1] On the spectrum of a weighted diagonal shift operator on the torus. *Vestsi Akad. Navuk BSSR Ser. Fiz.–Mat. navuk*, 1976, N3, p.48–53 (Russian).

W.B. Arveson
[1] A theorem on the action of abelian unitary groups. *Pacif. J. Math.*, 1966, v.16, N2, p.205–212.
[2] Operator algebras and measure preserving automorphisms. *Acta. Math.*, 1967, v.118, p.95–109.
[3] *An invitation to C*–algebras*. Springer–Verlag, 1976.

W.B. Arveson, K.B. Josephson
[1] Operator algebras and measure preserving automorphisms. II. *J. Funct. Anal.*, 1969, v.4, N1, p.100–134.

M.F. Atiyah
[1] *K–theory*. Harvard Univ., Cambridge, Mass., 1965.

N.V. Azbelev, L.F. Rahmatullina
[1] Functional–differential equations. *Diff. uravnenija*, 1978, v.14, N5, p.771–797 (Russian).

Ch. Babbage
[1] An essay towards the calculus of functions. *Philos. Trans. London*, 1816, v.106, N2, p.179–256.

S. Banach
[1] *Functional analysis course*. Kiiv, Radjan'ska shk., 1948 (Ukrainian).

A.B. Batkhin
[1] On Noetherity of two dimensional singular integral operators with discrete groups of shifts in $L_2(R^2)$. *Linejnye operatory v funktsional'nykh prostranstvah: Tez. dokl. Sev.–kavkaz. reg. konf. Groznyi*, 1989, p.17 (Russian).
[2] An isomorphism theorem for C^*–algebras of multidimensional convolution type operators with discrete groups of shifts. *Dokl. rasch. zased. sem. Inst. prikl. mat. im. I.N. Vekua, Tbilisi*, 1990, v.5, N1, p.26–29 (Russian).
[3] Multidimensional convolution type operators with piece–wise continuous symbol and shifts, generating discrete group. The Noetherity in algebras of operators. *Dep. v Ukr. Zh. INTEI* 1992, N654–UK92 (Russian).
[4] The Noether theory of multidimensional operators of convolution type with discrete groups of shifts in the space $L_2(\mathbb{R}^n)$. Ph.D. diss., Odessa, 1992 (Russian).

E.F. Beckenbach, R. Bellman
[1] *Inequalities*. Springer–Verlag, 1961.

R. Bellman, K. Cooke
[1] *Differential Difference Equations*. Academic Press, 1963.

L.M. Berezanskij
[1] On the question of conditions of reducibility of functional differential equations to the canonical form. *Diff. uravnenija*, 1977, v.13, N10, p.1749–1756 (Russian).
[2] Linear functional equations and functional differential equations with a deviating argument. Ph.D. diss., Alma–Ata, 1979 (Russian).

A. Beurling
[1] Sur les integrales de Fourier absolument convergentes et leur application a une transformation functionelle. *IX Congres des Math. Scand., Helsingfors*, 1938, p.345–366.

R.A. Biktashev
[1] Spectra of pseudodifferential operator over C^*–algebras. *Vestnik Moskov Univ. Ser. I. Mat. Mekh.*, 1982, N4, p.36–38 (Russian).

R.A. Biktashev, A.S. Mischenko
[1] Spectra of elliptic unbounded pseudodifferential operators over C^*–algebras. *Vestnik Moskov Univ. Ser. I., Mat. Mekh.*, 1980, N3, p.56–58 (Russian).

S. Bochner, R.S. Phillips
[1] Absolutely convergent Fourier expansions for non–commutative rings. *Ann. Math.*, 1942, v.43, N3, 409–418.

A. Böttcher, B. Sillberman
[1] *Analysis of Toeplitz Operators*. Berlin, Akademie, 1989.

N. Bourbaki
[1] *Integration* (Eléments de mathématique, LIVRE VI). Hermann, 1952.
[2] *Theories spectrales* (Eléments de mathématique, Fascicule XXXII). Hermann, 1967.
[3] *Varietes differentielles et analytiques* (Eléments de mathématique, Fasc. XXXIII, XXXVI). Hermann, 1967, 1971.

M. De Brabanter
[1] The classification of rational rotation C^*–algebras. *Arch. Math.*, 1984, v.43, N1, p.79–83.
[2] A classificational theorem for C^*–algebras generated by weighted shifts. *Simon Stevin*, 1984, v.58, N3, p.267–270.

M. De Brabanter, H. Zettl
[1] C^*–algebras associated with rotation groups and characters. *Manuscr. math.*, 1984, v.47, N1–3, p.153–174.

B. Brebken
[1] Representations and automorphisms of the irrational rotation algebra. *Pacif. J. Math.*, 1984, v.111, N2, p.257–282.

G.E. Bredon
[1] *Introduction to compact transformation groups.* Academic Press, 1972.

V.V. Brenner
[1] On the spectral radius of operators with locally independent shifts. *Izv. AN BSSR*, 1981, N3, p.48–55 (Russian).
[2] Operators with locally independent shifts. Ph.D. diss., Minsk, 1981 (Russian).
[3] On spectral properties of operators with a shift. *Uspekhi Mat. Nauk.*, 1982, v.37, N1, p.139–140 (Russian).
[4] On symbolic sequences I, II. *Dep. VINITI*, 1983, N4399–83; *Dep. VINITI*, 1983, N4400–83 (Russian).

I.U. Bronstein
[1] *Nonautonomous dynamical systems.* Kishinev. Shtiintsa, 1984 (Russian).

M. Burnat
[1] Die Spectraldarstellung einiger differential Operatoren mit periodischen Koeffizienten in Raum der fastperiodischen Functionen. *Studia Math.*, 1964, v.25, p.33–64.

B.F. Bylov, R.E. Vinoghrad, D.M. Grobman, V.V. Nemytskij
[1] *The Lyapunov exponents theory and its applications to problems of stability.* M. Nauka, 1966 (Russian).

S.L. Campbele
[1] Linear operators for which T^*T and TT^* commute. I. *Proc. Amer. Math. Soc.*, 1972, 34, N1, p.177–180; II. *Pacif. J. Math.*, 1974, 53, N2, p.355–361.

C. Chicone, R. Swanson
[1] The spectrum of the adjoint representation and the hyperbolicity of dynamical systems. *J. Diff. Eqns.*, 1980, v36, p.28–39.

C.C. Cowen
[1] Composition operators on Hilbert spaces of analytic functions: A status report. *Oper. Th.: Oper. Algebras and Appl.: Proc. Summer Res. Inst., Durham, July 3–23, 1988. Providence (R.I), 1990, p.131–145.*

Yu. L. Daletskij, M.G. Krein
[1] *Stability of solutions of differential equations in Banach space.* M. Nauka, 1970 (Russian).

J. Dauns, K.H. Hofmann

[1] Representations of rings by sections. *Mem. Amer. Math. Soc.*, 1968, N83, 180p.

A.M. Davie

[1] Invariant subspaces for Bishop's operator. *Bull. London Math. Soc.*, 1974, N6, p.343–348.

J. Dixmier

[1] *Les C*-algébras et leurs representations*, 2nd edition. Paris, Gauthier–Villars, 1969.

[2] *Les algebres d'operateurs l'espace hilbertien*, 2nd edition. Paris, Gauthier–Villars, 1969.

D.P. O'Donovan

[1] Weighted shifts and covariance algebras. *Trans. Amer. Math. Soc.*, 1975, v.208, p.1–25.

R.G. Douglas

[1] *Banach algebras technique in operator theory.* Academic Press, 1972.

[2] Banach algebra techniques in the theory of Toeplitz operators. *Lectures given at a GBMS Regional Conference, Series in Mathematics*, 1973.

B.A. Dubrovin, S.P. Novikov, A.T. Fomenko

[1] *Modern geometry: Methods and applications.* M. Nauka, 1979 (Russian).

R.V. Duduchava

[1] Integral convolution equations with discontinuous presymbols, singular integral equations with fixed singularities and their applications to mechanics problems. *Tr. Tbilissk. mat. in–ta*, 1979, v.60, p.5–135 (Russian).

N. Dunford, Ja. T. Schwartz

[1] *Linear operators: general theory.* Interscience, New York, v.I, 1958.

L.E. El'sgholts, S.B. Norkin

[1] *Introduction into the theory and applications of differential equations with a deviating argument* (2nd edition). M. Nauka, 1971 (Russian).

Fr. Forelli

[1] The isometries of H^p. *Canad. J. Math.*, 1964, 16, p.721–728.

B. Fuglede, R.V. Kadison

[1] On determinants and a property of the trace in finite factors. *Proc. Nat. Acad. Sc. USA*, 1951, 37, p.425–431.

[2] Determinant theory in finite factors. *Ann. Math.*, 1952, 55, p.520–530.

F.D. Gahov

[1] *Boundary value problems* (3rd edition). M. Nauka, 1977 (Russian).

Th. W. Gamelin

[1] *Uniform Algebras.* Prentice–Hall, Englewood Cliffs, N.J., 1969.

I.M. Glazman
[1] *The direct methods of the qualitative spectral analysis of singular differential operators.* M., Physmatghiz., 1963 (Russian).

J. Glimm
[1] A Stone–Weierstrass theorem for C^*–algebras. *Ann. Math.*, 1960, v.72, p.216–244.

I. Ts. Gohberg, I.A. Feldman
[1] *Convolution equations and projection methods of their solution.* M. Nauka, 1971 (Russian).

I. Ts. Gohberg, N. Ja. Krupnik
[1] *An introduction to the theory of one-dimensional singular integral operators.* Shtiintsa, Kishinjev, 1973 (Russian).

V. Ja. Golodets
[1] Cross products of Neumann algebras. *Uspekhi Mat. Nauk*, 1971, v.26, N5, p.3–50 (Russian).

I. Ja. Gol'shejd, G.A. Margulis
[1] Lyapunov exponents of the product of random matrixes. *Uspehi Mat. Nauk*, 1989, v.44, N5, p.13–60 (Russian).

Fr. P. Greenleaf
[1] *Invariant Means on Topological Groups and Their Applications.* Van Nostrand–Reinhold Comp., 1969.

R.I. Grigorchuk
[1] Symmetrical wanderings on discrete groups. *Mnoghokomponent. slutchain. sistemy., M. Nauka*, 1978, p.132–152 (Russian).
[2] On the group growth Milnor problem. *DAN SSSR*, 1983, v.271, N1, p.30–33 (Russian).

A.L. Gromov
[1] On invariant subspaces of weighted composition operators. *Funkts. anal. i egho prilozhen.*, 1988, v.22, N2, p.75–76 (Russian).
[2] On spectral properties of unitary composition operators with a weight. *Funkts. anal. i egho prilozhen.*, 1990, v.24, N3, p.74–75 (Russian).
[3] Spectral classification of certain types of unitary composition operators with a weight. *Alg. i anal.*, 1991, v.3, N5 p.62–87 (Russian).
[4] Spectral analysis of operators of an irrational rotation with a weight in $L^2(T)$. Ph.D. diss., Sankt–Peterbourgh, 1991 (Russian).

D.K. Gupta, B.S. Komal
[1] Fredholm composition operators on weighted sequence spaces. *Indian J. pure and appl. Math.*, 1983, v.14, N3, 293–296.

J. Hale
[1] *Theory of Functional Differential Equations.* Springer–Verlag, 1977.

P.R. Halmos
[1] *Lectures in ergodic theory.* Math. Soc., Japan, Tokyo, 1956.

[2] *A Hilbert Space Problem Book.* D. Van Nostrand Comp. Inc, Princeton, N.J., 1967.

P. de la Harpe, G. Scandalis
[1] Déterminant associé á une trace sur unt algébre de Banach. *Ann. Inst. Fourier*, 1984, 34, N1, p.241–26.

H. Helson
[1] Cocycles on the circle. *J. Oper. Theory*, 1986, v.16, p.183–199.

[2] The spectral theorem. *Lect. Notes Math.*, Springer–Verlag, 1986, v.1227, p.87–91.

A.B. Hevelev
[1] On invertibility of equations with two shifts on the circle. *Dep. VINITI*, N1305–78, 1978, 7p. (Russian).
[2] On invertibility and Noetherity of certain classes of operators with non Carleman shift. Ph.D. diss., University Rostov–na–Donu, 1979 (Russian).

E. Hewitt, K.A. Ross
[1] *Abstract Harmonic Analysis* v.1. Springer–Verlag, 1963.
[2] *Abstract Harmonic Analysis* v.2. Springer–Verlag, 1970.

K. Hoffman
[1] *Banach spaces of analytic functions.* Prentice–Hall, Englewood Hills, N.J., 1962.

K.H. Hofmann
[1] Representations of algebras by continuous sections. *Bul. Amer. Math. Soc.*, 1972, N78, p.291–373.

L. Hörmander
[1] *The Analysis of Linear Partial Differential Operators* v.I–IV. Springer–Verlag, 1983, 1983, 1985, 1985.

D. Husemoller
[1] *Fibre Bundles.* McGraw–Hill Book Company, 1966.

R.A. Johnson, K. Palmer, G.R. Sell
[1] Ergodic properties for linear differential systems. *SIAM J. Math. Anal.*, 1987, v.18, N1, p.1–33.

R.V. Kadison
[1] Strong continuity of operator functions. *Pacif. J. Math.*, 1968, v.26, p.121–129.

R.V. Kadison, J.R. Ringrose
[1] *Fundamentals of the theory of operator algebras* v.I, II. Academic Press, 1986.

H. Kamowitz
[1] On compact multipliers of Banach algebras. *Proceeding of AMS*, 1981, v.81, N1, p.79–80.

N.K. Karapetyants

[1] On a class of discrete convolution operators with oscillating coefficients. *DAN SSSR*, 1974, v.216, N1, p.28–31; (Russian).

[2] On a certain class of shift operators. *Izv. Sev.–Kavkaz. nautch. tzentr. vysch. shk.*, 1976, N3, p.11–12 (Russian).

N.K. Karapetyants, S.G. Samko

[1] On functional equation $\psi(x + \alpha) - b(x)\ \psi(x) = g(x)$. *Izv. AN Arm. SSR.*, 1971, v.5, N5, p.441–445 (Russian).

[2] On discrete Wiener–Hopf operators with oscillating coefficients. *DAN SSSR*, 1971, v.200, N1, p.17–20 (Russian).

Yu. I. Karlovich

[1] Locally–trajectorial method of investigation of invertibility in C^*–algebras of operators with discrete groups of shifts. *DAN SSSR*, 1988, v.299, N3, p.546–550 (Russian).

Yu. I. Karlovich, V.G. Kravchenko

[1] The Noether theory of a singular integral operator with a shift having periodic points. *DAN SSSR*, 1976, v.231, p.277–280 (Russian).

[2] On a certain algebra of singular integral operators with a non–Carleman shift. *DAN SSSR*, 1978, v.239, N1, p.38–41 (Russian).

[3] On systems of functional and integro–functional equations with a non–Carleman shift. *DAN SSSR*, 1977, v.236, N5, p.1064–1067 (Russian).

[4] Systems of singular integral equations with a shift. *Mat. Sborn.*, 1981, N1, p.87–110 (Russian).

Yu. I. Karlovich, V.G. Kravchenko, G.S. Litvinchuk

[1] The Noether theory of singul integral operators with a shift. *Izv. Vuzov. Matem.*, 1983, N4, p.3–27 (Russian).

Sh. Kawamura, J. Tomiyama

[1] Properties of topological dynamical systems and corresponding C^*–algebras. *Tokyo J. Math.*, 1990, v.13, N2, p.251–257.

A.K. Kitover

[1] On the spectrum of an automorphism with a weight, and Kamowitz–Sheinberg theorem. *Funktsion. Anal. i Prilozhen*, 1979, v.13, N1, p.70–71 (Russian).

V.B. Kolmanovskij, V.R. Nosov

[1] *Stability and periodic regimen of controlled systems with an aftereffect.* M. Nauka, 1981 (Russian).

[2] Systems with an aftereffect of neutral type. *Avtomatica i telemehanika*, 1984, N1, p.5–35 (Russian).

I.P. Kornfel'd, Ja. G. Sinaj

[1] The general ergodic theory of transformation groups with an invariant measure. *Dinamicheskie sistemy. Sovr. probl. matem. Fundam. inapravl.*, 1985, v.2, p.5–106 (Russian).

I.P. Kornfel'd, Ja. G. Sinaj, S.V. Fomin

[1] *Ergodic theory.* M. Nauka, 1980 (Russian).

A.V. Kozak
[1] Local principle in the theory of projection methods. *DAN SSSR*, 1973, v.212, N6, p.1287–1289 (Russian).

S.M. Kozlov, M.A. Shubin
[1] On the coincidence of spectra of random elliptic operators. *Mat. sb.*, 1984, v.123, N4, p.460–476 (Russian).

V.G. Kravchenko
[1] On a singular integral operator with a shift. *DAN SSSR*, 1974, v.215, N6, p.1301–1304 (Russian).

V. Krengel
[1] Ergodic theorems. 1985.

N. Ja. Krupnik, I.A. Feldman
[1] On invertibility of certain Fredholm operators. *Izv. AN MSSR. Ser. fiz.–tehn. i mat. nauk.*, 1982, N2, p.8–14 (Russian).

A. Kumar
[1] Fredholm composition operators. *Proc. AMS*, 1980, v.79, N2, p.233–236.

K. Kuratowski
[1] *Topology* v. I. Academic Press, 1966.

V.G. Kurbatov
[1] On invertibility of differential–difference operators in various topologies. *Dep. VINITI*, 1984, N764 (Russian).
[2] *Linear differential-difference equations.* Voronegsh, Izdat. Voronezhskogo Univer., 1990 (Russian).
[3] On estimate of the spectral radii of delaying operators in the space of continuous and bounded on the axis functions. *Funkts. analis i egho prilozhen.*, 1975, v.9, N3, p.56–60 (Russian).
[4] Linear functional–differential equations of neutral type and the delaying spectrum. *Sibirsk. Mat. Zh.*, 1975, v.16, N3, p.536–548 (Russian).

M.B. Landstad
[1] Duality theory for covariant systems. *Trans. Amer. Math. Soc.*, 1979, v.248, p.223–267.

B.V. Langhe, V.S. Rabinovich
[1] Algebra of generalized discrete convolution operators with oscillating coefficients. *Dep. VINITI*, 1982, N3661–82 (Russian).

Yu. D. Latushkin, A.M. Stepin
[1] Weighted shift operators, spectral theory of linear extensions and multiplicative ergodic theorem. *Mat. Sborn*, 1990, v.181, N6, p.723–792 (Russian).
[2] Weighted shift operators and linear extensions of dynamical systems. *Uspekhi Mat. Nauk.*, 1991, v.46, N2, p.85–143 (Russian).

A.V. Lebedev
[1] On the spectrum of weighted shift operators. *IV Shkola po teorii operat. v funkts. prostr., Tez. dokl Minsk.*, 1978, p.78 (Russian).

[2] On the invertibility of elements in C^*–algebras generated by dynamical systems. *Uspekhi Mat. Nauk.*, 1979, v.34, N4, p.199–200 (Russian).
[3] On the spectrum of weighted shift operators generated by a Morse–Smale diffeomorphism. *Vestsi Akad Navuk BSSR, Ser. Fiz–Mat. Navuk*, 1979, N3, p.24–28 (Russian).
[4] On certain operator algebras. *Godnensk. konfer. molod. utchen., Tez. dokl part III Grodno*, 1979, p.25–27 (Russian).
[5] On the spectrum of operators which can be decomposed into a direct integral. *Vestsi Akad. Navuk BSSR, Ser. Fiz–Mat. Navuk*, 1979, N5, p.57–62 (Russian).
[6] On the decomposition of weighted shift operators into integrable families of operators. *Vestsi Akad. Navuk BSSR, Ser. Fiz.–Mat. Navuk*, 1980, N3, p.25–32 (Russian).
[7] Weighted shift type operators. Ph.D. diss., Minsk, 1980 (Russian).
[8] On the spectrum of intransitive weighted shift operator on the circle. *Voprosy funkts. analiza (sb. statej), Petrozavodsk*, 1980, p.8–18 (Russian).
[9] Invertibility of elements in algebras of operators with a shift. I. *Vestsi Akad Navuk BSSR Ser. Fiz.–Mat. Navuk*, 1982, N6, p.36–43 (Russian).
[10] On invertibility of operators with a shift. *Dokl. AN BSSR*, 1983, v.27, N9, p.773–775 (Russian).
[11] On Wiener type theorems in cross–products and algebras associated with automorphisms. *Dokl. AN BSSR*, 1986, v.30, N6, p.493–495 (Russian).
[12] On invertibility and compactness of elements in algebras associated with automorphisms. *Dokl. AN BSSR*, 1986, v.30, N7, p.589–592 (Russian).
[13] On certain C^*–methods used while investigating algebras associated with automorphisms and endomorphisms. *Dep. VINITI*, 1987, N5351–B87 (Russian).
[14] On Szego–Kolmogorov–Krein theorem and Fuglede–Kadison determinant. *Dep. VINITI*, 1990, N2188–B90 (Russian).
[15] On certain algebras related to C^*–dynamical systems. *Dep. VINITI*, 1991, N1281–B91 (Russian).
[16] On certain ideas arising when searching for the conditions of invertibility of elements in algebras associated with automorphisms. *Dep. VINITI*, 1991, N3290–B91 (Russian).
[17] On hyperbolicity of operators in algebras associated with automorphisms. (To appear).

A.V. Lebedev, Dangh Suan Tkhan'
[1] Local method – C^*–scheme. *Dep. VINITI*, 1991, N1796–B1, 25p. (Russian).
[2] C^*–algebras generated by local type operators and their automorphisms. *Dep. VINITI*, 1991, N2449–B91, 29p. (Russian).

G.S. Litvinchuk
[1] The Noether theory for a system of singular integral equations with the Carleman shift and complex conjugate unknowns. *Izv. AN SSSR. Ser. mat.*, 1967, v.31, N3, p.563–586 (Russian).
[2] *Boundary value problems and singular integral equations with a shift.* M. Nauka, 1977 (Russian).

J.L. Massera, J.J. Sheffer
[1] *Linear differential equations and function spaces.* Academic Press, 1966.

J. Mather
[1] Characterization of Anosov diffeomorphisms. *Indag. Math.*, 1968, v.30, p.479–483.

S.G. Michlin, S. Pröbdorf
[1] *Singulare Integral Operatoren.* Akademic–Verlag, 1980.

A.S. Mischenko
[1] *Vector bundles and their applications.* M. Nauka, 1984 (Russian).

A.S. Mischenko, F. Sharipov
[1] Independence of spectrum of an elliptic operator with random coefficients. *Vestn. Mosk. Univ., Ser. I, Matem. Mekh.*, 1983, N6, p.51–56 (Russian).

Yu. A. Mitropol'skij, D.I. Martynjuk
[1] *Periodic and quasiperiodic oscillations of systems with delay.* Kiev, Vischa. shk., 1979 (Russian).

Ae. Mukhamadiev, B.N. Sadovskij
[1] On a spectral radius estimate for a certain operator related to neutral type equations. *Mat. zametki*, 1973, v.13, N1, p.61–78 (Russian).

P.S. Muhly
[1] Imprimitive operators. Preprint, 1972.

N.I. Muskhelishvili
[1] *Singular integral equations* M. Nauka, 1968 (Russian).

F.J. Murray, J. von Neumann
[1] On rings of operators. I. *Ann. Math.*, 1936, v.37, p.116–229; II. *Trans. Amer. Math. Soc.*, 1937, v.41, p.208–248.

A.G. Myasnikov, L.I. Sazonov
[1] On singular integral operators with a non–Carleman shift. *DAN SSSR*, 1977, v.237, N6, p.1289–1292 (Russian).

A.D. Myshkis
[1] *Linear differential equations with a delay in an argument.* M. Nauka, 1972 (Russian).
[2] On certain problems of the theory of equations with a deviating argument. *Uspekhi Mat. Nauk.*, 1977, v.32, N2, p.173–202 (Russian).

M.A. Naimark
[1] *Normed rings.* M. Nauka, 1968 (Russian).

J. Von Neumann
[1] On rings of operators. III. *Ann. Math.*, 1949, v.41, p.94–161.

N.K. Nikolskij
[1] *Lectures on the shift operator.* M. Nauka, 1980 (Russian).

Z. Nitecki
[1] *Differentiable dynamics.* M.I.T. Press, 1971.

E.A. Nordgren
[1] Composition operators. *Canad. J. Math.*, 1968, v.20, p.442–449.
[2] Composition operators in Hilbert spaces. *Lect. Notes Math.*, Springer–Verlag, 1978, v.693, p.37–64.

E.A. Nordgren, P. Rosental, F.S. Wintrobe

[1] Composition operators and the invariant space problem. *Math. Repts. Acad. Sci. Can.*, 1984, v.6, N5, p.279–283.

D. Olesen, G.K. Pedersen

[1] Applications of the Connes spectrum to C^*–dynamical systems I, II, III. *J. Funct. Anal.*, 1978, v.30, p.179–197; 1980, v.36, p.18–32; 1982, v.45, p.357–390.

V.I. Oseledets

[1] Multiplicative ergodic theorem. Lyapunov characteristic exponents. *Trudy Mosk. mat. obschest.*, 1968, v.19, p.179–210 (Russian).

R.S. Palais

[1] *Seminar on the Atiyah-Singer index theorem.* Princeton University Press, 1965.

S. Parrot

[1] Weighted translation operators. Ph.D. diss., Univ. of Michigan, Ann Arbor, Mich., 1965.

L.T.A. Paterson

[1] Amenability. *Math. Surv. and Monogr. N29, Amer. Math. Soc.*, 1988.

G.K. Pedersen

[1] *C^*–algebras and their Automorphisms Groups.* Academic Press, 1979.

Ja. B. Pesin

[1] General theory of smooth hyperbolic dynamical systems. *Itogi nauki i tehniki. Sovr. problemi matematiki. Fund. naprovl.*, 1985, v.2, p.123–171 (Russian).

K. Petersen

[1] The spectrum an commutant of a certain weighted translation operator. *Math. Scand.*, 1975, v.37, N2, p.295–306.

R.R. Phelps

[1] *Lectures on Choquet's Theorem.* Van Nostrand, Princeton, N.J., 1966.

M. Pimsner, D. Voiculescu

[1] Imbedding the irrational rotation C^*–algebras into AF–algebras. *J. Oper. Theory*, 1980, N4, p.93–118.

B.A. Plamenevskii

[1] *Algebras of pseudodifferential operators* M. Nauka, 1986 (Russian).

L.S. Pontryagin

[1] *Continuous groups* (3rd edition). M. Nauka, 1973 (Russian).

D. Przeworska–Rolewicz

[1] *Equations with transformed argument an algebraic approach* Warsawa, PWN, 1973.

I. Putnam, K. Schmidt, Ch. Skau
[1] C^*–algebras associated with Denjoy homeomorphisms of the circle. *J. Oper. Theory.*, 1986, 16, N1, p.99–126.

N.I. Radbel'
[1] Decomposition of spectrum and normal points of linear bundle of operators in B–spaces. *Vestu. Khar'kov. univ. Ser. Mekh. Mat.*, 1974, N39, p.17–20 (Russian).

C.E. Rickart
[1] *General theory of Banach algebras.* Van Nostrand Company Inc., 1960.

W.C. Ridge
[1] Composition operators. Thesis, Indiana University, 1969.
[2] Spectrum of a composition operator. *Proceedings AMS*, 1973, v.37, N1, p.121–127.

N. Riedel
[1] Classification of C^*–algebras associated with minimal rotations. *Pacif. J. Math.*, 1982, v.101, N1, p.153–161.
[2] Almost Mathieu operators and rotation C^*–algebras. *Proc. London Math. Soc.*, 1988, v.56, N8, p.284–302.

M. Rieffel
[1] C^*–algebras associated with irrational rotations. *Pacif. J. Math.*, 1981, v.93, N2, p.415–429.

F. Riesz, B. Sz.–Nagy
[1] *Lecons d'analyse fonctionnelle* (2nd ed.). Budapest, Akad. Kiado, 1972.

E. Yu. Romanenko, A.N. Sharkovskij
[1] An asymptotics of solutions of linear differential–functional equations. *Asimpt. poved, reshen. differents.-funktsion, uravn., Kiev Inst. Mat. AN USSR*, 1978, p.3–39 (Russian).

W. Rudin
[1] *Functional Analysis.* McGraw–Hill Book Company, 1973.

D. Ruelle
[1] Characteristics exponents and invariant manifolds in Hilbert space. *Ann. Math.*, 1982, v.115, p.243–290.

S. Sakai
[1] *C^*-algebras and W^*-algebras.* Springer–Verlag, 1971.

R.J. Sacker, G.R. Sell
[1] Existence of dichotomies and invariant splitting for linear differential systems. I. *J. Diff. Eqns.*, 1974, v.15, p.429–458.
[2] A spectral theory for linear differential systems. *J. Diff. Eqns.*, 1978, v.27, p.320–358.

A.M. Samoilenko
[1] *Elements of mathematical theory of multifrequency oscillations. Invariant tori.* M. Nauka, 1987 (Russian).

B.–W. Schulze
[1] Pseudodifferential operators and asymtotics on manifolds with corners I–IX. *Rept. Akad. Wiss DDK. Karl–Weierstrass–Inst. Math.* 1989, N7, I–IV, p.1–78; 1990, N2, I–IV, p.1–60; 1990, N4, I–V, p.1–70; 1990, N6, I–IV, p.1–61; 1990, N7, I–VI, p.1–70; 1990, N8, I–V, p.1–71; 1991, N1, I–VI, p.1–123; 1991, N2, I–VII, p.1–91; 1991, N3, p.1–121.

R.T. Seeley
[1] Integro–differential operators on vector bundles. *Trans. AMS*, 1965, v.117, N5, p.167–204.

G.R. Sell
[1] Linearization and global dynamics. *Proc. Intern. Congress of Math.*, Aug. 16–24, 1983, Warsawa, v.2, p.1283–1296.

V.N. Semenjuta
[1] On singular operator equations with a shift on the circle. *DAN SSSR*, 1977, v.237, N6, p.1301–1302 (Russian).
[2] Local approach to problems with a shift. *Izv. Sev.–Kavkaz. nautch. tzentra vysch. schk. Estestv. n.*, 1985, N4, p.28–31 (Russian).

B.V. Shabat
[1] *An introduction to complex analysis, part I, functions of one variable.* M. Nauka, 1976 (Russian).

F. Sharipov
[1] Independence of spectrum of an elliptic operator over C^*–algebra. *Vestn. Mosk. Univ., Ser. I, Matem. Mekh.*, 1985, N1, p.87–88 (Russian).

A.L. Shields
[1] Weighted shift operators and analytic function theory. *Top. oper. theory, Mathem. Surv.*, 1974, v.13, p.49–128.

M.A. Shubin
[1] *Pseudodifferential operators and spectral theory.* M. Nauka, 1978 (Russian).
[2] Pseudodifference operators and their inversion. *DAN SSSR*, 1984, v.276, N3, p.567–570 (Russian).
[3] Pseudodifference operators and their Green function. *Izv. AN SSSR. Ser. mat.*, 1985, v.49, N3, p.652–671 (Russian).

I.B. Simonenko
[1] A new general method of investigation of linear operator integral equations. I. II. *Izv. AN SSSR, Ser. mat.*, 1965, v.29, N3, p.567–586; N4, p.775–782 (Russian).
[2] On multidimensional discrete convolutions. *Mat. issled. Kishinev*, 1968, v.3, N1, p.108–122 (Russian).

I.B. Simonenko, Chin' Ngok Min'
[1] *Local method in the theory of one-dimensional singular integral equations with piece-wise continuous coefficients. Noetherity.* Izdatel'stvo Rostovskogho univ., 1986 (Russian).

R.K. Singh
[1] Compact and quasinormal composition operators. *Proceedings AMS*, 1974, v.45, N1, p.80–82.

R.K. Singh, N.S. Dharmdhikari
[1] Compact and Fredholm composite multiplication operators. *Acta. sci. math.*, 1988, 52, N3–4, p.437–441.

R.K. Singh, A. Kumar
[1] Compact composition operators. *J. Austral. Math. Soc.*, 1979, v.28, N3, p.309–314.

R.K. Singh, R. Kumar
[1] Compact weighted composition operators on L_2. *Acta. Sci. Math.*, 1985, v.49, p.339–344.
[2] Weighted composition operators on functional Hilbert spaces. *Bull. Austral. Math. Soc.*, 1985, v.31, p.117–126.

R.K. Singh, R. Kumar, David Chandra
[1] Compact weighted composition operators. *Acta. sci. math.*, 1985, v.49, N1–4, p.339–344.

R.K. Singh, S.D. Sharma
[1] Non–compact composition operators. *Bull. Austral. Math. Soc.*, 1980, v.21, p.125–130.

R.K. Singh, T. Veluchamy
[1] Spectra of normal composition operators on L^2 spaces. *Indian J. pure and appl. Math.*, 1985, v.16, N10, 1123–1131.
[2] Non–atomic measure spaces and Fredholm–composition operators. *Acta. sci. math.*, 1987, v.51, N3–4, p.461–465.

S.L. Sobolev
[1] An example of a correct boundary value problem for the string oscillation equation with the conditions set on the whole of the boundary. *DAN SSSR*, 1956, v.109, N4, p.707–709 (Russian).

A.P. Soldatov
[1] Algebra B of singular operators. *DAN SSSR*, 1985, v.282, N6, p.1312–1316 (Russian).
[2] *One dimensional singular operators and boundary value problems of the theory of functions.* M. Vysch. schk., 1991 (Russian).

M.E. Taylor
[1] *Pseudodifferential operators.* Princeton Univ. Press, Princeton, N.J., 1981.

P. Thieullen
[1] Fibres dynamiques asymptotiquement compact, exposants de Lyapunov. Entropie. Dimension. *Ann. Inst. H. Poincare*, 1987, v.4, N1, p.49–97.

F. Treves
[1] *Introduction to Pseudodifferential and Fourier Integral Operators.* v.1–2. Plenum Press, New York and London, 1982.

K. Vala

[1] On compact sets of compact operators. *Ann. Acad. Sci. Fenn. A.I.*, 1964, N351, p.2–9.

[2] Sur les éléments compact d'une algébre normée. *Ann. Acad. Sci. Fenn. A.I.*, 1967, N407, p.2–8.

J. Varela

[1] Duality of C^*–algebras. *Mem. Amer. Math. Soc.*, 1974, N148, p.97–108.

N.L. Vasilevskii

[1] Local principles in operator theory. *Lineinyie operatory v funktzionalnyikh prostranstvah. Tez. dokl. Severo-Kavkaz. reg. konf., Groznii*, 1989, p.32–33 (Russian).

A.M. Vershik

[1] Trajectorial theory ('Dynamical systems' 2, ch. 5). *Sovrem. problemy matematiki. Fundament. napravl*, v.2. *VINITI*, 1985 (Russian).

A.M. Vershik, A.L. Fedorov

[1] Trajectorial theory. *Sovrem. problemy matematiki. Noveisch. dostizh.*, v.26. *VINITI*, 1985, p.171–212 (Russian).

A.M. Vershik, S.V. Kerov

[1] Locally semisimple algebras. Combinatorial theory and K_0–functor. *Sovrem. problemy matematiki Noveisch. dostizh*, v.26. *VINITI*, 1985, p.3–56 (Russian).

A.M. Vershik, I.P. Kornfel'd

[1] Periodic approximations and their applications. Ergodic theorems, spectral and entropy theory for general group actions ('Dynamical systems' – 2, ch.4). *Sovrem. problemy matematiki. Fundament. napravl*, v.2. *VINITI*, 1985 (Russian).

P. Walters

[1] An introduction to ergodic theory. *Graduate texts in Math.*, 1982, v.79.

N. Wiener

[1] *The Fourier integral and certain of its applications.* Dover publ., N.Y., 1956.

K. Ylinen

[1] Compact and finite–dimensional elements of normed algebras. *Ann. Acad. Sci. Fenn. A.I.*, 1968, N428, p.6–37.

Index